AF464381

Data Communications and their Performance

IFIP – The International Federation for Information Processing

IFIP was founded in 1960 under the auspices of UNESCO, following the First World Computer Congress held in Paris the previous year. An umbrella organization for societies working in information processing, IFIP's aim is two-fold: to support information processing within its member countries and to encourage technology transfer to developing nations. As its mission statement clearly states,

> IFIP's mission is to be the leading, truly international, apolitical organization which encourages and assists in the development, exploitation and application of information technology for the benefit of all people.

IFIP is a non-profitmaking organization, run almost solely by 2500 volunteers. It operates through a number of technical committees, which organize events and publications. IFIP's events range from an international congress to local seminars, but the most important are:

- the IFIP World Computer Congress, held every second year;
- open conferences;
- working conferences.

The flagship event is the IFIP World Computer Congress, at which both invited and contributed papers are presented. Contributed papers are rigorously refereed and the rejection rate is high.

As with the Congress, participation in the open conferences is open to all and papers may be invited or submitted. Again, submitted papers are stringently refereed.

The working conferences are structured differently. They are usually run by a working group and attendance is small and by invitation only. Their purpose is to create an atmosphere conducive to innovation and development. Refereeing is less rigorous and papers are subjected to extensive group discussion.

Publications arising from IFIP events vary. The papers presented at the IFIP World Computer Congress and at open conferences are published as conference proceedings, while the results of the working conferences are often published as collections of selected and edited papers.

Any national society whose primary activity is in information may apply to become a full member of IFIP, although full membership is restricted to one society per country. Full members are entitled to vote at the annual General Assembly, National societies preferring a less committed involvement may apply for associate or corresponding membership. Associate members enjoy the same benefits as full members, but without voting rights. Corresponding members are not represented in IFIP bodies. Affiliated membership is open to non-national societies, and individual and honorary membership schemes are also offered.

Data Communications and their Performance

Proceedings of the Sixth IFIP WG6.3 Conference on Performance of Computer Networks, Istanbul, Turkey, 1995

Edited by

Serge Fdida
Laboratoire MASI
Paris
France

and

Raif O. Onvural
IBM
Research Triangle Park
USA

SPRINGER-SCIENCE+BUSINESS MEDIA, B.V.

First edition 1996

Originally published by Chapman & Hall in 1996
MyCopy version of the original edition 1996

DOI 10.1007/978-0-387-34942-8

A catalogue record for this book is available from the British Library

∞ Printed on permanent acid-free text paper, manufactured in accordance with ANSI/NISO Z39.48-1992 and ANSI/NISO Z39.48-1984 (Permanence of Paper).
www.springer.com/mycopy

CONTENTS

PREFACE

This is the sixth conference in the series which started in 1981 in Paris, followed by conferences held in Zurich (1984), Rio de Janeirio (1987), Barcelona (1991), and Raleigh (1993). The main objective of this IFIP conference series is to provide a platform for the exchange of recent and original contributions in communications systems in the areas of performance analysis, architectures, and applications.

There are many exiciting trends and developments in the communications industry, several of which are related to advances in Asynchronous Transfer Mode (ATM), multimedia services, and high speed protocols. It is commonly believed in the communications industry that ATM represents the next generation of networking. Yet, there are a number of issues that has been worked on in various standards bodies, government and industry research and development labs, and universities towards enabling high speed networks in general and ATM networks in particular.

Reflecting these trends, the technical program of the Sixth IFIP W.G. 6.3 Conference on Performance of Computer Networks consists of papers addressing a wide range of technical challenges and proposing various state of the art solutions to a subset of them. The program includes 25 papers selected by the program committee out of 57 papers submitted.

We would like to thank the members of program committee and the external reviewers for their meticulous and timely reviews. We also would like to thank Ms. Margaret Hudacko, Center for Advanced Computing and Communication, North Carolina State University and Ms. Barbara Sampair, IBM, Research Triangle Park, North Carolina for their help in every stage of organizing this event from the time we prepared the call for papers to the time the program was mailed out.

Serge Fdida, France
Raif O. Onvural, IBM

PCN'95

Program Committee Co-Chairs

Fdida Serge, France
Onvural Raif O., USA

Program Committee

Blondia Chris, Belgium
Bonatti Mario, Italy
Bruneel Herwig, Belgium
Budrikis Zygmuntas, Australia
Bux Werner, Switzerland
Casals Olga, Spain
De Moraes Luis Felipe, Brazil
Fratta Luigi , Italy
Iversen Villy, Denmark
Kelly F.P., UK
Körner Ulf, Sweden
Kühn Paul, Germany
Kurose Jim, USA
Le Boudec Jean-Yves, Switzerland
Lubacz Josef,.Poland
Mark Jon, Canada
Mason Lorne, Canada
Mitrou Nikolas, Greece
Perros Harry, USA
Puigjaner Ramon, Spain
Pujolle Guy , France
Roberts Jim, France
Spaniol Otto , Germany
Takagi Hideaki, Japan
Takahashi Yutaka, Japan
Tran-Gia Phuoc, Germany

Publicity Chairs

Atmaca Tulin , Europe
Halici U., Turkey
Viniotis Ioannis , USA
Yamashita Hideaki, Japan

Local Arrangements

Goker Arica, Turkey
Onvural A. , Turkey

PCN'95
List of Reviewers

Abdulmalak R.
Aydemir M.
Baguette O.
Becker M.
Blaabjerg S.
Blondia C.
Boel R.
Bruneel H.
Casals O.
Chang J-F.
Chen K.
Claudé J-P.
De Vleeschauwer D.
De Moraes L. F.
Diot C.
Duarte O.
Elsayed K.
Fayet C.
Fdida S.
Ferrari D.
Fratta L.
Gagnaire M.
Garcia F.J.
Garcia J.
Gay V.
Gelenbe E.
Genda K.
Guillemin F.
Gun L.
Hamel A.
Henk T.
Horlait E.
Hou X.
Hutchison D.
Kaltwasser J.
Kawashima K.
Kesidis G.
Korezlioglu H.
Körner U.
Kouvatsos D.
Kurose J.
Kvols K.
Labetoulle J.
Le Boudec J-Y.
Lee B.J.
Lubacz J.
Mark J.
Meteescu M.
Matragi W.
Melen R.
Mitrou N.
Musumeci L.
Nilsson A.
Ohsaki H.
Onvural R.
Owen S.
Panken F.
Parmentier P.
Perros H.
Pioro M.
Potlapalli Y.
Puigjaner R.
Pujolle G.
Raghavan S. V.
Ramaswami R.
Ren J. F.
Roberts J.
Rolin P.
Rosenberg C.
Rossello F.
Rouskas G.
Spaniol O.
Stavrakakis I.
Steyaert B.
Streck J.
Schatzmayr R.
Takagi H.
Takahashi Y.
Towsley D.
Tran-Gia P.
Tripathi S. K.
Ventre G.
Vinck B.
Viniotis I.
Virtamo J.
Vishnu M.
Vu C.D.
Wittevrongel S.
Wolisz A.
Yamashita H.

PART ONE

ATM Multiplexing

1

An exact model for the multiplexing of worst case traffic sources[1]

J. García, J.M. Barceló, O. Casals

Polytechnic University of Catalonia
Computer Architecture Department
c/ Gran Capitán, Módulo D6
E-08071 Barcelona, Spain
tel : + 34 3 4015956
fax : + 34 3 4017055
e-mail : (jorge,joseb,olga)@ac.upc.es

Abstract

In this paper we analyse a multiplexer handling a number of identical and independent *Worst Case Traffic (WCT)* sources. Each WCT source produces a periodic stream of cells consisting of a constant number of back-to-back cells followed by a silent period of constant duration. The WCT can model the traffic produced by a "malicious" user who sends an ON/OFF traffic where a burst of back-to-back cells whose length is the largest compatible with the tolerance introduced in the control function alternates with an idle period whose length is the smallest compatible with the policed peak cell rate. WCT can also model, for example, the traffic produced by some ATM Adaptation Layer multiplexing schemes in the Terminal Equipment.

Exact results are obtained, both for the discrete and the fluid-flow model. The numerical examples show the dramatic impact that WCT can have on the multiplexer buffer requeriments. The model presented can be useful to assess the convenience of using a traffic shaping device at the entry point of the ATM network.

Keywords

Worst Case Traffic, ATM, Traffic Management, Congestion Control.

[1]This work was supported in part by the CIRIT, Grant no. GRQ93-3.008

1 INTRODUCTION

The B-ISDN which will be based on the ATM technique, is designed to transport a wide variety of traffic classes with different transfer capacity needs and Network Performance objectives. The traffic flow present in such networks will be subject to unpredictable statistical fluctuations which will cause congestion. During a congestion state, the network will not be able to meet the negociated Network Performance objectives for the already established connections.

It is generally assumed that for real time services Traffic Control for ATM networks will be done in three steps:

- The user requests the set up of a connection characterized by a declared *Traffic Descriptor.*
- The network determines by means of a *Connection Admission Control (CAC)* function whether this connection can be accepted while maintaining the agreed Quality of Service (QoS). If the connection can be accepted, the network allocates the necessary resources.
- The network controls the established connection by means of a *User Parameter Control (UPC)* algorithm to verify that the negociated parameters of the Traffic Contract are not violated.

The Traffic Contract at the Public UNI consists of a Connection Traffic Descriptor, a requested QoS class and the definition of a compliant connection The ATMForum, 1993. The Connection Traffic Descriptor consists of:

- The Source Traffic Descriptor which can include parameters like *Peak Cell Rate, Sustainable Cell Rate, Burst Tolerance* and/or source type.
- The *Cell Delay Variation (CDV) Tolerance.* CDV refers to the random perturbation on the interarrival time of consecutive cells of a given connection produced by cell multiplexing and other functions of the *ATM Layer.* CDV Tolerance represents a bound on the cell clumping phenomenon due to CDV and it is defined according to the UPC algorithm used.
- The Conformance Definition based on one or more applications of the *Generic Cell Rate Algorithm (GCRA).* The GCRA is a Virtual Scheduling Algorithm or

a Continuous-state Leaky Bucket Algorithm which is used to specify the conformance at the public or private UNI to declared values of CDV tolerance and of traffic parameters Peak Cell Rate, Sustainablc Cell Rate and Burst Tolerance.

An important consequence of the introduction of a CDV Tolerance is that the UPC algorithm will allow that a certain number of cells violate the Peak Cell Rate declared at connection set up. In fact this tolerance makes possible that a burst of a certain number of back-to-back cells (i.e. emitted at link rate) are viewed as *conforming cells* by the UPC algorithm.

As an example, let us assume that at connection set up a user declares a *Constant Bit Rate (CBR)* connection with a certain Peak Cell Rate and a certain CDV Tolerance. The UPC algorithm will declare as conforming a maximum of N back-to-back cells if this burst of cells is preceded by a silence state long enough. In fact the user could send a periodic flow of cells consisting of N back-to-back cells followed by a silence state and the UPC will declare this flow of cells as conforming. We will call this kind of periodic traffic *Worst Case Traffic (WCT)*. This name comes from the fact that the multiplexing of this kind of traffic requires by far more resources than the required by a periodic connection. In fact WCT is not the "worst" traffic that can be declared as conforming by the UPC algorithm (see Aarstad, 1993). However WCT as defined above is more tractable and the results obtained are not far from the "worst" case.

The study of the effect of WCT in ATM multiplexers is important for several reasons: A misbehaving customer can try to take advantage of the UPC function tolerance to send traffic with different parameters than the negociated during the Connection set up phase. In our example the network cannot relay on the assumption that the user is actually sending CBR traffic as declared because it does not have means of checking that this is really happening. Therefore in order to ensure a certain QoS the CAC has to assume WCT to decide whether a connection can be accepted or not.

Besides the "tricky user" some AAL schemes may generate WCT in a natural way. As an example (Boyer, 1992), let us assume that a multimedia workstation involves several AALs each of them generating CBR traffic at different peak bit rates. Each AAL accumulates data in a private queue. The queues are periodically emptied at a rate of 150 Mbps with scanning period set to the lowest involved peak emission period.

A possible solution to the problem caused by clumps of cells is to use a *Traffic Shaping function* together with the UPC/NPC in order to retrieve as far as possible the negociated Peak Cell Rate of the connection. A device which performs this shaping function (together with a UPC function) known as *Spacer/Controller* has been proposed (see Boyer, 1992, Wallmeier, 1992). The model presented in the paper can be used to study the trade-off between the cost of introducing a shaping device such as the Spacer/Controller and the low utilization or large buffers needed to cope with WCT.

In this paper we present an exact model for the multiplexing of N identical and independent WCT sources. In García an approximate model for the multiplexing of WCT together with geometric or VBR traffic is developed. Approximate solutions of the model presented here are found in Roberts, 1993 (fluid approach) and in Kvols, 1992 (discrete time). In both cases the authors use what is known as the *Benes bound* which gives an upper bound for the queue length distribution. The problem of obtaining the queue length distribution using the Benes approach which can be reduced to computing the ith-fold convolution of a pulse, is solved approximately in Roberts, 1993 by using the saddle-point method and in Kvols, 1992 by means of an FFT for the discrete time case. In appendix A and B we give closed formulas for these convolutions for both the discrete and fluid case. In Ramamurthy, 1991 an exact formula for the mean queue length is given and they suggest an approximation for the queue length distribution.

The model developed is used to demonstrate the decrease of network utilization when considering WCT or, alternatively, the increase of buffer length needed to maintain a certain QoS. Roughly, allowing clumps of length b in the network means to increase the required buffer capacity by a factor of b if no traffic shaping is used.

The paper is structured as follows: In section 2 we present a discrete time model of the multiplexing of N identical and independent WCT sources in a slotted queue. In section 3 we see how to extend the model for a fluid approach. In section 5 we compare the exact results we obtain with some approximation suggested in the literature. In section 5 we use the model developed to study the impact of WCT in the dimensioning of ATM networks. Finally conclusions are drawn in section 6 .

2 MULTIPLEXING WORST CASE TRAFFIC SOURCES IN A SLOTTED QUEUE

In this section we consider a multiplexer of capacity one cell per time slot loaded with N identical and independent *Worst Case Traffic (WCT)* sources. Each WCT source produces a periodic stream of cells, of period T, with the following pattern: It emits a constant number, b, of back-to-back cells and then it remains silent during a constant time $T-b$ (Figure 1). The time slots where each source becomes active are uniformly and independently distributed within the period. In order to have a stable queue we assume that $\frac{Nb}{T} < 1$. We consider that arrivals take precedence on departures (i.e. first we have cell arrivals (if any), then the service (if any), and finally we observe the system).

We use the following definitions: $N(t)$ is the number of arrivals at slots $-(t-1), \ldots, 0$; $\phi(t) = N(t) - t$, B_t is the number of sources which become active at slots $-(t-1), \ldots, 0$;

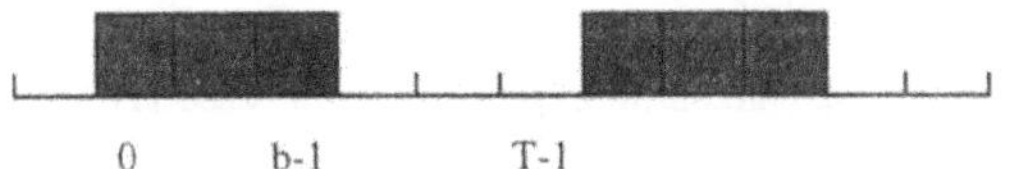

Figure 1: Traffic offered by a WCT source

γ_t is the number of sources which were active at $-t$, and L_t is the queue length at $-t$.

Following the Benes approach for the analysis of a slotted statistical multiplexer with a periodic input of period T (see Roberts, 1991) we express $p\{L_0 > x\}$ as:

$$p\{L_0 > x\} = \sum_{t=1}^{T} p\{\phi(t) = x\} p\{\phi(u) < x, t < u \leq T | \phi(t) = x\}. \tag{1}$$

Introducing γ_t and B_t, (1) can be written as:

$$p\{L_0 > x\} =$$
$$\sum_{t=1}^{T}\sum_{i=1}^{N} p\{\phi(t) = x, \gamma_t = 0, B_t = i\} p\{\phi(u) < x, t < u \leq T | \phi(t) = x, \gamma_t = 0, B_t = i\}.$$

2.1 The term $p\{\phi(t) = x, \gamma_t = 0, B_t = i\}$.

The term $p\{\phi(t) = x, \gamma_t = 0, B_t = i\}$ can be expressed as

$$p\{\phi(t) = x, \gamma_t = 0, B_t = i\} = p\{\phi(t) = x | \gamma_t = 0, B_t = i\} p\{\gamma_t = 0, B_t = i\}. \tag{2}$$

To derive an expression for these probabilities, we distinguish between three cases depending on the values of t:

- *Case (I):* $t = 1, \ldots, b-1$

  We have:

  $$p\{\gamma_t = 0, B_t = i\} = \binom{N}{i} \frac{t^i (T-b-t)^{N-i}}{T^N}. \tag{3}$$

  and:

  $$p\{\phi(t) = x | \gamma_t = 0, B_t = i\} = \frac{1}{t^i} q_t^{(i)}(t+x), \tag{4}$$

  where $q_t(x)$ is a discrete-time unitary pulse in $[1, t]$ and $q_t^{(i)}(x)$ is its i-th discrete-time convolution. A simple expression for $q_t^{(i)}(x)$ is derived in Appendix A.

- *Case (II):* $t = b, \ldots, T - b$

  $p\{\gamma_t = 0, B_t = i\}$ has the same expression as in case (I) (equation (3)). For the other term we get:

$$p\{\phi(t) = x | \gamma_t = 0, B_t = i\} = \sum_{j=0}^{i} \binom{i}{j} \frac{(t-b+1)^j}{t^i} q_{b-1}^{(i-j)}(t + x - bj). \tag{5}$$

- *Case (III):* $t = T - b + 1, \ldots, T$

  For $i = 1, \ldots, N - 1$ this term vanishes. For $i = N$ we have:

$$p\{\gamma_t = 0, B_t = N\} = (\frac{t}{T})^N. \tag{6}$$

and

$$p\{\phi(t) = x | \gamma_t = 0, B_t = N\} = \sum_{j=0}^{N} \binom{N}{j} \frac{(t-b+1)^j}{t^N} q_{b-1}^{(N-j)}(t + x - bj). \tag{7}$$

2.2 The term $p\{\phi(u) < x, t < u \leq T | \phi(t) = x, \gamma_t = 0, B_t = i\}$.

This term can be written as:

$$p\{\phi(u) < x, t < u \leq T | \phi(t) = x, \gamma_t = 0, B_t = i\} = \\ = \frac{p\{\phi(u) < x, t < u \leq T | \phi(t) = x, B_t = i\}}{p\{\gamma_t = 0 | \phi(t) = x, B_t = i\}}$$

A simple expression for the numerator can be derived by means of a similar argument as in section III of Roberts, 1991:

We note that the event $\{\phi(u) < x, t < u \leq T | \phi(t) = x, B_t = i\}$ corresponds to the arrival patterns that would result in an auxiliar queue loaded with periodic arrivals of period $T - t$ being empty at time $-t$.

The periodic arrivals at this auxiliar queue belong to two classes (Figure 2):

- A batch arrival at time $-(T-1)$ consisting of $ib - t - x$ cells. These cells correspond to the ones that were emitted in the original system by the i sources that had become active at slots $-(t - 1 - T), \ldots, -T$.

- $N - i$ independent WCT sources of period $T - t$ emitting b back-to-back cells.

If the original system is stable, the same will occur with the auxiliar queue ($Nb < T$ implies $ib - t - x + (N-i)b < T - t$) . Therefore we do not need to take into account the contribution of the first batch arrival, obtaining:

$$p\{\phi(u) < x, t < u \leq T | \phi(t) = x, B_t = i\} = 1 - \frac{(N-i)b}{T-t}. \tag{8}$$

The denominator is : $p\{\gamma_t = 0 | \phi(t) = x, B_t = i\} = (\frac{T-t-b}{T-t})^{N-i}$.

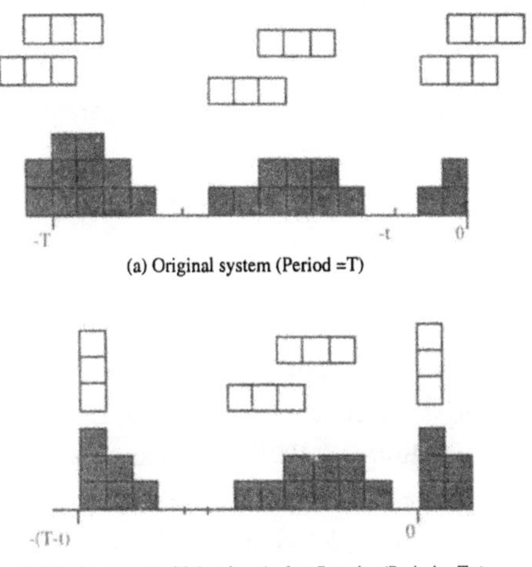

(a) Original system (Period =T)

(b) Equiv. system with batch arrival at first slot (Period = T-t)

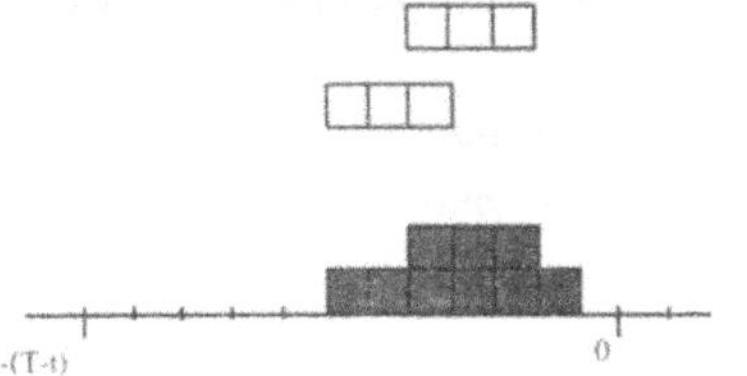

(c) Equiv. system without batch arrival at first slot (Period = T-t)

Figure 2:

3 THE FLUID WCT MODEL

A similar study can be done for a system using a fluid-flow approximation. In this case the WCT is defined as a periodic source which produces h information units per time slot during a constant time b and which remains silent during a constant time $T - b$. The multiplexer is able to serve at rate c. We asume $h \geq c$. The analogous to (1) in the fluid case is (see Roberts, 1993):

$$p\{L_0 > x\} = c\int_0^T p(\psi(t) = x)p(\psi(u) < x, t < u \leq T|\psi(t) = x\}dt, \tag{9}$$

where $W(t)$ is the work arriving in the interval $(-t, 0)$, and $\psi(t) = W(t) - ct$.

The derivation of the final formula in the case of WCT fluid sources follows the same steps as in the discrete-time case: We also have to distinguish between three cases depending on the values of t. (Namely: case (I): t < b; case (II): b < t < T-b and case (III): T-b < t). The formula for the i-th convolution of a continuous-time pulse is derived in appendix B. The resulting integrals can be evaluated, for example, by means of the Gauss method.

4 ERRORS AND BOUNDS

It is well known that the queue length distribution of a $nD/D/1$ queueing system can be approximated by the queue length distribution of an $M/D/1$ queueing system when the period of the sources is large (see Roberts, 1991). In Figure 3 we make a comparison between the $nWCT/D/1$ system and a slotted queue loaded with Poisson batch arrivals (curve with the points). The batchs have a deterministic distribution of $b = 2$ cells. We observe that when the period becomes large, the two systems have a similar distribution. The queue with Poisson batch arrivals gives an upper bound of the buffer length distribution of the $nWCT/D/1$ system.

In Ramamurthy, 1991 the following approximation is suggested: Let $L^{(b)}$ be the queue length of a multiplexer loaded with N WCT sources of period $T' = bT$ for a given T. Then

$$p\{L^{(b)} > x\} \approx p\{L^{(1)} > \frac{x}{b}\}. \tag{10}$$

Figure 4 compares this approximation (curves with the points) with the results obtained from the exact model. As expected, the differences with the exact model are more important for larger values of b.

5 RESULTS

In this section we present some results obtained for the discrete-time case.

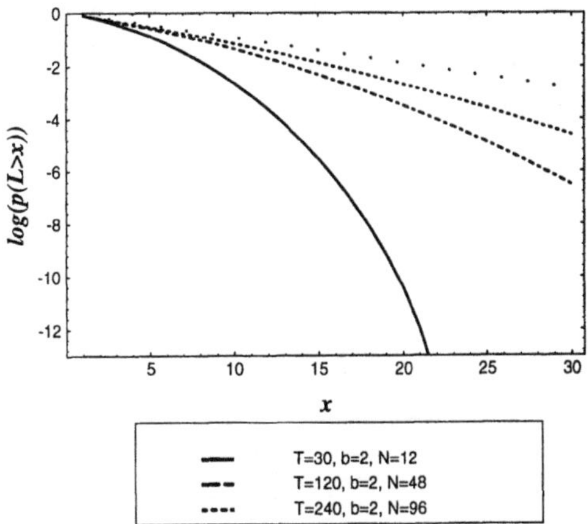

Figure 3:

Figure 5 shows, for different values of b, the complementary queue length distribution of a multiplexer handling 12 WCT traffic. The ratio $\frac{b}{T} = \frac{1}{15}$ is maintained constant. Even for $b = 2$ and $b = 3$ we observe a dramatic impact on the queue length distribution: For example, if we have CBR connections ($b = 1$) the probability of having more than 10 cells in the multiplexer is 3.277×10^{-11}. In the case of WCT with $b = 2$ the probability of this event is incremented about 6 orders of magnitude (7.823×10^{-4}) while with $b = 3$ we obtain a value of 1.737×10^{-2}. On the other hand, to have a quantile of the buffer length probability lower than 1×10^{-10} we need for $b = 1$ a buffer length $L = 10$, for $b = 2$ we need $L = 20$ while for $b = 3$, $L = 28$.

Figure 6 shows the admissible load to have a quantile of the buffer queue length probability lower than 1×10^{-10}. The traffic parameters are the same as above. We can again observe the impact of the value of b: For $b = 1$ the admissible load is 0.80 while for $b = 2$ it decreases to 0.33 and for $b = 3$ to 0.27.

A similar experiment is performed when $\frac{b}{T} = \frac{1}{100}$. We observe again the strong impact of b on the complementary buffer length distribution for a multiplexer loaded up to 0.8 (figure 7) and on the admissible load when the buffer has a capacity of 24 cells (figure 8).

Now we study an example in which the user produces WCT. Let us assume that we scan each 480 time slots the AAL buffers of a multimedia workstation with different CBR connections. The user generates traffic at a total rate of 10 Mbps, and the physical link rate is 150 Mbps. This means that clumps of 32 back-to-back cells will enter the ATM network. Figure 9 shows the complementary buffer queue length distribution of a multiplexer handling such traffic sources (load 0.80). To have a quantile of the buffer length

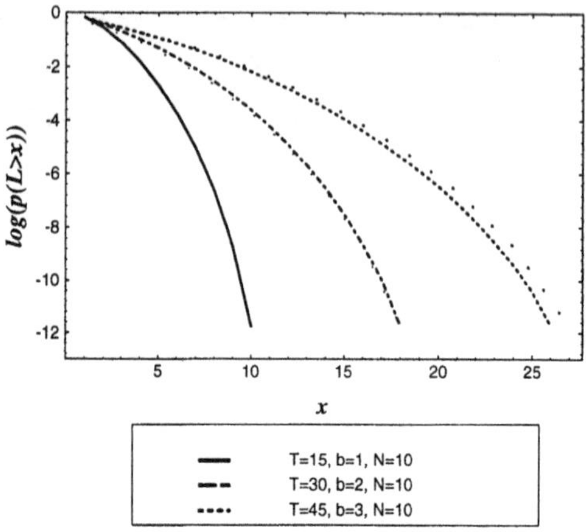

Figure 4:

probability lower than 1×10^{-10} we need a value of $L = 288$. Using $L = 128$ means to reduce the load to 0.33 while for $L = 64$ the admissible load is 0.13.

6 CONCLUSIONS

In this paper we have developed an exact model for a multiplexer loaded with a number of identical and independent Worst Case Traffic (WCT) sources.

The model can be used to demonstrate the decrease of network utilization when considering WCT or, alternatively, the increase of buffer length needed to maintain a certain QoS. This analysis is important to assess the convenience of using traffic shaping devices at the ATM entry points.

7 ACKNOWLEGDMENT

We would like to thank Pierre Boyer for useful discussions and suggestions.

APPENDIX A : I-TH CONVOLUTION OF A PULSE IN DISCRETE TIME

Let $q_m(t)$ be a discrete-time pulse of amplitude 1 in $[1, m]$ and let $q_m^*(z)$ be the z-transform of such pulse. We are interested in finding a simple expression for the i-th discrete-time convolution of $q_m(t)$. This is equivalent to find a simple expression for the coefficients of

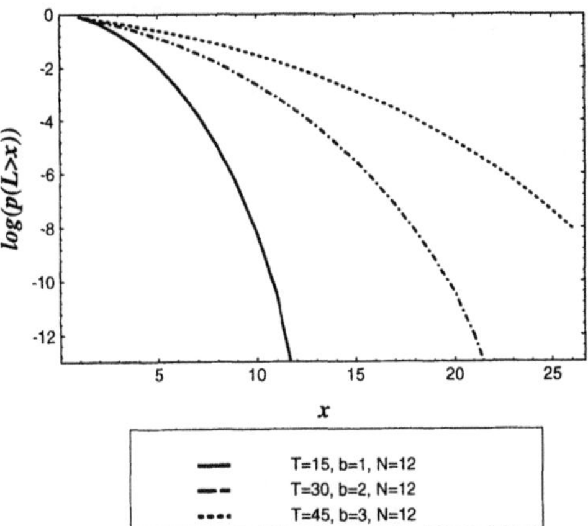

Figure 5:

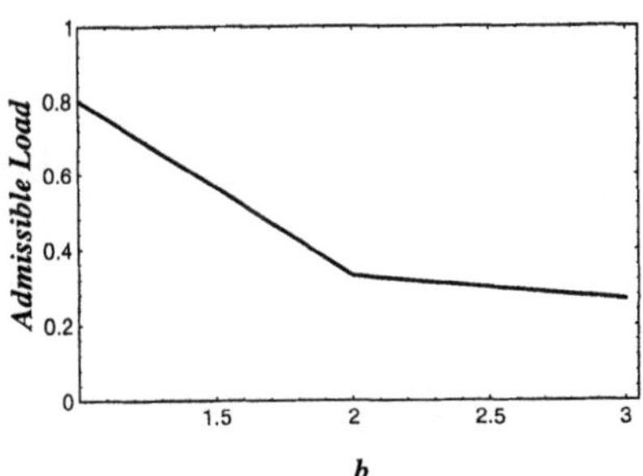

Figure 6:

the polynomial:

$$(q_m^*(z))^i = (\sum_{n=1}^{m} z^n)^i. \tag{11}$$

We can express $q_m^*(z)$ as:

$$q_m^*(z) = (1 - z^m)p_\infty^*(z). \tag{12}$$

We can easily derive (from example, using the convolution algorithm, Buzen, 1973 that:

$$(q_\infty^*(z))^i = z^i \sum_{n=0}^{\infty} \begin{pmatrix} i+n-1 \\ n \end{pmatrix} z^n. \tag{13}$$

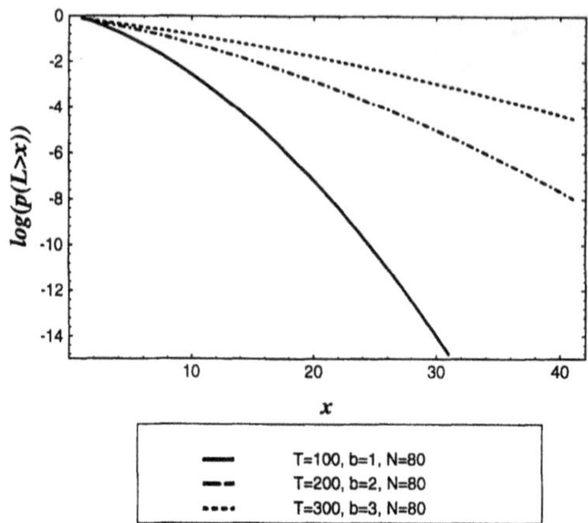

Figure 7:

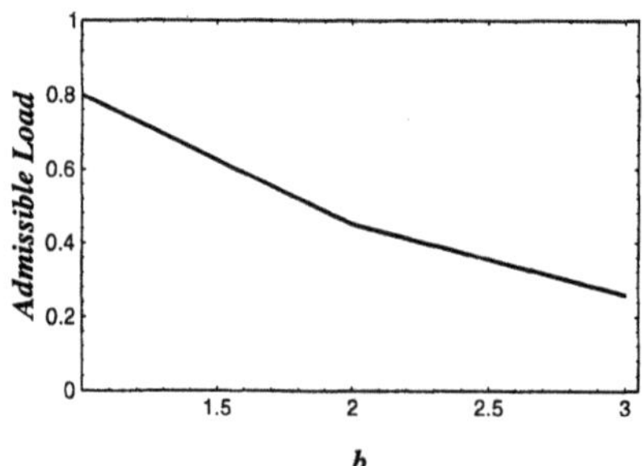

Figure 8:

Hence, we obtain ($i > 0$):

$$(q_m^*(z))^i = z^i \sum_{n=0}^{i(m-1)} \sum_{s=0}^{[\frac{n}{m}]} (-1)^s \begin{pmatrix} i+n-sm-1 \\ n-sm \end{pmatrix} \begin{pmatrix} i \\ s \end{pmatrix} z^n. \tag{14}$$

From that we deerive a formula for $q_m^{(i)}(t)$ ($i > 0$ and $t = i, \ldots, im$):

$$q_m^{(i)}(t) = \sum_{s=0}^{[\frac{t-i}{m}]} (-1)^s \begin{pmatrix} i \\ s \end{pmatrix} \begin{pmatrix} t-sm-1 \\ i-1 \end{pmatrix}. \tag{15}$$

and $q_m^{(i)}(t) = 0$ for other values of t.
For $i = 0$ we define

$$q_m^{(0)}(t) = \delta(t). \tag{16}$$

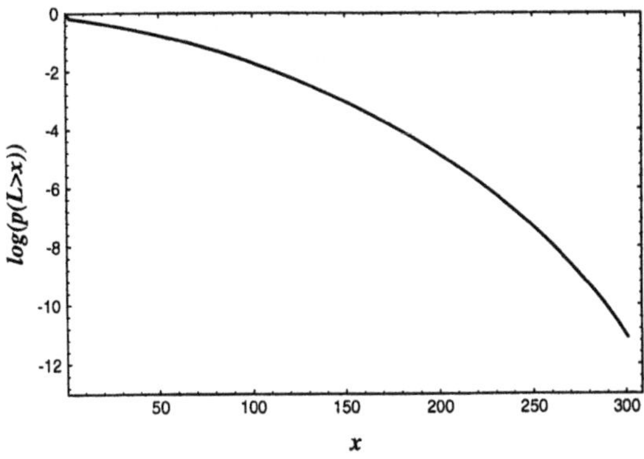

Figure 9:

APPENDIX B : I-TH CONVOLUTION IN CONTINUOUS TIME

Let $p_y(x)$ be a pulse with unitary amplitude in $(0, y)$. In order to get an explicit formula for its i-th convolution, we obtain first the Fourier transform of $p_y(x)$:

$$P_y(\omega) = \int_{-\infty}^{+\infty} e^{-j\omega x} p_y(x) dx = j\frac{e^{-j\omega y} - 1}{\omega}. \tag{17}$$

Hence, the transform of $p_y^{(i)}(x)$ is:

$$P_y^i(\omega) = j^i (\frac{e^{-j\omega y} - 1}{\omega})^i. \tag{18}$$

and $p_y^{(i)}(x)$ can be expressed as $(i > 0)$:

$$p_y^{(i)}(x) = \frac{1}{2\pi} \int_{-\infty}^{+\infty} e^{j\omega x} P_y^i(\omega) d\omega = \frac{j^i}{2\pi} \sum_{k=0}^{i} \left(\begin{array}{c} i \\ k \end{array} \right) (-1)^{i-k} \int_{-\infty}^{+\infty} \frac{e^{-j\omega(ky-x)}}{\omega^i} d\omega. \tag{19}$$

We now consider the integral:

$$\int_{-\infty}^{+\infty} \frac{e^{j\omega(x-ky)}}{\omega^i} d\omega = \int_{-\infty}^{+\infty} \frac{cos(\omega(x-ky))}{\omega^i} d\omega + j \int_{-\infty}^{+\infty} \frac{sin(\omega(x-ky))}{\omega^i} d\omega. \tag{20}$$

If i is even the imaginary part of the rigth hand side of equation (20) vanishes. For the real part we have Papoulis, 1962, Prudnikov):

$$\int_{-\infty}^{+\infty} \frac{cos(\omega(x-ky))}{\omega^i}d\omega = \frac{(x-ky)^{i-1}}{(i-1)!}cos(i\frac{\pi}{2})\int_{-\infty}^{+\infty}\frac{sin(\omega(x-ky))}{\omega}d\omega = \pi\frac{|x-ky|^{i-1}}{(i-1)!}cos(i\frac{\pi}{2})$$

If i is odd the real part vanishes while for the imaginary part we have:

$$\int_{-\infty}^{+\infty} \frac{sin(\omega(x-ky))}{\omega^i}d\omega = \frac{(x-ky)^{i-1}}{(i-1)!}cos((i-1)\frac{\pi}{2})\int_{-\infty}^{+\infty}\frac{sin(\omega(x-ky))}{\omega}d\omega = \pi\frac{(x-ky)^{i-1}}{(i-1)!}cos((i-1)\frac{\pi}{2})sgn(x-ky).$$

Finally we obtain the following expression for $p_y^{(i)}(x)$ $(i>0)$:

$$p_y^{(i)}(x) = \frac{1}{2}\sum_{k=0}^{i}\begin{pmatrix} i \\ k \end{pmatrix}(-1)^k\frac{(x-ky)^{i-1}sgn(x-ky)}{(i-1)!}. \qquad (21)$$

For $i=0$ we define:

$$p_y^{(0)}(x) = \delta(x), \qquad (22)$$

where $\delta(x)$ denotes the Dirac delta in continuous-time.

8 REFERENCES

Aarstad E. A comment on Worst Case Traffic, COST 242 TD, 1993.

Boyer, P. Guillemin F. M., Servel M. J. and Coudreuse J. P. Spacing Cells At The ATM Network Entry Points. Special Issue of IEEE Network Magazine on Switching and Congestion Control in ATM Networks, September 1992.

Buzen J. P. Computational Algorithm for Closed Networks with Exponential Servers", Comm. Assoc. Comput. Mach., 1973, 16, 527-531.

García J., Casals O, A Discrete Time Queueing Model to Study the Cell Delay Variation in an ATM Network", Performance Evaluation 21 3-22, 1994.

Kvols K., Blaabjerg S. Bounds and Approximations for the Periodic On/Off Queue with Applications to ATM Traffic Control, IEEE Infocom'92, Florence, May 1992.

Papoulis A. The Fourier Integral and its Applications, McGraw-Hill, 1962, New York.

Prudnikov A. P., Brychkov Y. A. and Marichev O. I. Integrals and Series, Volume 1. Gordon and Breaach Science Publishers, New York.

Ramamurthy G., Dighe R. S. A Distributed Source Control: A Network Access Control for Integrated Broadband Packet Networks. IEEE JSAC, vol. 9, No. 7, pp 990-1002, September 1991.

Roberts J. W., Virtamo J. The Superposition of Periodic Cell Arrival Streams in an ATM Multiplexer, IEEE T. on Comm., Vol. 39, No. 2, Feb. 1991.

Roberts J. W., Bensaou B., Canetti Y. A Traffic Control Framework for High-Speed Data Transmission, Modelling and Performance Evaluation of ATM Technology, La Martinique, January 1993.

The ATM Forum, ATM User-Network Interface Specification, Version 3.0, September, 1993.

Wallmeier E., Worster T. The Spacing Policer, an Algorithm for Efficient Peak Bit Rate Control in ATM Networks, Proc. of ISS'93, Yokohama (Japan), October 1992.

9 BIOGRAPHY

Jorge García graduated and received his Ph.D. in Telecommunications Engineering from Polytechnic University of Catalonia (UPC) in 1988 and 1992, respectively. He joined UPC in 1988 and currently is Associate Professor of the Computer Architecture Department. In 1992-93 he was Visiting Scientist at the Systems and Industrial Department of the University of Arizona with a NATO fellowship. He has ben involved in several RACE project and currently is involved in COST-242 project.

José María Barceló graduated in Telecommunications Engineering from UPC in 1992. He has been involved in RACE project EXPLOIT and currently he is Ph. D. student in the Computer Architecture Department of UPC.

Olga Casals graduated and received his Ph. D. from UPC in 1983 and 1986 respectively, both in Telecommunications Engineering. She joined UPC in 1983 where she became Full Professor in 1994 and she is head of a research group on traffic in B-ISDN communications systems. She has been working on ATM networks since 1988 with her participation in the RACE project R1022. She has been also involved in the RACE projects EXPLOIT and BAF and currently she is involved in ACTS project EXPERT and in COST-242.

2

On per-session performance of an ATM multiplexer with heterogeneous speed links

R.I. Balay and A.A. Nilsson
Center for Advanced Computing and Communication
North Carolina State University, Raleigh, NC 27695-7914, USA.
Phone: (919) 515-5130. Fax: (919) 515-2285.
email: {rajesh,nilsson}@eceyv.ncsu.edu

Abstract

Our aim in this paper is to provide insight on a session's performance in the presence of different speed links at an ATM multiplexer. We describe a queueing model for such scenarios considering a finite capacity buffer and bursty traffic with correlated arrivals. We use Markov Modulated Bernoulli Process (MMBP) as the traffic model for the session of interest and model the cross traffic as an MMBP with batch arrivals. We present exact analysis using an evolution method to obtain the steady-state performance measures of the session. To enhance its efficiency, a closed form, partial approximation method is also described. Using numerical results we compare a sessions performance in different scenarios of link speeds, and also study the influence of a few network and traffic characteristics.

Keywords

ATM, quality of service, high-speed networks, performance modeling, per-session performance, discrete-time analysis, Markov Modulated Bernoulli Process, finite capacity queue

1 INTRODUCTION

It is anticipated that Broadband Integrated Services Digital Networks (B-ISDNs) will support data, voice, video and multimedia applications using Asynchronous Transfer Mode (ATM) technology. Most applications will traverse different speed links and will have stringent quality of service (QoS) requirements to be guaranteed by the network. In enabling and efficiently utilizing such networks, per-session performance study becomes a necessity for admission control, network design and for the service provider, who desires to know the behavior of a session given the traffic characteristics and network topology. In this paper we address the above issue by providing discrete-time queueing analysis for per-session performance of a single switching node.

Discrete-time queueing models have received much attention due to increasing focus on ATM networks where packets are transmitted as fixed size units called 'cells'. In literature,

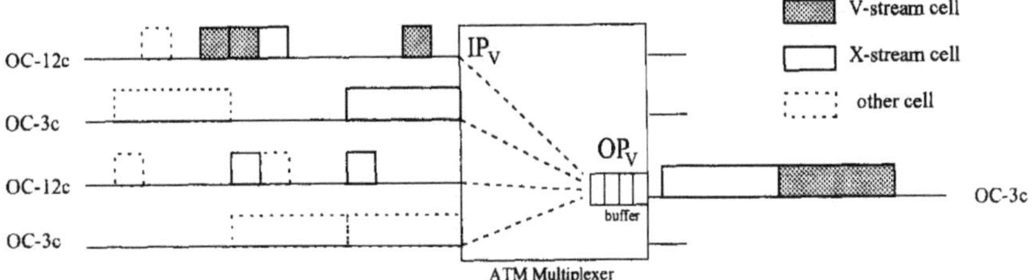

Figure 1 An example ATM switching scenario with SONET links.

most performance studies deal with a multiplexer as a whole and only a few consider per-session performance (Murata, *et al.* (1990), Herrmann (1993)). To model the bursty and correlated nature of ATM traffic many models (e.g. MMBP) have been proposed which are complex and non-renewal. Since closed form analytic solutions are difficult to obtain for systems with such traffic characteristic, algorithmic methods such as matrix-geometric techniques developed by Neuts (1981) are commonly used. We present an alternate algorithmic approach using evolution method which is computationally efficient and has the following advantages: (1) the flexibility to trade computational time for accuracy of results using a closed form, partial approximation and (2) transient behavior of a session can be studied (Balay and Nilsson, 1996).

This paper is organized as follows: In section 2, using a switching scenario we define our analytical model. In section 3, by exploiting the Markov renewal property of an MMBP, we present an exact analysis to derive a session's steady state performance measures such as cell loss probability and delay distribution. In section 4, we discuss the complexity of our method and also compare it with matrix-geometric methods. An approximation technique to enhance the efficiency of our method is described in section 5. We present some numerical results in section 6 and make our concluding remarks in section 7.

2 MODEL DESCRIPTION

The switching node we model is shown in figure 1, where IP_V and OP_V are the arrival and exit ports respectively for the session of interest (V-stream). All other sessions which are routed through OP_V constitute the cross traffic (X-stream). In modeling such scenarios for per-session study, a normal trend is to consider a single server queue of deterministic service time of one slot and assume at most one arrival to occur in a slot; the reasoning being cell transmission time on a link is constant. This model is a gross approximation if the session of interest traverses different speed links (which will be evident when we look at some numerical results). For an accurate model, it is important to capture the nature of cell arrivals with respect to the cell transmission; we do so using two parameters:

N_A: The constant number of slots for a cell arrival from V-stream. It means that two consecutive arrivals from V-stream are at least N_A slots apart, and a cell arrival

Input Link (IP_V)	Output Link (OP_V)	Arrival slots (N_A)	Service slots (N_S)
OC-3c	OC-3c	1	1
OC-12c	OC-3c	1	4
OC-12 (4 x OC-3c)	OC-3c	1	1
OC-3c	OC-12c	4	1
OC-48 (4 x OC-12c) OC-48 (4 x 4 x OC-3c)	OC-3c	1 1	4 1

Table 1 Link speeds and corresponding model parameters

event is considered to occur in the last of the cell arrival slots (i.e. N_Ath slot). Note that this does not apply to cross-traffic cells since it constitutes cells from many sessions which may arrive on different speed links; hence we consider cross-traffic cells to take one slot to arrive.

N_S: The deterministic service time for all cells at the multiplexer.

In our discussion, a 'slot' has a new meaning; it is an interval of time which is the same as the least common multiple of the time to transport a cell on the input and output links traversed by V-stream. To make it more clear, for a few SONET speed links we list the corresponding model parameters in Table 1. Notice that the parameters depend only on the underlying concatenated SONET protocol used, since it determines the rate at which a cell is transported on a link. Some modeling assumptions are: buffer capacity is finite and is of size K cells, a cell cannot receive service in its arrival slot, arrivals occur before the departure event in a slot, an arrival is lost if buffer is full and a cell in service does not occupy buffer space. If a simple cyclic polling scheme is used to transfer the incoming cells to their respective output buffers, the performance of a session depends on the position of the arrival port. To obtain bounds on performance (which is more informative than average) we consider two slot policies: $\mathcal{P}_V$ and $\mathcal{P}_X$, which denote that in a slot the V-stream arrival is considered for buffer allocation before a cross-traffic arrival, and vice-versa, respectively.

2.1 Traffic model

We model V-stream as a Markov Modulated Bernoulli Processes (MMBP), the discrete-time analogue of the Markov Modulated Poisson Process (MMPP) (Fischer and Meier-Hellstern, 1993). These processes capture the notion of burstiness and correlation properties of an arrival stream. An MMBP is a Bernoulli process where the arrival rate is varied by a multiple-state Markov chain. It is characterized by a state transition probability matrix $\mathbf{P}$ and arrival probability descriptor $\mathbf{\Lambda}$, where $[\mathbf{P}]_{i,j}$ denotes the probability of transition from state 'i' to state 'j' and $[\mathbf{\Lambda}]_i$ denotes the probability of arrival in a slot when in state 'i'. We use $\mathbf{\Lambda_V}$ and $\mathbf{P_V}$ to describe the V-stream MMBP. For our analysis we use $a_v(v', j)$, the conditional probability density function (p.d.f.) of inter-arrival time

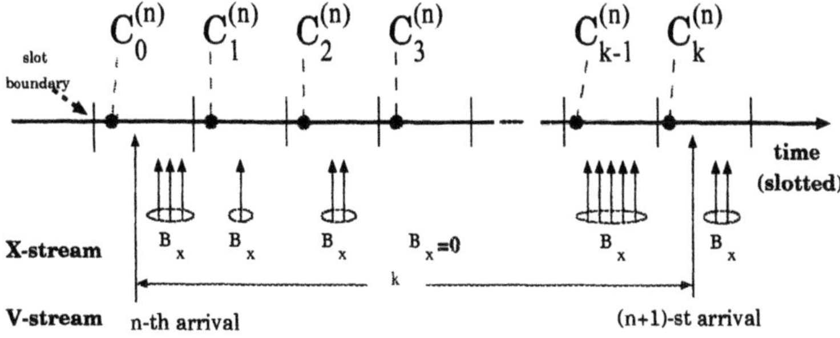

Figure 2 Relating system states observed by two consecutive V-stream arrivals.

of V-stream arrivals, which is defined as

$$a_v(v', j) = P[\textit{next Vstream arrival occurs in state } v' \textit{ and after } j \textit{ slots} \mid \textit{a Vstream arrival occurred in state } v].$$

For the case where an arrival takes more than a slot ($N_A > 1$), a modified MMBP is considered (Balay and Nilsson,1996). The X-stream traffic is modeled as an MMBP with batch arrivals (B-MMBP), described by $\mathbf{P_X}$, $\mathbf{\Lambda_X}$ and a general distribution for the batch size. For our analysis we use:

$$b_x(j) = P[\textit{Xstream cells arriving in a slot} = j \mid \textit{MMBP state} = x],\ j \geq 0.$$

3 EXACT ANALYSIS

The basis of our queueing analyses is the identification of "embedded points" (observation instants) of the system such that it is "regenerative" (Wolff,1989) at these instants. We follow the evolution path along the embedded points to obtain steady-state information. A high level description of the 'evolution method' is:

1. Identify embedded points and define a regenerative system state at these points.
2. *Evolution step*: obtain the state distribution at the next embedded point using the current state information and the arrival characteristics.
3. Repeat the above evolution step until the state distribution at two consecutive embedded points is the same, i.e., evolve the system until it reaches steady-state.

We consider the V-stream arrival slots as the embedded points. For a regenerative system state, the MMBP state at a V-stream arrival instant should also be a part of the state descriptor, since MMBP is not a *renewal process* but is a *Markov renewal process* (Cinlar,1975); the time interval between consecutive arrivals depends on the MMBP states at these arrival instants. We define $C_k^{(n)}$ as the random variable (r.v.) for the system state at

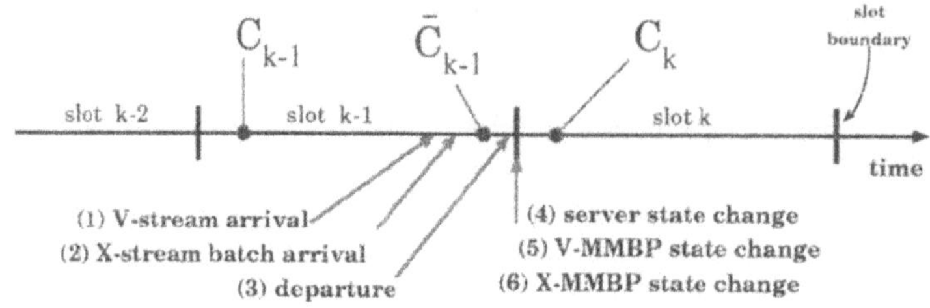

Figure 3 Relative order of events and observations in a slot (policy $\mathcal{P}_V$)

the beginning of the kth slot following the nth V-stream arrival (i.e. in the nth evolution step) given that the $(n+1)^{st}$ V-stream arrival does not occur in the preceding $k-1$ slots. $C_k^{(n)}$ is a 4-tuple, $C_k^{(n)} \equiv (V_k^{(n)}, X_k^{(n)}, S_k^{(n)}, J_k^{(n)})$ (p.d.f. given by $c_k^{(n)}(\cdot)$), $k \geq 0, n \geq 1$, where

$V_k^{(n)}$: Arrival state of V-stream MMBP (V-MMBP) in the nth V-stream arrival slot. Note that $V_k^{(n)}$ does not change with k, i.e. $V_k^{(n)} = V_0^{(n)}$, $\forall k > 0$. It's state space is denoted by $\mathcal{V}$.

$X_k^{(n)}$: State of X-stream MMBP (X-MMBP) in a slot. Its state space is denoted by $\mathcal{X}$.

$S_k^{(n)}$: State of the server. $S_k = i$ implies that the current slot is the ith service slot of cell in service; $S_k = 0$ when server is idle. A departure occurs at the end of slot k if $S_k = N_S$.

$J_k^{(n)}$: Number of cells in system (i.e. queue and server), $(0 \leq J_k \leq K+1)$.

From a known system state at an embedded point, to obtain the system state at the next embedded point we use the statistical distribution of the V-stream arrival process and the system states at each of these possible V-stream arrival slots (see figure 2), for which slot-to-slot system dynamics are required. The events which influence the change in system state between two consecutive slots, say *k-1* and *k*, are shown in figure 3. For ease of derivation we consider them in two phases: **phase(1)**: arrivals which occur in slot *k-1*, and **phase(2)**: state transitions at the slot boundary and the departure event. $\bar{C}_{k-1}^{(n)} = (\bar{V}_{k-1}^{(n)}, \bar{X}_{k-1}^{(n)}, \bar{S}_{k-1}^{(n)}, \bar{J}_{k-1}^{(n)})$ (p.d.f. represented as $\bar{c}_{k-1}^{(n)}(\cdot)$) is the system state after phase (1), where
$\bar{V}_{k-1}^{(n)} = V_{k-1}^{(n)}, \quad \bar{X}_{k-1}^{(n)} = X_{k-1}^{(n)}, \quad \bar{S}_{k-1}^{(n)} = S_{k-1}^{(n)}$, and

$$\bar{J}_{k-1}^{(n)} = min(J_{k-1}^{(n)} + B_x + \bar{u}(k-1), K + u(S_{k-1}^{(n)})) \tag{1}$$

for $k \geq 1$, $n > 0$. Here $\bar{u}(n) \stackrel{\text{def}}{=} 1 - u(n)$ and

$$u(n) = \begin{cases} 0 & if\ n = 0 \\ 1 & if\ n \neq 0. \end{cases} \tag{2}$$

Equation (1) accounts for the fact that a V-stream arrival needs to be considered only in slot *0*, and the maximum cells that can be accommodated in the system is K if the server is idle or else it is $K+1$. The relations to derive $C_k^{(n)}$ from $\bar{C}_{k-1}^{(n)}$ are

$$\begin{aligned} V_k^{(n)} &= \bar{V}_{k-1}^{(n)}, \;\; J_k^{(n)} = \bar{J}_{k-1}^{(n)} - u(N - S_{k-1}^{(n)}), \\ X_k^{(n)} &= x \; with \; probability \; [\mathbf{P_X}]_{i,x} \; where \; \bar{X}_{k-1}^{n} = i, \; x \in \mathcal{X} \end{aligned} \tag{3}$$

$$S_k^{(n)} = \begin{cases} \bar{S}_{k-1}^{n} + 1 & if \;\; 0 < \bar{S}_{k-1}^{n} < N \\ 1 & if \;\; \bar{S}_{k-1}^{n} = 0, \; \bar{J}_{k-1}^{(n)} > 1 \;\; or \;\; \bar{S}_{k-1}^{n} = N, \; \bar{J}_{k-1}^{(n)} > 0 \\ 0 & if \;\; \bar{S}_{k-1}^{n} = 0, \; \bar{J}_{k-1}^{(n)} = 1 \;\; or \;\; \bar{S}_{k-1}^{n} = N, \; \bar{J}_{k-1}^{(n)} = 0 \end{cases} \tag{4}$$

for $k \geq 1$ and $n \geq 1$. Let $\tilde{m}_k \stackrel{\text{def}}{=} K + u(S_k) - \bar{u}(k)$. In terms of p.d.f's the relations are

$$\bar{c}_k^{(n)}(v,x,s,j) = \begin{cases} \sum_{i=0}^{j-\bar{u}(k)} c_k^{(n)}(v,x,s,i) \, b_x(j-i-\bar{u}(k)) & if \;\; j < \tilde{m}_k \\ \sum_{l=\tilde{m}_k}^{\infty} \sum_{i=0}^{j-\bar{u}(k)} c_k^{(n)}(v,x,s,i) \, b_x(l-i-\bar{u}(k)) & if \;\; j = \tilde{m}_k \end{cases} \tag{5}$$

$$c_k(v,x,s,j) = \begin{cases} \sum_{x'} [\mathbf{P_X}]_{x',x} (\bar{c}_{k-1}(v,x',0,0) + \bar{c}_{k-1}(v,x',N_S,1)) \\ \qquad\qquad if \;\; s=0, \;\; j=0 \\ \sum_{x'} [\mathbf{P_X}]_{x',x} (\bar{c}_{k-1}(v,x',0,j) + \bar{c}_{k-1}(v,x',N_S,j+1)) \\ \qquad\qquad if \;\; s=1, \;\; 1 \leq j \leq K \\ \sum_{x'} [\mathbf{P_X}]_{x',x} \bar{c}_{k-1}(v,x',s-1,j+1) \\ \qquad\qquad if \;\; 2 \leq s \leq N_S, \;\; 1 \leq j \leq K+1 \\ 0 \qquad\qquad otherwise \end{cases} \tag{6}$$

where $k \geq 1$, $n \geq 1$, $x, x' \in \mathcal{X}$, $v \in \mathcal{V}$. The evolution step to obtain the p.d.f. of system state at the next embedded point from the p.d.f. of system state at the nth embedded point is given as

$$c_0^{(n+1)}(v',x,s,j) = \sum_k \sum_v a_v(v',k) \, c_k^{(n)}(v,x,s,j), \tag{7}$$

where $k \geq 1$, $v, v' \in \mathcal{V}$, $x \in \mathcal{X}$, $0 \leq s \leq N_S$ and $0 \leq j \leq K+1$. If steady state exists then the steady state p.d.f. of the system state at an embedded point is given as

$$c_0(v,x,s,j) = \lim_{n \to \infty} c_0^{(n)}(v,x,s,j). \tag{8}$$

$c_0(\cdot)$ denotes the steady-state p.d.f. of system state at the beginning of a V-stream arrival slot. We use it obtain the V-stream performance measure in the next section; X-stream performance measures can also be obtained (Balay and Nilsson,1994).

3.1 Per-session performance (V-stream)

Let $C^v = (V^v, X^v, S^v, J^v)$ be the r.v. for the system state seen by a V-stream arrival at steady state and let $c^v(\cdot)$ be its p.d.f.
Case 1: $\mathcal{P}_V$ arrival policy. V-stream arrival in a slot is considered first for buffer allocation. Hence, we have $C^v \equiv C_0$.
Case 2: $\mathcal{P}_X$ arrival policy. The system state seen by a V-stream arrival is after the X-stream cells of the same slot. We obtain its p.d.f. as follows:

$$V^v = V_0, \quad X^v = X_0, \quad S^v = S_0, \quad J^v = min(J_0 + B_x, K + u(S_0)). \tag{9}$$

$$c^v(v,x,s,j) = \begin{cases} \sum_{i=0}^{j} c(v,x,s,i)\, b_x(j-i) & if \quad j < K + u(s) \\ \sum_{l=K+u(s)}^{\infty} \sum_{i=0}^{j} c(v,x,s,i)\, b_x(j-i) & if \quad j = K + u(s). \end{cases} \tag{10}$$

Some of the interesting performance measures of V-stream are found as follows:

$$P[buffer\ occupancy = j] = \sum_{v,x} \left[c^v(v,x,0,j) + \sum_{s=1}^{n} c^v(v,x,s,j+1) \right], \quad 0 \le j \le K. \tag{11}$$

$$P[cell\ loss] = \sum_{v,x} \left[c^v(v,x,0,K) + \sum_{s=1}^{N_S} c^v(v,x,s,K+1) \right] \tag{12}$$

$$P[delay = j\ slots] = \begin{cases} \sum_{v,x} c^v(v,x,0,l-1) + \sum_{v,x} c^v(v,x,N_S,l) & if\ j \bmod N_S = 0 \\ \sum_{v,x} c^v(v,x,s,l) \quad if\ j \bmod N_S \neq 0\ and\ N_S \le j < (K+1)N_S \\ 0 \qquad otherwise \end{cases} \tag{13}$$

where $s = (N_S - j) \bmod (N_S)$, $l = \lfloor j/N_S \rfloor$ and delay is the time spent in the system.

4 DISCUSSION ON TIME COMPLEXITY

Let N denote the size of the system state; the main component is the buffer size. Computational time to obtaining slot-to-slot system dynamics is proportional to the system size N and the maximum arrivals which can occur in a slot; we consider the maximum arrivals in a slot to be a constant since it is limited by the number of input links at the multiplexer. Each evolution step requires M_I slot-to-slot operations where M_I denotes the maximum inter-arrival time for the session of interest; it is large for bursty processes. The third factor of computational complexity is the number of evolutions required for convergence which we denoted by N_C. The computational complexity of the arrival-wise method is $O(NM_IN_C)$ to obtain exact steady-state results. Using the approximation technique described in the next section the latter two factors of complexity are reduced to constants and the complexity of the method is $O(N)$.

For a matrix-geometric solution for the considered system, the transition probability matrix $\mathbf{P}$ of size N^2 is required where each element (i,j) denotes the transition probability of the system from state i to state j at consecutive embedded points. The computational time to obtain each element is the same as that of an evolution step $(O(NM_I))$. The complexity of generating the transition probability matrix $(O(N^3M_I))$ prohibits using a matrix-geometric technique for per-session analysis of a multiplexer while an evolution method is much more efficient.

5 APPROXIMATION TECHNIQUE

In this section, we describe an approximate characterization for the change between system states at slots which are N_S slots apart, as a slow changing function. This approximation is applied to partially reduce the evolution step (equation (7)) to a closed-form relation. Similarly, we can approximate the steady-state relation (equation (8)) by considering this approximation for the change between system states at consecutive embedded points.

Let $c_k(q)$ $(= P[C_k = q])$,where $q = (q_v, q_x, q_s, q_j)$ be the p.d.f. of system state at the beginning of slot k. Let $\delta_k(q)$ represent the difference in p.d.f. of system states N_S slots apart, and be defined as $\delta_k(q) = c_{k+N_S}(q) - c_k(q)$, $\forall q, k$. For some finite value $T > 0$, the system behavior can be approximately characterized by the following relation

$$\delta_{k+N_S}(q) \;=\; \alpha\, \delta_k(q) \qquad \forall k \geq T \tag{14}$$

where α is a non-negative constant and satisfies $0 < \alpha < 1$. For a valid approximation we choose T such that $0 \leq c_T(q) + \frac{\delta_T(q)}{(1-\alpha)} \leq 1$, $\forall q$. We consider the system states in the tail (i.e.$\geq T$) as N_S geometric streams i.e., $\{c_{T+N_S i}(q)\}$, $\{c_{T+1+N_S i}(q)\}$, $\cdots$, $\{c_{T+N_S-1+N_S i}(q)\}$. Each stream (j) is treated separately with respect to the approximation and hence we have a constant α for each stream which is denoted as α_j where $0 \leq j < N_S$. Using this approximation we can now represent the system state at the beginning of any slot $(\geq T)$ as

$$c_{k+j+N_S i}(q) \;=\; c_{k+j}(q) \;+\; \frac{(1-\alpha_j^i)}{(1-\alpha_j)}\,\delta_{k+j}(q), \quad \forall q,\ \forall k \geq T,\ i \geq 0,\ 0 \leq j < N_S \tag{15}$$

For any m-state MMBP with one state as completely bursty (i.e. prob. of arrival in this state is 1), the following is true of its p.d.f.

$$a_v(v', k+1) \;=\; \beta\, a_v(v', k) \qquad \forall k \geq m$$

where β is a constant $(0 < \beta < 1)$. Using this property of an MMBP and the above stated approximation we can reduce and rewrite the evolution step (equation (7)) as

Traffic type	V-stream (offered load (λN_S)=0.2)				X-stream (offered load $(\lambda N_S \bar{B})$=0.4)					
	bursty	CV^2	correlated	ψ_1	bursty	CV^2	correlated	ψ_1	Batch	$\bar{B}$
A	√	16	√	0.1	√ (low)	1	no	0	Geo	3
B	√	16	no	0	no	1	no	0	constant	1
C	no	1	no	0	no	1	no	0	constant	1

Table 2 Traffic parameters for V-stream and X-stream

$$c_0^{(n+1)}(q') = \sum_{k=1}^{T-1} \sum_{q,v'} a_{q_v}(v',k)\, c_k^{(n)}(q) + \sum_{q,v'} \sum_{j=0}^{N_S-1} \frac{a_{q_v}(v',T+j)}{(1-\beta_S^N)} \left(c_{T+j}^{(n)}(q) + \frac{\beta_S^N\, \delta_{T+j}^{(n)}(q)}{(1-\alpha_j^{(n)}\beta_S^N)} \right) \quad (16)$$

For each evolution step we compute $\delta_{T+j}^{(n)}(q)$ and $\alpha_j^{(n)}$ from $c_{T+j}^{(n)}(q)$, $c_{T+j+N_S}^{(n)}(q)$ and $c_{T+j+2N_S}^{(n)}(q)$ (these are p.d.f.s of system states obtained using exact relations) as $\delta_{T+j}^{(n)}(q) = c_{T+j+N_S}^{(n)}(q) - c_{T+j}^{(n)}(q)$ and $\alpha_j^{(n)}(q) = \frac{c_{T+j+2N_S}^{(n)}(q) - c_{T+j+N_S}^{(n)}(q)}{\delta_{T+j}^{(n)}(q)}$. We consider α_j to be a weighted average which is calculated as $\alpha_j^{(n)} = \sum_q c_{T+j}^{(n)}(q)\, \alpha_j^{(n)}(q)$.

6 NUMERICAL RESULTS

In this section we study the effect of link speeds, traffic characteristics and buffer sizing on the performance of a session. We also validate the accuracy of the approximation technique by comparing results with those obtained from simulation. We consider a scenario where a session's offered load is 20% of an output link capacity while it is 40% for the cross traffic. We compare the session's performance for three cases of link speeds:

1. $N_A = 1, N_S = 1$: both input and output links of V-stream are of same speed.
2. $N_A = 4, N_S = 1$: input link of V-stream is four times slower than the output link.
3. $N_A = 1, N_S = 4$: input link of V-stream is four times faster than the output link.

We use the traffic descriptors $(\lambda N_S, CV^2, \psi_1)$ for V-stream and $(\lambda N_S \bar{B}, CV^2, \psi_1, \bar{B})$ for X-stream where, λ is the mean arrival rate of an MMBP, CV^2 is the burstiness defined as the squared coefficient of variation, ψ_1 is the autocorrelation of lag 1 slot, of the arrival process and $\bar{B}$ is the mean batch size. Given the traffic descriptors we obtain 2-state MMBPs as described by Balay and Nilsson (1994). The traffic descriptors used for numerical results are tabulated in table 2.

6.1 Effect of different speed links

Notice that we use 'offered load' as one of the traffic descriptors; it helps us compare the three systems. In the figures 4(a), 5(a), 5(b), 6(a) and 6(b), we plot some of the performance measures (buffer occupancy distribution, mean delay and loss probability) of a session. We see that the performance measures are distinctly different for the three systems considered. These results clearly indicate the importance of considering the link speeds for modeling. One way of interpreting these results is as follows: the results for the case of the homogeneous system ($N_A = 1, N_S = 1$) are the same as the results of a heterogeneous system ($N_A \neq N_S$), if modeled ignoring the different speed links. We notice that ignoring the speed of links is a gross approximation and results in very erroneous performance predictions.

6.2 Validating the accuracy of results

To validate the accuracy of the analytical results obtained using approximation, we compare them with simulation results. Figure 4(a) shows the p.d.f. of buffer occupancy seen by V-stream and X-stream, obtained using exact analysis for systems with type A traffic, policy $\mathcal{P}_X$ and buffer size of $K = 64$ cells. For the case of ($N_A = 1, N_S = 4$) system we also plot the results obtained using approximation and simulation. Notice that there is no visible difference between the simulation and analytical results; for a better view of the comparison, we plot the absolute error (= *analytical - simulation*) along with the 95th % confidence regions of simulation results in figure 4(b). For this example, in obtaining exact results the maximum interarrival time for V-stream cells was 2060 slots with an accuracy of 10^{-6}, i.e., if $a(j)$ is the interarrival time p.d.f., $\sum_{j=1}^{2060} a(j) > 1 - 10^{-6}$. We see that, approximation with $T > 150$ slots yields near exact results; a reduction of computational time by a factor of 13. For the case where $N_S = 1$ we obtain results within confidence intervals with $T = 25$ and a gain by a factor of 23. These results shows that by using the approximation technique we can obtain results efficiently while maintaining accuracy. Some observations are that it is computationally much better than simulation, and is more efficient when $N_S = 1$ or when the system size is large; further work is required to engineer the minimum value of T for a given system.

6.3 Effect of burstiness and correlation

First, we illustrate an important phenomenon which occurs in systems with a slower output link and bursty traffic. Notice in figure 4(a), the curve for ($N_A = 1, N_S = 4$) system exhibit an oscillatory behavior for smaller buffer occupancy values; the probability values for occupancies which are multiples of ($N_S - 1$) are higher than its neighboring values. It can be explained using an example scenario: Consider a busy period of the system where the first n ($n > N_S$)arrivals are from V-stream and all of which occur in consecutive slots. The first cell arrives in slot 1, begins service in slot 2, and departs at the end of slot 5 (since $N_S = 4$). The arrivals which occur in slot 5 and 6 respectively see the buffer occupancy to be 4. Also, for each departure i which occurs at the end of slot $iN_S + 1$ in this interval of n arrivals, the occupancy seen by an arrival in slots $iN_S + 1$ and $iN_S + 2$ is $i(N_S - 1)$ cells; hence a higher probability value for $i(N_S - 1)$. Such scenarios are very likely to occur when both streams are bursty. This behavior will also be reflected in the delay distribution and departure process; it introduces periodic (or negative) correlation between departures which makes the characterization of the departure process difficult. In figures 5(a) and 5(b) we plot the loss probability of V-stream for change in V-stream's burstiness and correlation values respectively. We observe that a source which is initially

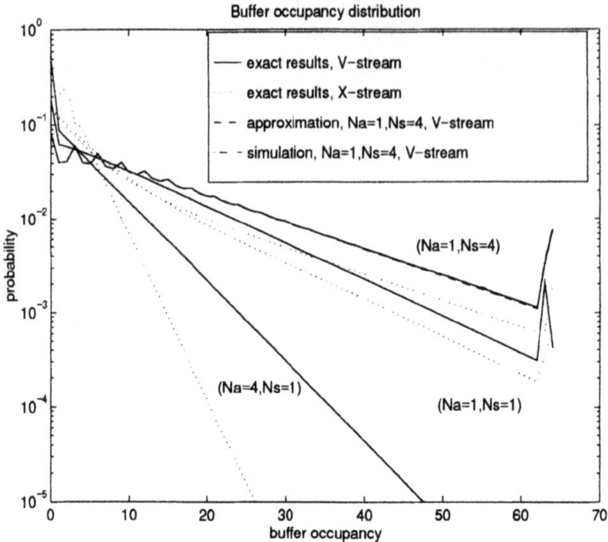

(a) Buffer occupancy distribution

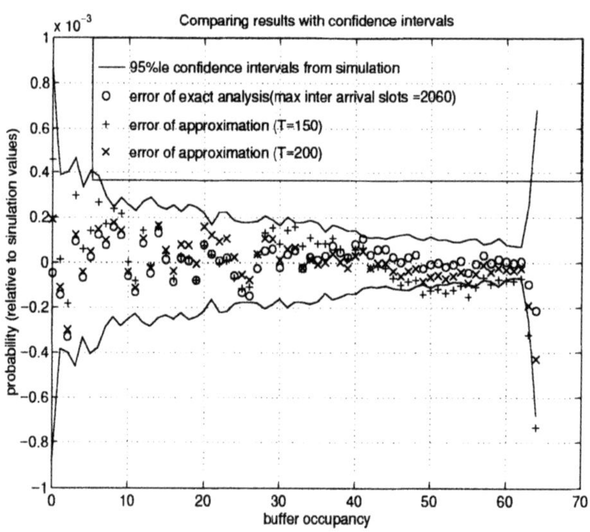

(b) Comparing results with 95 % confidence intervals of simulation: ($N_A = 1, N_S = 4$) system

Figure 4 Buffer occupancy seen by V-stream and X-stream in systems with K=64, Type A traffic and policy $\mathcal{P}_X$.

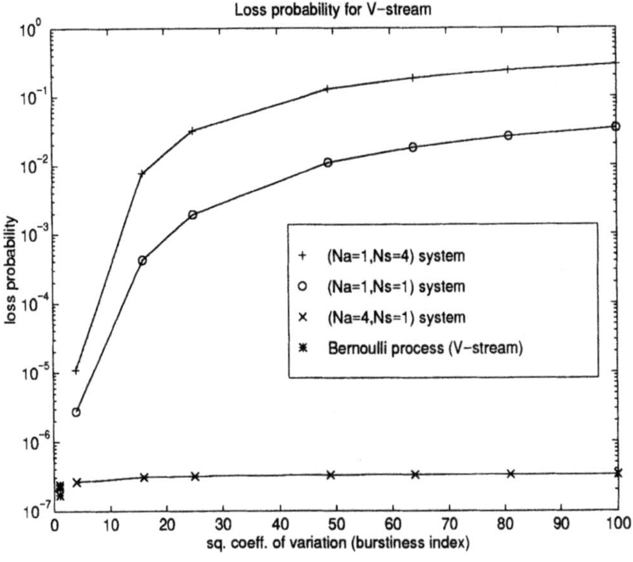

(a) Effect of change in burstiness

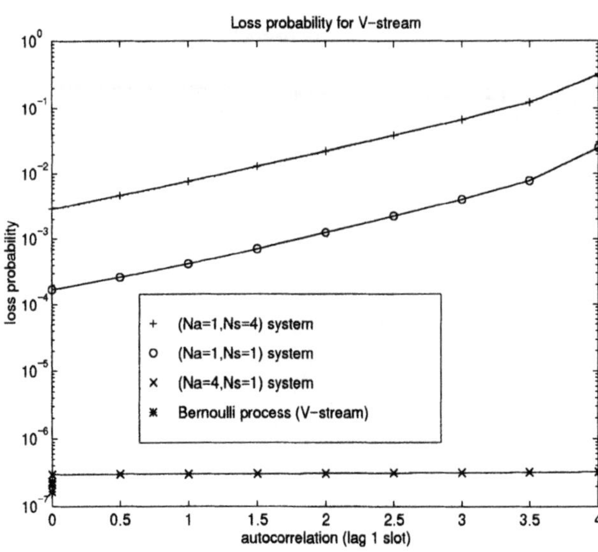

(b) Effect of change in autocorrelation

Figure 5 Effect of change in burstiness and autocorrelation of V-stream on its loss probability: systems with K=64, policy $\mathcal{P}_X$ and type A traffic.

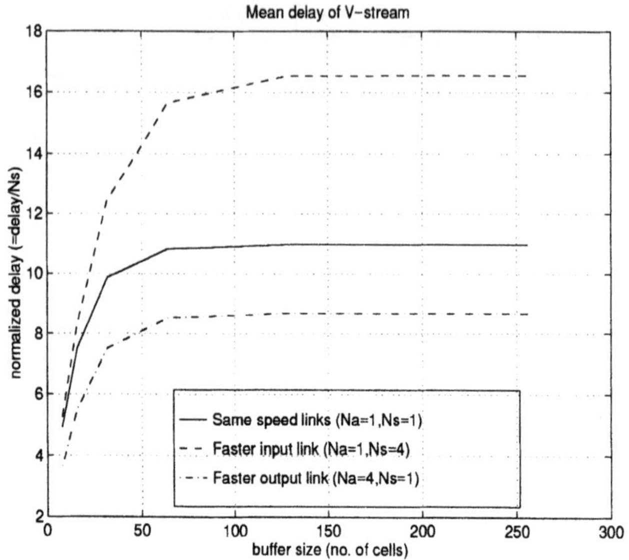

(a) Effect of change in buffer size on mean delay of V-stream (type A traffic)

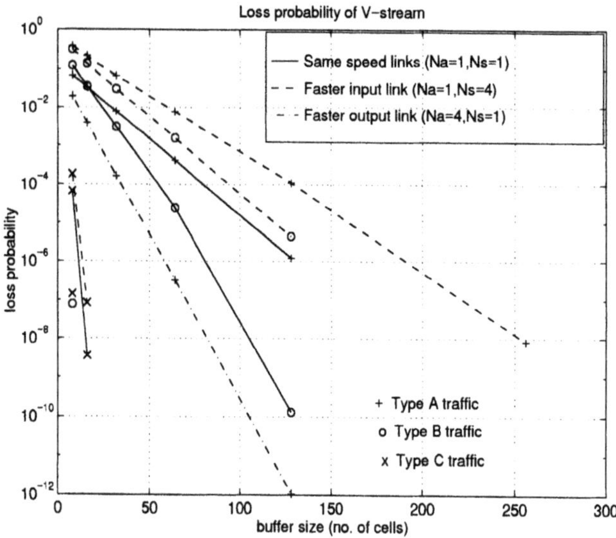

(b) Effect of change in buffer size on loss probability of V-stream

Figure 6 Influence of change in buffer size on mean delay and loss probability of V-stream: systems with policy $\mathcal{P}_X$ and all types of traffic.

less bursty is more sensitive to change in burstiness. The converse seems true for the effect of autocorrelation, i.e., correlated streams are more sensitive than non-correlated streams, for change in autocorrelation. We observe that the performance variation due to change in traffic characteristics is less in a ($N_A = 4, N_A = 1$) system compared to the other systems.

6.4 Influence of slot policy ($\mathcal{P}$) on performance

Consider the curves of figures 5(a) and 5(b) for systems where $N_S = 1$. The loss probabilities plotted are for systems with policy $\mathcal{P}_X$. V-stream cells are not lost when the policy is $\mathcal{P}_V$, since in each slot there is always space for one cell in the buffer which is made available at the beginning of a slot when a cell starts its service. This means the sessions loss probability lies between '0' and the values plotted in the figures, depending on the location of the arrival port. The port of arrival at a switching element is an important factor which influences the performance of a session if $N_S = 1$ when a simple round-robin policy is used for switching the incoming cells. Such variations in performance can be reduced by randomizing the polling policy or randomly distributing the incoming cells to the input ports for switching.

6.5 Influence of buffer size

An important aspect of designing a switch is buffer dimensioning since loss guarantees may be of the order $< 10^{-10}$. Obtaining accurate results of such small magnitude through simulation techniques is clearly a difficult (time consuming) task; one reason why discrete-time analysis is important in studying ATM systems. To illustrate the buffer requirements to provide delay and loss guarantees for systems with different speed links and different traffic characteristic, we plot the mean delay and loss probability for the three systems for change in buffer size in figures 6(a) and 6(b) The values plotted in figure 6(b) are for the three traffic types, buffer size of 2^i, $3 \leq i \leq 8$, and when the loss probability is more than 10^{-12}. We observe that the loss probabilities change almost linearly (on a logarithmic plot) with buffer size while the buffer requirements are very different for the considered system.

7 CONCLUSIONS

In this paper we have shown the importance of considering the speed of links traversed by a session, for studying its performance at an ATM multiplexer. The numerical results presented indicate that a sessions performance is distinctly different when it traverses different speed links; they also show that ignoring the link speeds is a gross approximation and will result in incorrect results.

For studying a session's performance in the presence of different speed links, we have described a queueing model of an ATM multiplexer. For our analysis we considered finite capacity buffer and modeled the session of interest and the cross-traffic as Markov renewal traffic streams. We used an evolution method to obtain the steady-state performance measures, and presented an approximation technique; together they make our analysis much more efficient than a matrix-geometric method. Using numerical results we have also illustrated the effects of traffic characteristics, port of arrival of a session and buffer size on the steady-state performance measures of a session. The presented analysis also contributes in studying the transient behavior of a session (Balay and Nilsson, 1996). Future work includes characterization of per-session departure process and end-to-end analysis.

8 REFERENCES

Balay, R.I. and Nilsson, A.A. (1994) Analyses of heterogeneous tandem links, part 1: Per-session performance of an ATM multiplexer. Technical Report TR-94/20, Center for Advanced Computing and Communication, NCSU Raleigh, NC.

Balay, R.I. and Nilsson, A.A. (1996) Transient performance of a session at an ATM multiplexer with heterogeneous speed links. *INFOCOM'96*, Submitted for publication.

Blondia, C. and Casals, O. (1992) Performance analysis of statistical multiplexing of VBR sources. *Proceedings, INFOCOM'92*, 828-838.

Cinlar, E. (1975) *Introduction to stochastic processes.* Prentice Hall Inc., NJ.

Fischer, W. and Meier-Hellstern, K. (1993) The mmpp cookbook. *Performance Evaluation*, 18(2):149-171.

Herrmann, C. (1993) Correlation effect on per-stream QoS parameters of ATM traffic superpositions relevant to connection admission control. *Proceedings, ICC*, 1027-1031.

Murata, M. and Oie, Y. and Suda, T. and Miyahara, H. (1990) Analysis of discrete-time single server queue with bursty inputs for traffic control in ATM networks. *IEEE Journal on Selected Areas in Communications*, 8(3):447-458.

Neuts, M. (1981) *Matrix-geometric solution in stochastic models: An algorithmic approach.* John Hopkins University Press.

Wolff, R.W. (1989) *Stochastic Modeling and the Theory of Queues.* Prentice Hall Inc., NJ.

9 BIOGRAPHY

Rajesh I. Balay received an M.Sc.(Tech) degree in Computer Science from the Birla Institute of Technology and Science (BITS Pilani), India, in 1989, and an M.S. degree from the University of Alabama at Birmingham, USA, in 1991. He is currently a doctoral candidate in Computer Science at North Carolina State University, Raleigh, USA. Since 1991 he has been with the Center for Advanced Computing and Communication, at the university, engaged in research on performance evaluation of high speed networks. His research interests include design of efficient network protocols, performance modeling of networks and communication protocols, and computer games of strategy.

Arne A. Nilsson received an M.S. degree in Electrical Engineering in 1968 and a Ph.D degree in Telecommunication Systems in 1976, from Lund University of Technology, Sweden. He is currently a Professor of Electrical and Computer Engineering, and the Technical Director of the Center for Advanced Computing and Communication, at the North Carolina State University, Raleigh, USA. His areas of research include performance modeling of local area networks and computer systems, routing and flow control in computer networks, computer communication synthesis and analysis, medical image networking and packet radio networks.

3

Effect of the on-period distribution on the performance of an ATM multiplexer fed by on/off sources : an analytical study

S. Wittevrongel and H. Bruneel
SMACS Research Group, Laboratory for Communications Engineering, University of Ghent
Sint-Pietersnieuwstraat 41, B-9000 Gent, Belgium

Abstract

In this paper, we consider a statistical multiplexer in an ATM network, which is fed by a finite number of independent bursty on/off traffic sources. Most of the previous work assumes a two-state on/off source model, where both the on-periods and the off-periods are geometrically distributed. We assume a *general* distribution for the on-period lengths, and we study the effect of the on-period distribution on the multiplexer performance. Both the homogeneous and the heterogeneous traffic case are considered. An analytical method is presented to analyze the system, which basically is a generating-functions approach and uses an *infinite-dimensional* state description. Exact closed-form expressions are obtained for the mean and the tail distribution of the system contents. The numerical evaluation of the derived formulas is simple and not CPU time and/or memory space consuming, whatever the on-period distribution is. Numerical results indicate that, for a given fixed on-period length, the multiplexer performance strongly depends on the actual distribution of the on-period.

Keywords

ATM, multiplexer performance, on/off sources, general on-period distribution

1 INTRODUCTION

The ATM (Asynchronous Transfer Mode) is regarded as the most promising transfer technique for various types of information in future broadband integrated services digital networks. An ATM network is expected to support various services with widely different traffic characteristics, such as voice, data, video. Hence, for the design of these networks it is essential to assess the impact of the traffic characteristics on the performance.

In this paper, we consider a statistical multiplexer in an ATM network, which is modeled as a discrete-time single-server queueing system with infinite storage capacity. In ATM networks, time is divided into fixed-length slots and the transmission time of a cell is one slot. The multiplexer supports a finite number N of independent bursty traffic sources. These sources belong to T traffic types and for traffic type t, $1 \leq t \leq T$, there are N_t sources. Each source is modeled as an on/off source, that is, each source alternates between on-periods (active) and off-periods (passive). During an on-period, a source generates one cell per slot. No cells are generated during an off-period. We assume that for a source of type t, the length of the off-period is geometrically distributed with mean value $1/(1-\beta_t)$. Furthermore, the lengths of the on-periods are assumed to be i.i.d. random variables with probability generating function (pgf) $A_t(z)$ and probability mass function (pmf) $a_t(i)$. Finally, it is assumed that the lengths of the on-periods and the off-periods are independently distributed.

The traffic model considered here allows us to study the impact of the on-period distribution on the multiplexer performance. Similar discrete-time models have been investigated in (Bruneel, 1988), (Steyaert, 1995) and (Xiong, 1992). In these papers homogeneous on/off sources (T=1) were considered, and the on-periods were purely geometrically distributed, were distributed according to a mixture of 2 geometric distributions, or consisted of a geometrically distributed multiple of fixed-length intervals, respectively. The present paper can hence be viewed as an extension of these studies, in the sense that the distribution of the on-periods is *general* here. The study in this paper is also related to (Sohraby, 1993). However, the analysis presented there only leads to approximate results for the tail distribution of the system contents, whereas our analysis is *exact* and leads to *closed-form* expressions for both the mean and the tail distribution (both coefficient and decay rate), whose numerical evaluation is not limited by the traffic characteristics. A general on-period distribution is also considered in (Elsayed, 1994). The system is analyzed there by numerically solving a set of balance equations, and hence the analysis is limited by the huge state space and the computational complexity of the algorithms. A heuristic approximation for the distribution of the system contents is derived in (Simonian, 1994).

The remainder of the paper is organized as follows. First, the homogeneous traffic case is considered. In Section 2, the analytical model of the multiplexer under study is described, and a functional equation is established which characterizes the behavior of the system under study. Section 3 concentrates on the steady-state cell arrival process. The mean and the tail distribution of the system contents are derived in Sections 4 and 5 respectively. In Section 6, the heterogeneous traffic case is considered. Some numerical examples are given in Section 7.

2 SYSTEM EQUATIONS AND FUNCTIONAL EQUATION

First, we will present the analysis for the case of a single traffic type (T=1), and for the clearness of the explanation, we omit the t-dependence in the above described source model.

As mentioned before, we assume that each source will alternately be passive (state 0), or active. An active source is in state n, $n \geq 1$, if it is in the nth slot of an on-period. Hence, each source can be characterized by an infinite-dimensional Markov chain with states n, $n \geq 0$, and transition probabilities as shown in Figure 1, where $p(n\text{-}1)$ is the probability of having an on-period of at least n slots, given that the on-period consists of at least n-1 slots, i.e.,

$$p(n-1)=\left(1-\sum_{i=1}^{n-1}a(i)\right)\left(1-\sum_{i=1}^{n-2}a(i)\right)^{-1} . \tag{1}$$

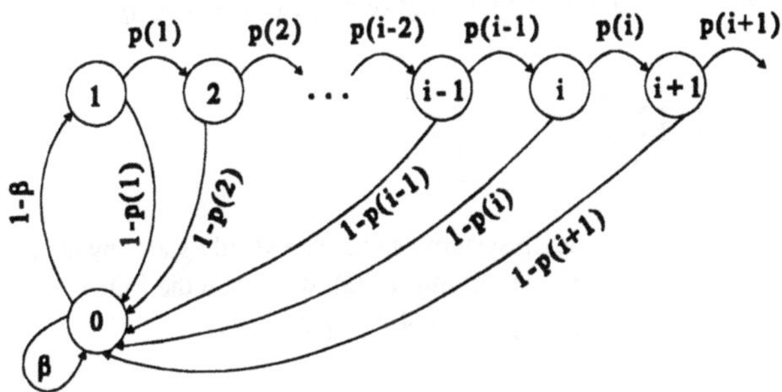

Figure 1 State transition diagram of an inlet.

Let us now define the random variables $d_{n,k}$ $(n \geq 1)$ as the number of sources in the nth slot of an on-period during slot k. Then, in view of Figure 1, the following relationships hold :

$$d_{1,k}=\sum_{i=1}^{N-\sum_{n=1}^{\infty}d_{n,k-1}} b_i \quad ; \qquad d_{n,k}=\sum_{i=1}^{d_{n-1,k-1}} c_{n-1,i} \ , \quad n>1 \ . \tag{2}$$

Here the b_i's are i.i.d. random variables with pgf

$$B(z) \triangleq E\left[z^{b_i}\right]=\beta+(1-\beta)z \ . \tag{3}$$

For given n, the $c_{n\text{-}1,i}$'s are i.i.d. random variables with pgf

$$C_{n-1}(z) \triangleq E\left[z^{c_{n-1,i}}\right]=1-p(n-1)+p(n-1)z \ , \quad n>1 \ . \tag{4}$$

Moreover, the b_i's and the $c_{n\text{-}1,i}$'s are mutually independent. Also, let e_k be the total number of cell arrivals during slot k. This random variable can be expressed as

$$e_k=\sum_{n=1}^{\infty}d_{n,k} \ . \tag{5}$$

Next, let s_k represent the system contents (i.e., the number of cells stored in the multiplexer buffer, including the potential cell in transmission) at the beginning of slot k, i.e., just after slot (k-1). Then the evolution of the system contents is described by the following system equation,

$$s_{k+1} = (s_k - 1)^+ + e_k \ , \tag{6}$$

where $(.)^+$ denotes max(., 0). Equations (1)-(6) imply that the set $\{(d_{n,k-1}\ (n \geq 1),\ s_k)\}$ is a Markov chain. If we now define the joint pgf of $d_{n,k-1}$ $(n \geq 1)$ and s_k as

$$P_k(x_1, x_2, \ldots, z) = E\left[\left(\prod_{n=1}^{\infty} x_n^{d_{n,k-1}}\right) z^{s_k}\right] , \tag{7}$$

and if we assume that the queueing system can reach a steady state, by using equations (2)-(6), in a similar way as described e.g. in (Xiong, 1992), we obtain the following functional equation for the steady-state version $P(x_1, x_2, \ldots, z)$ of $P_k(x_1, x_2, \ldots, z)$:

$$P(x_1, x_2, \ldots, z) = \frac{[B(x_1 z)]^N}{z} \left\{ P\left(\frac{C_1(x_2 z)}{B(x_1 z)}, \frac{C_2(x_3 z)}{B(x_1 z)}, \ldots, z\right) + (z-1)p_0 \right\} , \tag{8}$$

where the quantity p_0 indicates the probability of having an empty buffer at the beginning of an arbitrary slot in the steady state.

Next, let s be the system contents at the beginning of a slot in the steady state. Unfortunately, we are not able to derive from (8) an explicit expression for $P(x_1, x_2, \ldots, z)$ or not even for the pgf $S(z)$ of s. However, as shown in the following, it is possible to derive the moments and the tail distribution of s, if we now consider in (8) only those values of x_n $(n \geq 1)$ and z for which the arguments of the P-functions on both sides of (8) are equal to each other, i.e., $x_n = C_n(x_{n+1}z)/B(x_1 z)$, $n \geq 1$. From this equation, x_n $(n \geq 1)$ can be solved in terms of z. It turns out that for a given value of z, there may be more than one set of solutions. Here, we only choose the set of solutions which has the additional property that x_n=1, $n \geq 1$, for z=1. Denoting this set of solutions by $\chi_n(z)$, we get

$$z\,\chi_n(z) = \left\{ \sum_{i=n}^{\infty} a(i) \left(\frac{B(\chi_1(z) z)}{z}\right)^{n-1-i} \right\} \left\{ 1 - \sum_{i=1}^{n-1} a(i) \right\}^{-1} , \quad n \geq 1 \ . \tag{9}$$

Note in particular that

$$z\,\chi_1(z) = A\left(\frac{z}{\beta + (1-\beta)\chi_1(z) z}\right) . \tag{10}$$

By choosing $x_n = \chi_n(z)$, $n \geq 1$, in (10), we then obtain the function $P(\chi_1(z), \chi_2(z), \ldots, z)$ as

$$P(\chi_1(z),\chi_2(z),\ldots,z)=\frac{(z-1)p_0G(z)}{z-G(z)}\ , \tag{11}$$

where

$$G(z)=\left[B(\chi_1(z)z)\right]^N=\left[\beta+(1-\beta)\chi_1(z)z\right]^N\ . \tag{12}$$

From the normalization condition $P(\chi_1(z), \chi_2(z), \ldots, z)\,|_{z=1}$, it follows that $p_0 = 1\text{-}p$, where p is the total load into the multiplexer, i.e.,

$$p=N\,A'(1)(1-\beta)/\left[1+A'(1)(1-\beta)\right]\ . \tag{13}$$

In the next sections, we will describe a technique to calculate exactly the moments and the tail distribution of the buffer occupancy without having to calculate the whole distribution of s.

3 THE CELL ARRIVAL PROCESS

In this section, we will derive the steady-state pgf of the cell arrival process. Let d_n denote the number of sources in the nth slot of an on-period during an arbitrary slot in the steady state. The joint pgf $D(x_1, x_2, \ldots)$ of the random variables d_n ($n \geq 1$) is given by $P(x_1, x_2, \ldots, 1)$. Putting z=1 in the functional equation (8), we get

$$D(x_1,x_2,\ldots)=\left[B(x_1)\right]^N D\left(\frac{C_1(x_2)}{B(x_1)},\frac{C_2(x_3)}{B(x_1)},\ldots\right)\ . \tag{14}$$

$D(x_1, x_2, \ldots)$ is an Nth degree polynomial in x_n, $n \geq 1$, and it can be verified that the above equation is satisfied if

$$D(x_1,x_2,\ldots)=(1-\sigma)^N\left[1+(1-\beta)\sum_{n=1}^{\infty}\sum_{i=n}^{\infty}a(i)\,x_n\right]^N\ , \tag{15}$$

where σ is the load of one source, i.e., $\sigma=p/N$. The marginal pgf $D_n(z)$ of d_n is then obtained by putting x_i=1 ($i \geq 1$, $i \neq n$) and x_n=z in (15). The mean value of d_n is given by

$$E\left[d_n\right]=D_n'(1)=\frac{N(1-\beta)}{1+(1-\beta)A'(1)}\left(1-\sum_{i=1}^{n-1}a(i)\right)\ , \tag{16}$$

i.e., the mean number of sources in the first slot of an on-period during a slot times the probability of having an on-period of at least n slots. From (16), the mean number of arrivals during a slot is found to be equal to p. Consequently, the equilibrium condition is $p < 1$.

4 MEAN SYSTEM CONTENTS

In this section, we derive an expression for the mean buffer occupancy $E[s]$. First, we define the burstiness factor K of the source as

$$K = A'(1)(1-\sigma) = \frac{\sigma}{1-\beta} \ , \tag{17}$$

where $1/(1-\beta)$ is the mean off-period length and σ is the average load of a source. Note that K equals the ratio of the mean on-period length in our model, to the mean length of an on-period in case of a Bernoulli arrival process. It is clear that σ describes the ratio of the mean lengths of the on/off periods, whereas K is a measure for the absolute lengths of these periods. Also we define the variance factor L of the source as the ratio of the variance of the on-period length in our model, to the variance of a geometrically distributed on-period with the same mean length, i.e.,

$$L = \frac{A''(1) + A'(1) - [A'(1)]^2}{A'(1)[A'(1)-1]} \ . \tag{18}$$

Next, by evaluating the first derivative of equation (11) with respect to z at z=1, in a similar way as explained in (Bruneel, 1988), we get

$$E[s] = p + \frac{(N-1)p^2}{2N(1-p)}\left[K + L(K-1) + \frac{p}{N}(L-1)\right] \ . \tag{19}$$

It has been checked that the above general result is in agreement with the results obtained in (Bruneel, 1988), (Steyaert, 1995) and (Xiong, 1992). The above formula clearly demonstrates that the multiplexer performance depends not only on the mean length of the on-period, but also strongly on the actual on-period-length distribution. First of all, we observe that for a given total load, the mean length of the on-period has a substantial influence on $E[s]$. The mean system contents namely linearly increases with the burstiness factor K of the sources. Next, for a given load and a given mean length of the on-period (given K), the mean system contents $E[s]$ linearly increases with L, i.e., $E[s]$ increases linearly with the variance of the on-period. Higher-order moments of the on-period distribution have no impact on the mean system contents.

5 TAIL DISTRIBUTION OF THE SYSTEM CONTENTS

From the inversion formula for z-transforms it follows that the pmf Prob[s=n] of s can be expressed as a weighted sum of negative powers of the poles of $S(z)$. Since the modulus of all these poles is larger than one, it is obvious that for large n, Prob[s=n] is dominated by the contribution of the pole having the smallest modulus. Let us denote this dominating pole by z_0. It is shown in (Bruneel, 1994) that in order to ensure that the tail distribution is nonnegative anywhere, z_0 must necessarily be real and positive. Furthermore, we assume here that z_0 has multiplicity one. Therefore, for n sufficiently large, Prob[s=n] can be approximated as

$$\mathrm{Prob}[s=n] \cong -\frac{\theta}{z_0}\left(\frac{1}{z_0}\right)^n , \tag{20}$$

where θ is the residue of $S(z)$ in the point z=z_0.

5.1 Calculation of z_0

As in (Xiong, 1992), it can be argued that z_0 is also the pole with the smallest modulus of $P(\chi_1(z), \chi_2(z), ..., z)$. Hence, in view of (11) and (12), z_0 is a real root of $z - G(z) = 0$, or

$$z-\left[\beta+(1-\beta)\chi_1(z)z\right]^N = 0 \ . \tag{21}$$

This can even be proved. As all sources are statistically independent, $G(z)$ is the Perron-Frobenius eigenvalue related to the aggregated arrival process to the multiplexer (Neuts, 1989). Hence, the dominant pole z_0 of $S(z)$ is determined by $z - G(z) = 0$ (Sohraby, 1993). It is obvious that $\chi_1(z) > 0$ for $z > 1$. From (10) and (21) , we have

$$\frac{z^{1/N}-\beta}{1-\beta} - A\left(\frac{z}{z^{1/N}}\right) = 0 \ . \tag{22}$$

Hence, the pole z_0 can be easily calculated exactly from equation (22) by using, for instance, the Newton-Raphson algorithm.

Next, in order to assess the impact of the traffic characteristics on the geometric decay rate, we derive 2 approximations for z_0 in the heavy traffic case, where the total utilization of the multiplexer approaches to one. In case of heavy traffic, it is expected that z_0 will be close to one. By expanding the equation $z = G(z)$ around z=1, we obtain

$$z = G(1) + G'(1)(z-1) + \frac{G''(1)}{2}(z-1)^2 + \frac{G'''(1)}{6}(z-1)^3 + O(z-1)^4 \ . \tag{23}$$

By keeping terms up to $(z\text{-}1)^2$ in equation (23) and neglecting higher-order terms, we get the following approximation for z_0 :

$$z_0 \cong z_1 \triangleq 1 + \frac{2(1-p)}{p(p-1) + p\left(1-\frac{p}{N}\right)\left[K + L(K-1) + \frac{p}{N}(L-1)\right]} \ . \tag{24}$$

A more accurate heavy-traffic approximation for z_0 is obtained by keeping terms up to $(z\text{-}1)^3$ in equation (23) and neglecting higher-order (≥ 4) terms. As a result, we find

$$z_0 \cong z_2 \triangleq 1 + \frac{-3G''(1) + 3\sqrt{[G''(1)]^2 + \frac{8}{3}(1-p)G'''(1)}}{2G'''(1)} , \tag{25}$$

where

$$G''(1) = N(N-1)\sigma^2 + N\sigma(1-\sigma)[(L+1)(K+\sigma-1) - 2\sigma] ;$$

$$G'''(1) = N(N-1)(N-2)\sigma^3 + 3N(N-1)\sigma^2[(1-\sigma)(L+1)(K+\sigma-1) - 2\sigma(1-\sigma)]$$
$$+ N\sigma(1-\sigma)\left[-3\sigma(L+1)^2(K+\sigma-1)^2 + 12\sigma^2(1-\sigma) - (1-\sigma)^2 + (1-\sigma)^3\frac{M}{K}\right]$$
$$+ N\sigma(1-\sigma)(L+1)(K+\sigma-1)3\left(4\sigma^2 - 2\sigma - 1\right) ,$$

and M is the third moment of the on-period distribution, i.e., $M = A'''(1) + 3A''(1) + A'(1)$. In Table 1, we compare the exact value of z_0 with the approximations z_1 and z_2, for N=8, a negative binomial distribution for the on-period length, i.e., $A(z) = (1-\gamma)^2 z/(1-\gamma z)^2$ and various values of the load p and the burstiness factor K. The results show that both approximations are accurate for *very* high values of p, and behave in the same way as the pole z_0 as the source characteristics vary. Moreover z_2 is also quite accurate for intermediate loads. Expression (24) indicates that the variance of the on-period length has a strong influence on the heavy-traffic tail behavior of the multiplexer. Furthermore, as the total utilization approaches to one, the geometric decay rate $1/z_0$ becomes almost independent of higher-order (≥ 3) moments of the on-period distribution. For intermediate to high values of the load however, an accurate performance evaluation can be obtained by taking into account the first 3 moments of the on-period distribution. Studies based only on the first 2 moments could in this case lead to inaccurate results, and should be applied with careful consideration.

Table 1 Exact and approximate results for the dominant pole z_0, for N=8 and a negative binomial distribution for the on-period length

p	K	z_0	z_1	z_2
0.6	2	1.435059	1.632583	1.463659
0.7	2	1.305059	1.395415	1.315785
0.8	2	1.191038	1.224442	1.193924
0.9	2	1.090056	1.097130	1.090392
0.6	10	1.067622	1.099755	1.072289
0.7	10	1.049399	1.064594	1.051180
0.8	10	1.032153	1.037955	1.032641
0.9	10	1.015725	1.016992	1.015783

5.2 Calculation of θ

Let us consider the case where the number of cells stored in the multiplexer buffer just after a given slot is sufficiently large (>> N). Then we may think that the number of cell arrivals during this slot (which cannot be larger than N) has almost no impact on the total buffer contents. Consequently, if n is sufficiently large ($n > T$), we may assume that the conditional probabilities Prob[$d_1 = i_1$, $d_2 = i_2$, ... | $s=n$] are almost independent of n, and approach to some limiting values for $n \rightarrow \infty$, denoted by $\omega(i_1, i_2, ...)$, i.e.,

$$\text{Prob}\left[d_1 = i_1, d_2 = i_2, \ldots | s = n\right] \cong \omega(i_1, i_2, \ldots) \ , \quad n > T \ , \tag{26}$$

with corresponding joint pgf $\Omega(x_1, x_2, ...)$.

Using equation (26), the joint pgf $P(x_1, x_2, ..., z)$ can now be approximately expressed as

$$P(x_1, x_2, \ldots, z) \cong \sum_{i_1}\sum_{i_2}\cdots\sum_{j=0}^{T} \text{Prob}\left[d_1 = i_1, d_2 = i_2, \ldots, s = j\right]\left(\prod_{n=1}^{\infty} x_n^{i_n}\right) z^j$$

$$+\Omega(x_1, x_2 \ldots)\left(S(z) - \sum_{j=0}^{T} \text{Prob}[s = j] z^j\right) .$$

Setting $x_n = \chi_n(z)$, we know that z_0 is a pole of both the P-function and $S(z)$. As T is finite, multiplying both sides of the above equation by $(z - z_0)$ and taking the $z \rightarrow z_0$ limit, we find

$$\theta = \frac{(z_0 - 1)(1 - p) z_0}{\left[1 - G'(z_0)\right] \Omega\left(\chi_1(z_0), \chi_2(z_0), \ldots\right)} \ . \tag{27}$$

In order to derive the pgf $\Omega(x_1, x_2, ...)$, we let $\pi(i_1, i_2, ... \mid j_1, j_2, ...)$ denote the one-step transition probability that there are i_n ($n \geq 1$) sources in the nth slot of an on-period, given that there were j_l ($l \geq 1$) sources in the lth slot of an on-period in the previous slot. From equations (20) and (26), we then get

$$z_0 \ \omega(i_1, i_2, \ldots) = \sum_{j_1}\sum_{j_2}\cdots \ \pi(i_1, i_2, \ldots | j_1, j_2, \ldots) \ \omega(j_1, j_2, \ldots) \ (z_0)^{\sum_{k=1}^{\infty} i_k} .$$

Hence, we obtain the following equation for the pgf $\Omega(x_1, x_2, ...)$:

$$z_0 \ \Omega(x_1, x_2, \ldots) = \left[B(x_1 z_0)\right]^N \Omega\left(\frac{C_1(x_2 z_0)}{B(x_1 z_0)}, \frac{C_2(x_3 z_0)}{B(x_1 z_0)}, \ldots\right) . \tag{28}$$

As can be expected intuitively, it is possible to show that the solution $\Omega(x_1, x_2, ...)$ of (28) has the same form of expression as the pgf $D(x_1, x_2, ...)$ of the unconditional cell arrival process. Specifically, $\Omega(x_1, x_2, ...)$ can be expressed as

$$\Omega(x_1, x_2, \dots) = \left(1 - \sum_{n=1}^{\infty} \sigma_n^* + \sum_{n=1}^{\infty} \sigma_n^* x_n\right)^N , \tag{29}$$

where σ_n^* $(n \geq 1)$ is the (conditional) probability of finding a source in the nth slot of an on-period, when the number of cells in the buffer is extremely large. From equations (9), (21), (28) and (29), an expression can be derived for $\Omega(\chi_1(z_0), \chi_2(z_0), ...)$. Also, from equations (10) and (12), we obtain an expression for $G'(z_0)$. Finally, after some algebra, we find

$$\theta = \frac{(1-p)z_0^2(z_0-1)^{N+1}}{\left[1 - A'\left(\frac{z_0}{z_0^{1/N}}\right)\frac{(N-1)(1-\beta)z_0}{z_0^{2/N}}\right]\left[A'\left(\frac{z_0}{z_0^{1/N}}\right)\frac{(1-\beta)z_0}{z_0^{2/N}} + 1\right]^{N-1}\left(z_0 - z_0^{1/N}\right)^N} . \tag{30}$$

Consequently, the two parameters z_0 and θ of the geometric tail approximation have been determined. It is then easy to calculate the probability Prob[$s > S$] that the buffer contents exceeds a certain threshold S, which is often used in practice to approximate the cell loss ratio in a finite buffer with a waiting room of size S, i.e., the fraction of the arriving cells that is lost upon arrival because of buffer overflow.

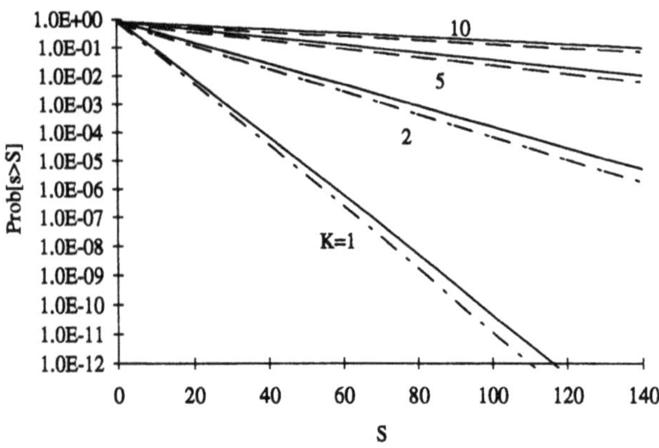

Figure 2 Prob[$s > S$] versus S : exact results (solid lines) and heavy-traffic approximation (dashed lines), for N=8, p=0.9, and a negative binomial distribution for the on-period length.

In Figure 2, we compare Prob[$s > S$] calculated from (22) and (30), with the heavy-traffic approximation calculated from (24) and (30), for N=8, p=0.9, a negative binomial distribution for the on-period length and various values of K. We see that for p=0.9, the heavy-traffic approximation based on the first 2 moments of the on-period distribution is quite accurate, and somewhat overestimates Prob[$s > S$]. Hence, we may conclude, that for *very* high utilization, the "overflow probability" is nearly independent of higher-order (≥ 3) moments of the on-period length.

6 THE HETEROGENEOUS TRAFFIC CASE

In this section, we consider the case of heterogeneous traffic sources. We assume that there are T traffic types, and for traffic type t, there are N_t sources. The mean and the tail distribution of the system contents can then be derived in a similar way as described above for the homogeneous traffic case. Specifically, the mean system contents $E[s]$ is given by

$$E[s] = p + \frac{1}{2(1-p)} \sum_{t=1}^{T} N_t \sigma_t (p - \sigma_t) \left[K_t + L_t (K_t - 1) + \sigma_t (L_t - 1) \right] . \tag{31}$$

Here p is the total load into the multiplexer and σ_t, K_t and L_t denote the load, the burstiness factor and the variance factor respectively, of a source of traffic type t. The dominant pole z_0 of $S(z)$ is determined by the set of equations

$$z - \prod_{t=1}^{T} \left[\beta_t + (1-\beta_t) \chi_{1,t}(z) z \right]^{N_t} = 0 \; ; \qquad z\chi_{1,t}(z) = A_t \left(\frac{z}{\beta_t + (1-\beta_t)\chi_{1,t}(z) z} \right) , \tag{32}$$

and hence z_0 can easily be obtained by means of the Newton-Raphson algorithm. The residue θ is given by

$$\theta = (1-p) z_0^2 (z_0 - 1)^{N+1} \left(1 - \sum_{t=1}^{T} \frac{N_t F_t(z_0)}{1 + F_t(z_0)} \right)^{-1} \prod_{t=1}^{T} \left[\left(z_0 - (z_0(t))^{1/N_t} \right) (1 + F_t(z_0)) \right]^{-N_t} , \tag{33}$$

where

$$z_0(t) = \left[\beta_t + (1-\beta_t) \, \chi_{1,t}(z_0) z_0 \right]^{N_t} \; ; \quad F_t(z_0) = (1-\beta_t) z_0 (z_0(t))^{-2/N_t} A_t' \left(\frac{z_0}{(z_0(t))^{1/N_t}} \right) . \tag{34}$$

7 NUMERICAL EXAMPLES

We will now illustrate the above analysis by means of some numerical examples. In order to show more clearly the influence of various traffic parameters of the sources on the multiplexer performance, let us first consider the case of a single traffic type. In Figure 3, we have plotted the mean system contents $E[s]$ in terms of the total load $p=N\sigma$, for N=16, K=5, and various values of the variance factor L of the sources. The figure shows that for given values of p and K, the variance of the on-period lengths strongly influences the mean system contents. The figure also indicates a considerable decrease in performance as L increases.

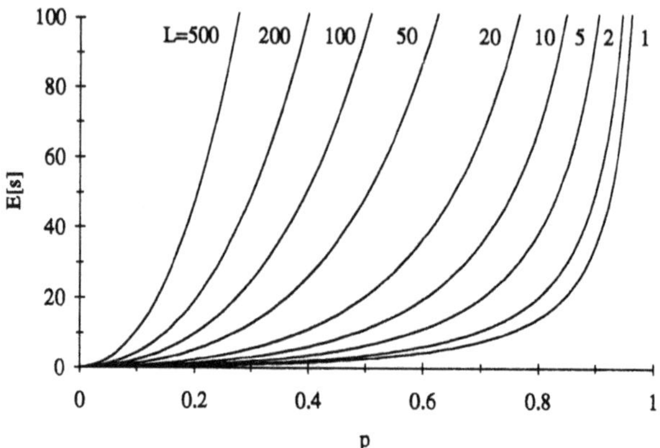

Figure 3 Mean system contents $E[s]$ versus the load p, for N=16 and K=5.

Next, we consider the following examples for the pgf $A(z)$:

$$A_1(z) = z^m \ ; \quad A_2(z) = \frac{(1-\gamma)^2 z}{(1-\gamma z)^2} \ ; \quad A_3(z) = \frac{(1-\alpha)z}{1-\alpha z} \ ,$$

i.e. constant-length on-periods, a negative binomial distribution and a geometric distribution respectively. In order to study the impact of the variance of the on-periods on the "overflow probability" Prob[$s > S$], we choose the parameters of these distributions such that the mean on-period length $A'(1)$ is equal to m in all cases, which corresponds to choosing

$$\alpha = \frac{m-1}{m} \quad \text{and} \quad \gamma = \frac{m-1}{m+1} \ .$$

The corresponding variances of the on-period lengths are then

$$\mathrm{var}_1 = 0 \ ; \quad \mathrm{var}_2 = \tfrac{1}{2}(m-1)(m+1) \ ; \quad \mathrm{var}_3 = m(m-1) \ .$$

In Figure 4, Prob[$s > S$] is plotted versus S, for N=16, p=0.8, K=5 and the above 3 distributions for the length of the on-period. The corresponding variances of the on-period lengths are then var_1=0, var_2=13.35 and var_3=22.44. The variance factors are given by L_1=0, L_2=0.595 and L_3=1. It is clear that for given values of p and m, the variance of the on-periods has a strong impact on the performance. We observe that the performance degrades with increasing variance of the on-period lengths.

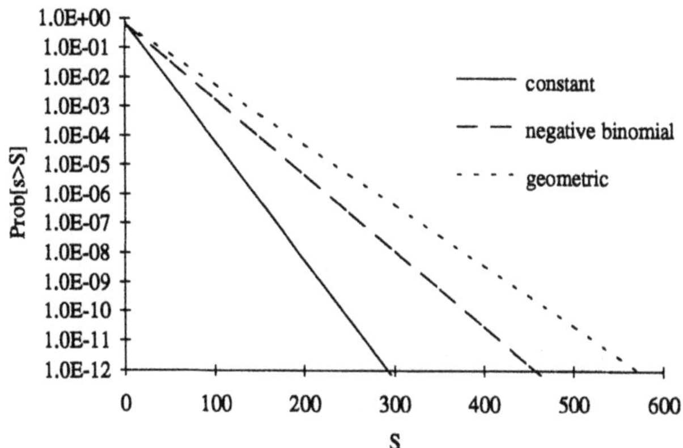

Figure 4 Prob[s > S] versus S, for N=16, p=0.8, K=5, and various on-period distributions.

Now, we consider a mixture of 2 geometric distributions for the on-period length, i.e.,

$$A(z)=\frac{q(1-\alpha_1)z}{1-\alpha_1 z}+\frac{(1-q)(1-\alpha_2)z}{1-\alpha_2 z} \quad .$$

In Figure 5, we have plotted Prob[s>S] in terms of S, for N=8, p=0.8, K=5, L=2 and various values of q. We see that in general, Prob[s>S] is not only determined by p, K and L, but also depends on higher-order moments of the on-period-length distribution. Therefore, as mentioned before in Section 5.1, the commonly used assumption that it suffices to take into account only the first 2 moments of the on-period distribution should be used with care.

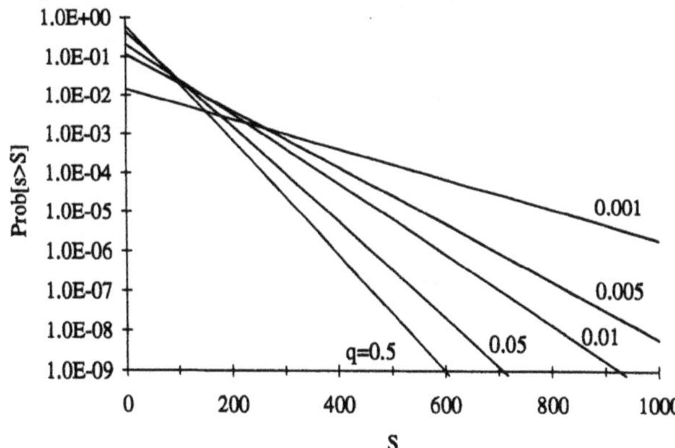

Figure 5 Prob[s > S] versus S, for N=8, p=0.8, a mixture of 2 geometric distributions for the length of the on-period, K=5, L=2 and various values of q.

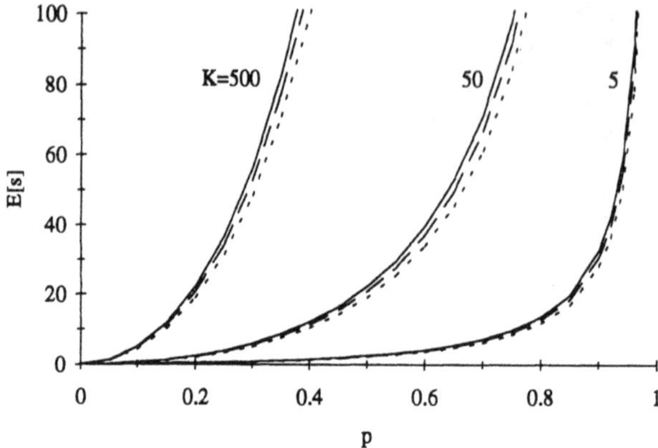

Figure 6 Mean system contents $E[s]$ versus the load p, for T=2, N_1=N_2=4, K_1=K_2=K, a geometric distribution for the on-period length, σ_1=a σ_2, K=5, 50, 500 and a=0 (dotted lines), a=0.2 (dashed lines), a=1 (solid lines).

Finally, we consider a multiplexer fed by 2 types of traffic sources, where N_1=N_2=4, K_1=K_2=K and the on-periods are geometrically distributed. In order to assess the impact of the heterogeneity of the loads of these traffic types, we consider the case that σ_1=a σ_2, for various values of a. In Figure 6, we have plotted $E[s]$ as a function of p, for various values of K and a. First of all, the figure reveals very clearly the strong impact of the burstiness factor K on $E[s]$, for all values of p. Hence, the influence of the absolute lengths of the on-periods and the off-periods on $E[s]$ is far from negligible, even when the ratio of these lengths is fixed. The performance deteriorates considerably as K increases. Secondly, for a given value of K, $E[s]$ decreases as the difference between σ_1 and σ_2 increases. This is also intuitively clear. The extreme cases are a=1, i.e., 8 homogeneous sources each with load p/8, and a=0, i.e., 4 homogeneous sources each with load p/4. Since for a=1, a maximum number of 8 cells can arrive during the same slot, whereas for a=0, this number is limited to 4, it is expected that the case a=1 will lead to a higher $E[s]$. However, the impact of the heterogeneity of the loads is limited as compared to the strong influence of the burstiness of the sources.

8 ACKNOWLEDGEMENT

The authors wish to thank the Belgian National Fund for Scientific Research (N.F.W.O.) for support of this research.

9 REFERENCES

Bruneel, H. (1988) Queueing behavior of statistical multiplexers with correlated inputs. *IEEE Transactions on Communications*, **36**, 1339-1341.

Bruneel, H. and Kim, B. G. (1993) Discrete-time models for communication systems including ATM. Kluwer Academic Publishers, Boston.
Bruneel, H.; Steyaert, B.; Desmet, E. and Petit, G. (1994) Analytic derivation of tail probabilities for queue lengths and waiting times in ATM multiserver queues. *European Journal of Operational Research*, **76**, 563-572.
Elsayed, K. (1994) On the superposition of discrete-time Markov renewal processes and application to statistical multiplexing of bursty traffic sources. *Proceedings of IEEE GLOBECOM '94*, San Francisco, 1113-1117.
Neuts, M. (1989) Structured stochastic matrices of M/G/1 type and their applications. Marcel Dekker Inc., New York.
Simonian, A. and Guibert, J. (1994) Large deviations approximation for fluid queues fed by a large number of on/off sources. *Proceedings of ITC 14*, Antibes Juan-les-Pins, 1013-1022.
Sohraby, K. (1993) On the theory of general ON-OFF sources with applications in high-speed networks. *Proceedings of IEEE INFOCOM '93*, San Francisco, 401-410.
Steyaert, B. and Bruneel, H. (1995) On the performance of multiplexers with three-state bursty sources : analytical results. *IEEE Transactions on Communications*, **43**, 1299-1303.
Xiong, Y. and Bruneel, H. (1992) Performance of statistical multiplexers with finite number of inputs and train arrivals. *Proceedings of IEEE INFOCOM '92*, Firenze, 2036-2044.

10 BIOGRAPHIES

Sabine WITTEVRONGEL was born in Gent, Belgium, in 1969. She received the M.S. degree in Electrical Engineering from the University of Ghent, Belgium, in 1992. Since September 1992, she has been with the SMACS Research Group, Laboratory for Communications Engineering, University of Ghent, first in the framework of various projects, and since October 1994, as a researcher for the Belgian National Fund for Scientific Research (N.F.W.O.). Her main research interests include discrete-time queueing theory, performance evaluation of ATM networks and the study of traffic control mechanisms.

Herwig BRUNEEL was born in Zottegem, Belgium, in 1954. He received the M.S. degree in Electrical Engineering, the degree of Licentiate in Computer Science, and the Ph.D. degree in Computer Science in 1978, 1979 and 1984 respectively, all from the University of Ghent, Belgium. Since 1979, he has been working as a researcher for the Belgian National Fund for Scientific Research (N.F.W.O.) at the University of Ghent, where he currently leads the SMACS Research Group of the Laboratory for Communications Engineering. He is also a Professor in the Faculty of Applied Sciences at the same university. His main research interests include stochastic modeling of digital communication systems, discrete-time queueing theory, and the study of ARQ protocols. He has published more than 80 papers on these subjects and is coauthor of the book H. Bruneel and B. G. Kim, "Discrete-Time Models for Communication Systems Including ATM" (Kluwer Academic Publishers, Boston, 1993).

PART TWO

High-Performance Protocols

4

General Bypass Architectures for High-Performance Distributed Applications

C. M. Woodside, G. Raghunath
Dept. of Systems and Computer Engineering
Carleton University, Ottawa, Canada K1S 5B6
cmw@sce.carleton.ca, raghu@scs.carleton.ca

Abstract

End-system performance limits are a barrier to achieving many of the planned applications for high-speed networks. We seek architectural principles for designing the end-system software which will make it easier to achieve high throughput and low delay. A bypass architecture, previously applied to protocol stacks, is generalized in this paper to any application which handles large volumes of information at a high rate. It is necessary to consider the applications as well as the protocol stack, to avoid having the processing bottleneck simply reappear at the higher level. The resulting architecture is modular, with attachments for bypass paths which traverse the system. The performance and workload impact of a bypass-based architecture may be dramatic.

Index terms: distributed computing, performance engineering, communications software, multimedia workstations.

1. Introduction

It is a commonplace observation that the use of high speed networks suffers from bottlenecks in the end-systems. Protocol execution alone is quite demanding, but various special optimizations have been found which make transfers in the range of some hundreds of Mbits/s feasible. Now similar optimizations are being investigated for the application design in particular kinds of end-systems, such as a video-on-demand server.

While specialized optimizations are useful in communications software, they often produce monolithic integrated designs that are very technology-dependent and hard to maintain. It would be better to find a general architectural principle with conventions which would make it easier to build end-systems from a variety of components, and also to apply optimizations across collections of components. The General Bypass Architecture is such a principle, which was developed originally for layered protocols in [1], [2], [3], [4]. This paper describes how it can be applied to application software of general modular structure, and explores its performance characteristics.

The Bypass Concept

The notion of the bypass begins from the conventional notion of a "fast path" in performance tuning. Figure 1 shows a segment of processing with a fast path. The segment has a general case which takes T sec to compute. Under "condition B", it reduces to a simpler computation F_B taking the smaller value T_B sec. The relative frequency of the condition being true is p_B, and the execution cost reduction factor is R_B:

$$R_B = \frac{p_B T_B + (1-p_B)T}{T} = p_B(T_B/T) + (1-p_B) = 1-p_B(1-T_B/T)$$

R_B is small when the improvement is large. We will call path F_B the bypass path and R_B the bypass workload ratio.

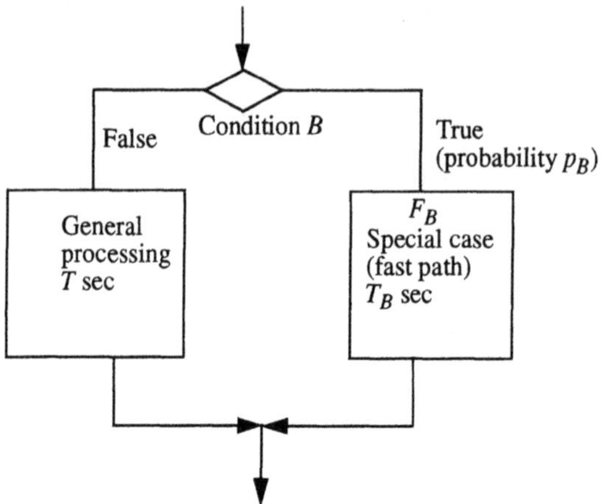

Figure 1. The Notion of a Fast Path or Bypass

In a protocol system the condition B and its special processing are somewhat complex because of protocol state. In [1] it was demonstrated that a combined bypass for several protocol layers can be constructed for bulk data transfers, based on a condition B depending on

- the input event and its arguments (e.g. data packets of a certain size)
- the protocol state (in "data transfer" state)
- the control state of the protocol entity (no data units being processed ahead of it).

The probability p_B depends on the size of the bulk data transfers (p_B =1 - 1/(data units per transfer)), provided there are no error packets or other exceptions in the flow. Thus p_B can be very close to 1. T_B also is very small, just a few machine instructions. The architecture of this protocol bypass is shown in Figure 2.

Related Work

Early work on efficient implementation of protocols was carried out by Clark, resulting in the widely used upcalls mechanisms [5]. Tuning a standard protocol stack such as TCP/IP was investigated by Jacobson et al., [6] culminating in the header-prediction algorithm. The protocol bypass is a generalization of the header-prediction algorithm that can be applied to both the sender and receiver stacks. More recently Clark and Tennenhouse [7] proposed an architecture for new generation protocols based on Integrated Layer Processing and Application Layer Framing. The main principle behind this architecture is the separation of data flows from control flows in the protocol processing. Other work, at the operating systems level has also addressed the issue of efficient

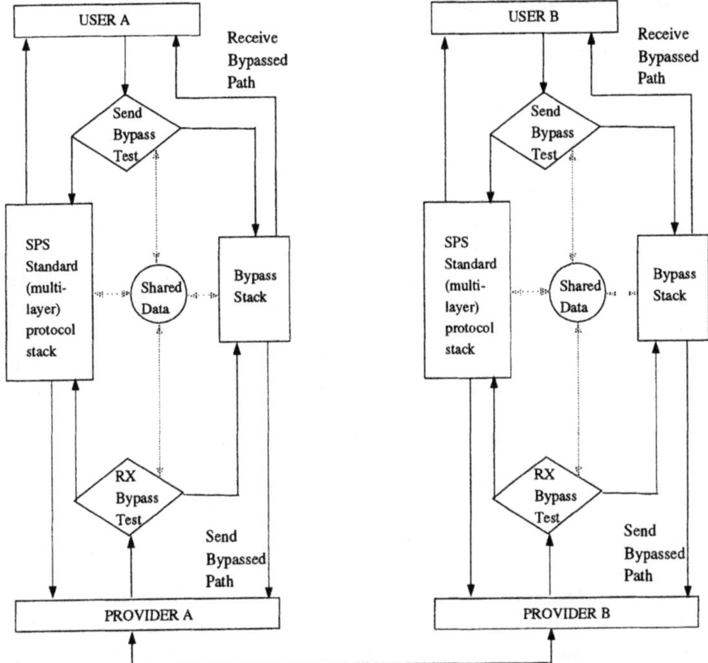

Figure 2. A Layered Protocol System with Bypass [1]

data movement, including the ideas of FBUFs of the X-kernel [8] and Container-shipping [9]. The protocol bypass architecture complements all these ideas and can be combined with them. It adds to performance improvements by identifying fast paths and allowing them to co-exist with the standard stack. One can view the protocol bypass as a design centered around the data path which is optimized for bulk data transfers.

The rest of the paper is organized as follows. Section 2 describes the further extension of the protocol bypass architecture to a general design principle for efficient implementation of high-performance distributed applications. It also provides an example of a Medical Consultation Application to illustrate these concepts. Section 3 discusses the various design options and their performance-cost trade-offs and provides expressions for quantifying them. It is important to consider the effect of bypass failures on the performance of systems, particularly in distributed applications requiring guaranteed bounds on delays. Section 4 discusses the causes of failures and develops analytical expressions for quantifying the delays. The discussion includes the three cases --- equal priorities, priority to bypass and priority among bypass streams.

2. Application Software Bypass

Layering and modularization are essential to deal with software complexity. The key to achieving performance gains in modularized software is to distinguish the data flows between modules from the control flows, and to improve the performance of data flows. There are three complementary ways to achieve these improvements -- customized monolithic implementations, general purpose mechanisms especially with support from operating systems and general design principles directly applicable to application components. The application bypass architecture falls under the last category. The application is analyzed to identify the modules and the data flows. Opportunities for bypassing are identified by finding relationships such as concatenated flows, common operations and common origins and destinations across layers. These data flows are reduced into simpler merged flows with minimal number of operations. The logical conditions under which these simpler flows can be used are also derived from the state machines of the interacting modules. Then a bypass path guarded by the bypass test is constructed. Further optimization of the bypass can then be done applying the operating systems mechanisms mentioned above.

Example of a Medical Consultation Application (MCA)

To illustrate these concepts consider a Medical Consultation Application (MCA). Medical consultation and remote diagnostics are some of the prime motivators for multimedia teleconferencing systems [10]. It has been recognized that teleconferencing is not merely an application but a *paradigm* of communication for many divergent applications. There are many differences in details between a multi-party business conference, a tele-lecture and a medical consultation application. However, all of them have much in common in terms of the structure of communications. Hence, it is reasonable to structure a medical consultation application on top of a teleconferencing service.

Application Modules of the MCA

A typical session of a medical consultation application may involve a patient, physician, radiologist and one or more specialists. The interactions may include text, voice, video and high-quality images. While the video and voice are live, images are usually stored in some server, possibly at the site of the radiologist. It is reasonable to partition the system into subsystems such as user-interface manager, conference manager, multicast transport service and image data base manager (Figure 3). If the underlying communication system provides multicast transport support, then that layer is not part of the application but the application interacts with that layer through a service interface.

Each of the modules provides a set of services. For example, the user-interface module may have such services as initiating a discussion, sending information to the other participants and presenting information received from the other participants in a manner suitable to the role of the participant. A patient, physician, or radiologist each may need to have different presentation styles. In addition, access rights may be associated with these roles and may have to be applied at run time.

The conference manager has a more general notion of the structure and roles of the consultation. It may introduce roles such as organizer (initiator), manager (chair), ordinary participant, special invited participant and so on. It may provide services such as initiating a consultation, the assignment or election of persons for roles, managing the entry and exit of participants in the conversa-

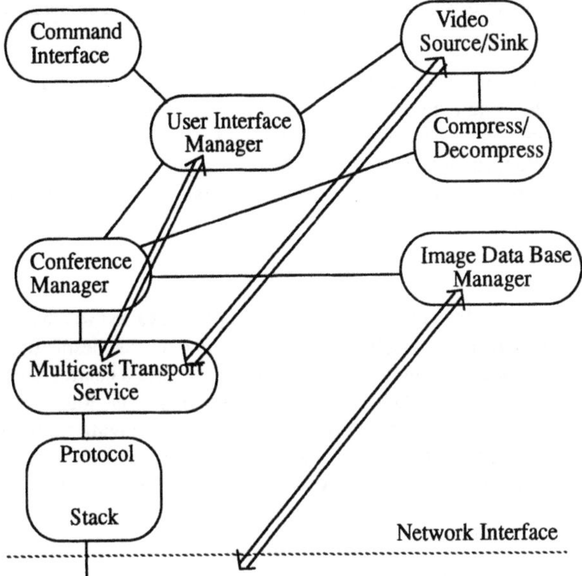

Figure 3. Modules for a Medical Consultation Application
(The double arrows represent paths which may be bypassable)

tion and floor control, apart from actual transmission of data [11].

The multicast transport layer provides services such as creation and destruction of groups, group membership management apart from transmission and reception of data.

If we treat the modules as independent while implementing them, it provides important flexibility in the design. However, it leads to redundant processing overheads. On the execution path of a single system operation, it often happens that the operations in one module are closely related to those in the next module along the path. Specifically, when an upper layer module is in data transfer state, it is normally the case that the lower layer modules are also in their data transfer states. These close relationships are exploited in a bypass path which efficiently combines the operations along the execution path.

Implementation Possibilities

Figure 3 contains several wide arrows representing data flows in MCA which provide opportunities for a bypass. They satisfy the criterion of high volume passing through several layers of applications, and simplification of processing in frequent but special cases. In the paths for video data for example the simplified processing could be a standard compression algorithm, and the bypass test could be passed by all video packets internal to a frame. Some bypass paths bridge several modules, which means they include functions from each one. These paths can be identified by considering the design trade-offs described in the next section. We can call these "customized bypasses."

It is also desirable to consider standardized bypasses. These would be standardized modules which provide a defined, commonly-used fast-path service, such as a transfer from CD-ROM to IP/TCP packets, or from compressed video to a stated video display-buffer format. These optimized paths could be plugged together with full-service modules which handle exceptions and establish paths and groupings, to give a building-block approach to software engineering which also has good performance.

Bypass Test Failures

A bypass designed to transfer bulk data, along the lines of [4], uses a test which requires that each data unit conforms to a standard description (e.g. in size, encoding, and destination storage address). In a stream of image-data units to the conference manager, a control data unit (for instance, for floor management) would have to go to the general module logic rather than a bypass. When this happens, a rather longer processing operation involving several modules may follow, introducing a transient interruption or degradation to the Quality of Service (QoS) of the image-data stream. These transients, called "bypass glitches", are considered in Section 4.

3. Execution Cost Trade-offs

Figure 4 illustrates the usual modular structure of software in an end-system. If a data-path lies through a sequence of several modules (illustrated in Figure 4), then bypass processing can be introduced in three different ways as described in Figure 5.

- In Figure 5(a) ("internal bypass") each module handles its own fast path, internally.
- In Figure 5(b) ("communicating bypasses") the fast path in one module accesses its continuation in the next directly, without a new test inside the second module.
- In Figure 5(c) ("integrated bypass") the fast path is removed from the modules entirely into a separate integrated bypass module.

In module i the normal execution cost is T_i, and the fast path cost is T_{Bi}.

The execution cost implications of these three options will be summarized for an operation traversing a set of n modules. It is assumed first that a single test condition governs the special processing in all the modules, and this test can be applied in any module. The data stream is such that the test is true with probability p_B as before. If we wish to add up the total execution cost it is useful to use the sums

$$T = \sum_1^n T_i, \quad T_B = \sum_1^n T_{Bi}$$

Also there is a cost of crossing a module boundary, of T_{OH}. This may represent procedure calling, object messaging or interprocess messaging overhead, depending on the design of the software. Finally there is a cost of applying a bypass test, of T_{TEST}.

By inspecting Figures 4 and 5 we see that without any bypass optimization the execution cost for n modules is $T + nT_{OH}$, and for the three options it is

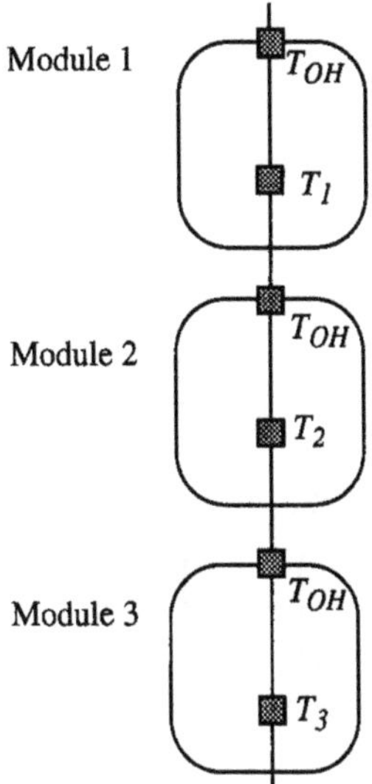

Figure 4. Flow Through a Modular Software System

- option (a), Cost = $T(1-p_B) + T_B p_B + n(T_{OH} + T_{TEST})$
- option (b), Cost = $T(1-p_B) + T_B p_B + T_{TEST} + nT_{OH}$
- option (c), Cost = $(T + nT_{OH})(1-p_B) + T_B p_B + T_{TEST} + T_{OH}$

For example, with the parameter values (normalized to the value of T) given by

$$T_B/T = T_{OH}/T = T_{TEST}/T = 0.05; \; n = 6$$

the values for Cost/T are 1.3 for the original case without any fast path, and for the three options the values are:

(a) Cost/T = $1.6 - 0.95p_B$

(b) Cost/T = $1.35 - 0.95p_B$

(c) Cost/T = $1.4 - 1.25p_B$

Figure 6 shows these values plotted against p_B, showing the marked advantage of option *(c)* as p_B approaches 1.0. With these parameters, as $p_B \to 1.0$ the workload cost in case *(c)* is only one quarter of that in case (a), and one eighth of that with no bypass. Also, the bypass has multiplied the workstation capacity by a factor ranging from 2 in case *(a)* to 8 in case *(c)*.

So far we have considered one bypass for one path threaded through a set of modules. There may be

- other paths through the same modules, with a different bypass condition and different pro-

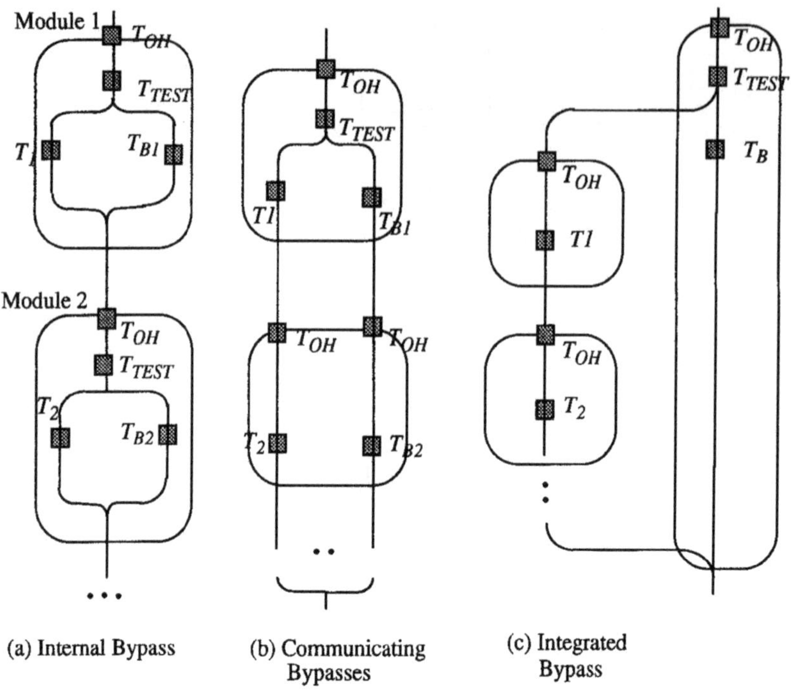

Figure 5. Bypass Paths in a Modular Software System

cessing (for example, a different encoding/decoding step). The tests for the various bypasses could be combined, adding somewhat to the cost T_{TEST},

- other bypass paths within the same module structure, with their own test points. (Figure 3, for example, indicates several distinct bypasses.)

Each bypass must be justified separately, and interaction between them is usually slight (consisting mainly of the possibility of combined bypass tests, as mentioned above).

4. Bypass Glitches

The main performance defect of a bypass is the "bypass glitch" which is a transient condition following a test failure. This places a transient in the otherwise smooth flow through the bypass, which may degrade the connection. When a bypass test fails, that data unit must be processed by the general version of the module or modules. A bypass test failure may be due to the end of the stage, to exceptional data, or to a communications error. We will assume that the following data unit, after waiting for the slow general processing, will pass the tests; that test failures are sufficiently separated in time that the resulting transients do not overlap, and that the processing cost of both paths are deterministic. The glitch is the response to the slowdown for one data unit, and

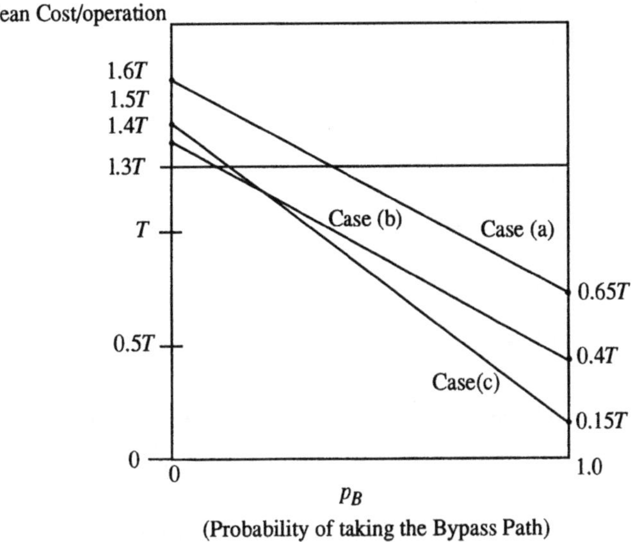

Figure 6. Execution costs vs p_B (for 6 modules and $T_{OH} = T_{TEST} = T_B = 0.05T$)

the backlog that builds up during that time. We will analyze the effect of glitches on quality-of-service (QoS) parameters.

We will consider communications processing as being carried out on several concurrent *streams*. In each stream some uniform operation (such as a video or graphics transfer) is carried out, for a period of time called a *stage*. We will consider one such stage, with bypassed operations, running perhaps in combination with other concurrent streams. Since a glitch gives a transient overload, a fluid approximation to the delay process [12] is appropriate for a first-order understanding of its effects. In a fluid approximation the total flow of work into the system for all streams, and the backlog of work to be done, is represented by a continuous deterministic flow. During a transient situation the delay values change with time. Consider a steady flow giving a total processor utilization of ρ, of which an amount T_B/τ is due to a particular bypassed stream with execution cost T_B and arrival period τ. At time 0, we have a pulse of work in this stream, of magnitude $(T - T_B)$, representing one bypassed operation of cost T_B being replaced by one non-bypassed operation of cost T. (We will assume that all the additional work is at the CPU). During the remainder of the transient the CPU is continuously busy and new work is still arriving at rate ρ units per sec, so the backlog declines at rate (1 - ρ)/sec, giving a work backlog transient as shown in Figure 7. Clearly the duration of the transient is $(T - T_B)/(1 - \rho)$. If the arrivals in our particular stream are spaced τ sec apart there are about $k = (T - T_B)/[\tau(1 - \rho)]$ operations during one transient.

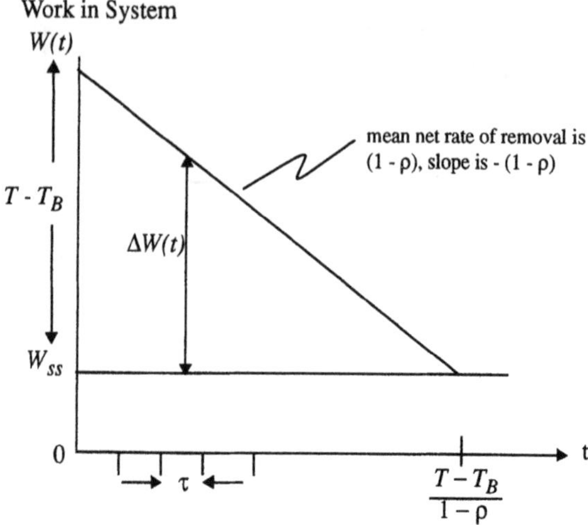

Figure 7. Transient of Work in System for One Glitch

Equal Priorities

Consider the response-time consequences, first for cases where all work is of equal priority. Then the response time for the given stream is just the work-in-system at the arrival instant [13] plus the service time. The *additional* response time due to the glitch, for an arrival at time *t*, is the amount $\Delta W(t)$ shown in Figure 7. As already noted, there are about $k = (T - T_B) / [\tau(1 - \rho)]$ responses during one transient, and from Figure 7 they have an average duration of $(T - T_B)/2$.

During a long period of *K* operations there are, on average, $(1\text{-}p_B)K$ glitches. The addition to the total response time due to glitches in the same stream is

$$(1 - p_B) Kk (T - T_B) / 2$$

and the addition to the mean response time of the stream is

$$\Delta R_G = (1 - p_B) (T - T_B)^2 / [2\tau(1 - \rho)] . \qquad \text{(Equal priorities)}$$

For example if $\rho = 0.7$, $(T - T_B) = 95$ msec, and $\tau = 15$ msec, then there are about 20 operations during one glitch and overall, $\Delta R_G \approx 1000 p_B$ msec.

Depending on how QoS is defined for a data stream, the important impact may either be an increased mean delay through the system, or operations which fail to meet deadlines. The effect ΔR_G on the mean delay as calculated above may be masked if the achieved end-to-end mean is normally less than the QoS target. However the figure calculated above serves as a pessimistic bound. Similarly, the effect on deadlines can be bounded by the number of operations affected by glitches, which during a long period of K operations is $N_G = (1\text{-}p_B)Kk$, giving

$P_G = \text{Prob}\{\text{operation misses deadline}\} \leq (1-p_B)k = p_B(T-T_B)/[\tau(1-\rho)]$

For the parameters given just above, this has value $21.3(1\text{-}p_B)$.

These equations show that the impact of bypass failures is considerably increased when the CPU is heavily loaded, in that the probability of a QoS impact is much larger than $(1\text{-}p_B)$ per operation, and the mean delay impact is much more than $(1-p_B)(T-T_B)$.

The above analysis is highly simplified, so it was compared to results for more realistic simulated response times. Figure 8 shows response times of packets following a glitch. The simulation has $\tau = 0.5$, $T = 1.0$, $T_B = 0.05$, a background workload sufficient to make $\rho = 0.7$, and equal priorities. The arrival processes are Erlang-10, so they have some limited degree of random variation, the bypassed stream has deterministic service, and the background workload has Erlang-10 service. The figure shows the mean of 100 transients, with 90% confidence intervals, plotted against the prediction shown as a dashed line. The agreement is excellent, and was equally close for other parameter values.

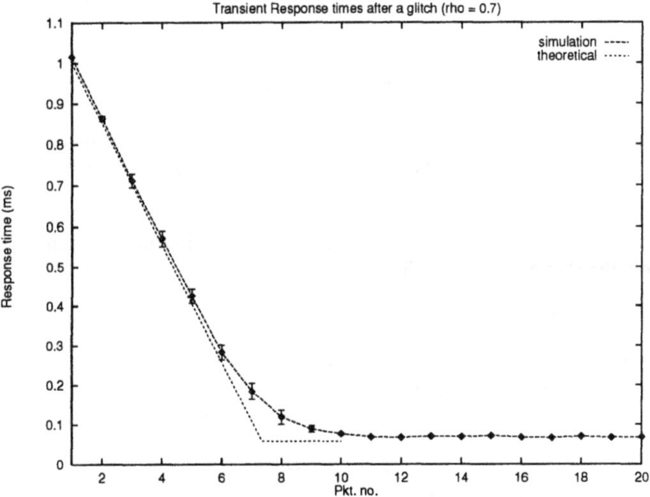

Figure 8. Transient Response times after a glitch ($\rho = 0.7$)
(Simulation vs the theoretical fluid approximation)

Priority to the Bypass

The impact of glitches on the same bypassed stream may be reduced if a part of the CPU workload other than this stream can be preempted (by giving priority to the communication-oriented processing), or by modifying the recovery from a glitch (perhaps by dropping some data). If the bypassed stream with partial utilization $\rho' = T_B/\tau$ has preemptive priority over the remaining "background" workload, then it does not "see" the background at all. Therefore, the same formulae are applied with ρ' replacing ρ. Thus

$$\Delta R_G = (1-p_B)\,(T-T_B)^2/[2\tau(1-\rho')] \quad \text{(Priority)}$$

$$P_G = (1-p_B)\,(T-T_B)/[\tau(1-\rho')] \quad \text{(Priority)}$$

Multiple Bypassed Streams: Glitch Effects

The above analysis only considers one bypassed stream, among all the various workload components of the CPU. If there are several bypassed streams running at the same priority, a glitch in any one will affect the others. The glitch rate will be the sum of the rates for all streams. Applying a subscript i for the ith stream, it has parameters p_{Bi}, t_i, T_i and T_{Bi}. When all work has the same priority, glitches in stream i contribute an amount which can easily be shown to be

$$\Delta R_{ij} = (1-p_{Bi})\frac{(T-T_{Bi})^2}{2\tau_i(1-\rho)}$$

to the mean response time of stream j, so for any one stream j the total addition is

$$\Delta R_j = \sum_i (1-p_{Bi})\frac{(T_i-T_{Bi})^2}{2\tau_i(1-\rho)}$$

One bypass stream with low p_B can negatively affect all of the processing on the end-system. This is the most serious problem with bypassing, and indeed with any kind of optimistic tuning of any component of work.

If the bypassed streams have priority over background processing and contribute a total of ρ'' to the utilization, the above expressions have ρ'' in place of ρ.

Priorities Between Bypasses

Priorities can reduce the impact of glitches on a high-priority stream, at the cost of worse impact on other bypassed streams. Consider two streams with the same bypass failure probability, arrival time and service times (p_B, τ, T, T_B), but stream 1 has pre-emptive priority over stream 2, and both have priority over the background. During a stream-1 glitch, stream 1 only waits for its own component of load, with the backlog $\Delta W_{11}(t)$ illustrated in Figure 9. It returns to normal after time $(T - T_B)/(1 - \rho')$, but during all this time stream 2 is stopped by the priority system. Remember that $\rho' = T_B/\tau$.

After a stream-1 glitch, stream-2 operations are totally blocked until the stream-1 backlog is cleared (i.e. until $(T-T_B)/(1-\rho')$). At this point there is a stream-2 backlog of $\rho'(T-T_B)/(1-\rho')$, which must be cleared; the backlog is reduced at rate $(1-2\rho')$. Thus the stream-2 backlog is $\Delta W_{12}(t)$ as shown in Figure 9.

For a low-priority stream like stream 2 the response-time is not the backlog at the arrival instant; it includes some high-priority arrivals that come after but leave first. If $\Delta W_{12}(t)$ is the backlog, an arrival at time t has response time

$$R_{12}(t) = \Delta W_{12}(t) + \rho' R_{12}(t) = \Delta W_{12}(t) / (1-\rho').$$

Thus for the right side of Figure 9, for $t < (T-T_B)/(1-\rho')$, this makes the response time of an arrival at time t equal to:

$$R_{12}(t) = \left[\rho' \frac{(T-T_B)}{1-\rho'} - (1-2\rho')\left(t - \frac{(T-T_B)}{1-\rho'}\right)\right] \frac{1}{1-\rho'}$$
$$= [(T-T_B) - (1-2\rho')t] / (1-\rho')$$

which reaches zero (ending the glitch) at $t = (T-T_B)/(1-2\rho')$.

For $t < (T-T_B)/(1-\rho')$, $R_{12}(t)$ is the time until the stream-1 backlog clears, plus a term due to the stream-2 backlog at time t, which is $\Delta W_{12}(t) = \rho' t$. Thus:

$$R_{12}(t) = \frac{T-T_B}{1-\rho'} - t + \frac{\Delta W_{12}(t)}{1-\rho'}$$
$$= [T-T_B - (1-2\rho')t] / (1-\rho')$$

which makes $R_{12}(t)$ a straight line from $(T-T_B)/(1-\rho')$ at $t = 0$ to zero at $(T-T_B)/(1-2\rho')$. Figure 10 shows simulation results for $R_{11}(t)$ and $R_{12}(t)$, for a case with $\rho' = 0.3$, $\tau = 0.167$, $T = 1.0$, $T_B = 0.05$.

The mean response-time increment to stream 2 can be found, after some manipulation, to be:

$$\Delta R_{G,12} = (1-p_B) \frac{(T-T_B)^2}{2\tau(1-\rho')^2(1-2\rho')}.$$

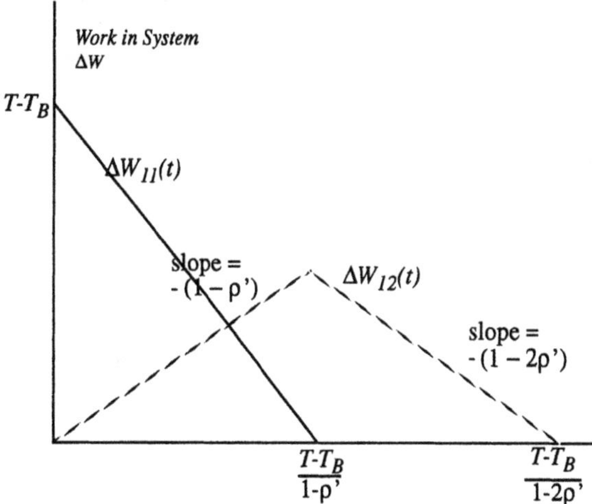

Figure 10. Response to a stream-1 glitch.

A stream-2 glitch gives a simpler analysis. Stream 1 is unaffected; stream 2 sees a backlog $\Delta W_{22}(t)$ starting at $(T - T_B)$ and dropping to zero at $t = (T - T_B) / (1 - 2\rho')$. So:

$$\Delta W_{22}(t) = (T - T_B) - t(1 - 2\rho')$$

$$R_{22}(t) = \Delta W_{22}(t) / (1 - \rho')$$

$$\Delta R_{G,22} = (1 - p_B)(T - T_B)^2 / [2\tau(1 - \rho')].$$

Overall then the effect on the mean response time is

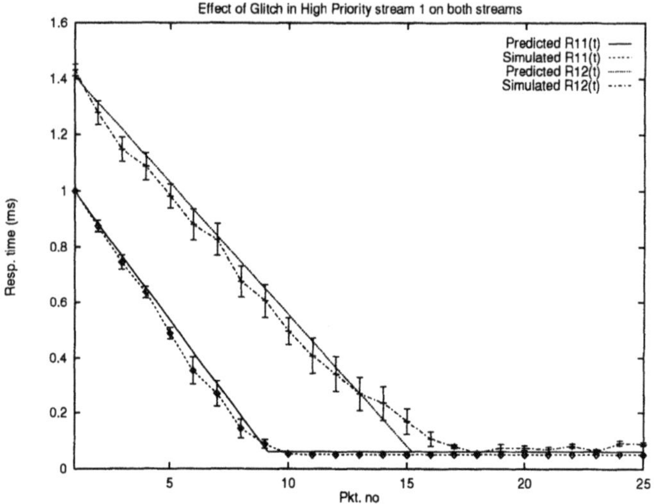

Figure 11. Response Times of Both Streams after Glitch in Stream-1

$$\Delta R_{G,1} = p_B (T - T_B)^2 / [2\tau(1 - \rho')]$$

$$\Delta R_{G,2} = \Delta R_{G,12} + \Delta R_{G,22}$$

$$= (1 - p_B)(T - T_B)^2 (2 - \rho') / [2\tau(1 - \rho')^2 (1 - 2\rho')].$$

As an example, suppose $\rho' = 0.3$, $\tau = 0.5$, $T - T_B = 0.95$, then $\Delta R_{G,1} = 1.29(1 - p_B)$, and $\Delta R_{G,2} = 7.83 p_B$. If the two streams have equal priority, however, $\Delta R_G = 4.512(1 - p_B)$.

5. Conclusions

We have identified the possibility for exploiting the bypass idea, previously described for layered protocols, in general modular communications software such as a medical consultation system. Calculations for the execution-cost advantage of a bypass were given, for various alternative bypass architectures. Substantial cost advantages, or one or more orders of magnitude, are possible.

A *glitch* in a data stream is a transient condition resulting from the need to apply the full processing logic to the data stream. If the end-system is heavily loaded, glitches may have significant impact on the quality of service parameters of a stream. We considered the effect on the mean end-system response, and the number of data units that are affected. The effect on certain streams can be reduced by the use of priorities, and this was investigated. When a bypass is designed the frequency and magnitude of glitches must be investigated as part of the assessment of its performance potential. If bypassing becomes important in practice it may be necessary to include QoS parameters that describe the relative immunity of the stream, as required by the user.

Acknowledgments

This research was supported by the Telecommunications Research Institute of Ontario, and by the Commonwealth Scholarships.

References

[1] C.M. Woodside, K. Ravindran and R.G. Franks, "The Protocol Bypass Concept for High Speed OSI Data Transfer", in *Proc. of Second IFIP International Workshop on Protocols for High-Speed Networks*, November, 1990 (Published by North-Holland as *Protocols for High-Speed Networks II*, 1991, pp. 107-122.)

[2] Y.H. Thia and C.M. Woodside, "High Speed OSI Protocol Bypass Algorithm with Window Flow Control" in *Proc. 3rd IFIP Workshop on Protocols for High Speed Networks*, Stockholm, May, 1992.

[3] Y.H. Thia and C.M. Woodside, "A Reduced Operation Protocol Engine (ROPE) for a Multiple-Layer Bypass Architecture" in *Proc. 4th IFIP Workshop on Protocols for High Speed Networks*, Vancouver, August, 1994, pp.203-218.

[4] C.M. Woodside and Y.H. Thia, "A Parallel Optimistic Bypass Architecture (POBA) for High-Speed Bulk Data Protocol Processing", Report SCE-94-03 of the Dept. of Systems and Computer Engineering, Carleton University, November, 1993.

[5] D.Clark "The structuring of systems using upcalls", In *Proc. 10th ACM Symposium on Operating System Principles*, pp. 171-180, December, 1985.

[6] D.D. Clark, V. Jacobson, J. Romkey and H. Salwen, "*An Analysis of TCP Processing Overhead*", IEEE Communications Magazine, vol. 27, no. 6, pp. 23-29, June, 1989.

[7] D.D.Clark and D.Tennenhouse, "Architectural Considerations for a New Generation of Protocols", in *Proc. ACM SIGCOMM*, pp. 200-208. 1990.

[8] P.Druschel, M.B.Abbot, M.M.Pagels and L.L.Peterson, *"Network Subsystem Design"*, IEEE Network,7(4) July, 1993.

[9] J.Pasquale, E. Anderson and P.Keith Muller, *Container Shipping - Operating System Support for I/O-intensive Applications*, IEEE Computer, March 1994, pp. 84-93.

[10] L.Orozco-Barbosa, A.Karmouch, N.D. Georganas and M.Goldberg, "*A Multimedia Inter-hospital Communications System for Medical Consultations*", IEEE Journal on Selected Areas in Communications, vol.10, No.7, September 1992, pp.1145-1157.

[11] ITU Study Group 8, *T.GCC: Generic Conference Control for Audiovisual and Audiographic Terminals and Multipoint Control Units*, May 23-5, 1994 (ITU Telecommunications Standardization Sector).

[12] G.F. Newell, "*Applications of Queueing Theory*", Chapman and Hall Ltd., London, England, 1971.

[13] L. Kleinrock, "*Queueing Systems*", vol. 1, John Wiley and Sons, 1976.

5

High Performance Protocol Architecture

W. Dabbous, C. Diot
INRIA Centre de Sophia Antipolis,
2004 Route des Lucioles, BP-93,
06902 Sophia Antipolis Cedex, FRANCE.
Tel: + 33 93 65 78 25, Fax: + 33 93 65 77 65
e-mail: {dabbous|cdiot}@sophia.inria.fr
http://www.inria.fr/rodeo/

Abstract

The performance enhancement of communication protocols is an essential step toward the development of high speed network applications. Application Level Framing (ALF) and Integrated Layer Processing (ILP) have been presented as two design principles for a new generation of protocols. In this paper, we study these high performance protocol design principles. We will first show the need for a new protocol architecture by presenting protocol optimization techniques and their limitations. We then describe ALF/ILP and study the impact of these principles on protocol design. Experiments with ALF and ILP are then presented. These experiments show that there is a performance gain when applying ALF/ILP to protocol design. We discuss the consequences of ALF and ILP based design on the way communication systems should be designed for more integration and more efficiency.

1 Introduction

The development of high speed networking applications requires efficient communication protocols. As networks proceed to higher speeds, the performance bottleneck is shifting from the bandwidth of the transmission media to the processing time necessary to execute higher layer protocols. In fact, the existing standard protocols were defined in the seventies. At that time, communication lines had low bandwidth and poor quality, and complex protocol functions were necessary to compensate for the transmission errors. The emergence of high speed networks has changed the situation, and these protocols would not be designed in the same way today.

At the same time, the application environment is changing. New applications (e.g. audio and video conferencing, collaborative work, supercomputer visualization etc.) with specific communication requirements are being considered. Depending on the application, these requirements may be one or more of the following: (1) high bit rates, (2) low jitter data transfer, (3) simultaneous exchange of multiple data streams with different "type of service" such as audio, video and textual data, (4) reliable multicast data transmission service, (5) low latency transfer for RPC based applications, etc.

The above requirements imply the necessity to revise the data communication services and protocols in order to fulfill the specific application needs. In fact, applications may "classically" choose either the connection-less or the connection oriented transport services. In both cases, the application needs are expressed in terms of quality of service (QoS) parameters such as e.g. the transit delay or the maximum throughput. However, the applications need more than a set of QoS parameters to control the transmission. The applications require to be involved in the choice of the control mechanisms and not only the parameters of a "standard" transport service. In fact, we argue that we need to define adaptation algorithms allowing a "Network Conscious Application" to govern the transmission. Placing most of the burden of network adaptation in the user equipment is in line with the "end to end argument", a key point of the Internet architecture. It contributes in keeping the network simple, and is very often the only way to scale up a complex system.

In this paper, we study and evaluate a new high performance protocol architecture. After a rapid survey on protocol optimization techniques in section 2, we discuss in some detail the need for a new protocol architecture and we present the ALF/ILP design principles in section 3. In section 4, we present experiments done in order to evaluate the performance gain obtained when the ALF/ILP principles are applied to the design of communication subsystems. The redesign of applications according to these principles may either be "manual" i.e. by applying the design rules directly resulting in a "network conscious application", or via a protocol compiler supporting theses design rules. Section 4.2 presents this "Network Conscious Applications" concept and section 4.3 shows how to perform automatic integration of communication systems in an efficient way according to the ALF/ILP principles. In section 5, we conclude the paper and present future work in this context.

2 High performance protocols

Early work on high performance protocols concentrated on the optimization of each layer separately. Concerning the transport protocols, several approaches have been studied such as the tuning of standard general purpose protocols [Wat87, Col85], the definition of new protocols or the work on enhanced implementations. In fact, good implementation techniques represent one of the most important factors in determining the performance of a given protocol [Cla89]. These techniques depend on the environment more than on the protocol itself. The proposed solutions focused on the enhancement of the protocol implementation performance in a given software or a hardware environment: outboard protocol processors (e.g. [Kan88], [Coo90]), hardware protocol implementations (early work on XTP/Protocol Engine [Ches89]) or parallel implementations of transport protocols [Brau92], [Rüt92], [Lap92], [Bjö93]. A detailed survey of protocol implementation optimization techniques can be found in [Dab91] and [Fel93].

Early work in the domain of enhanced transport *service* proposed the design of light weight special purpose protocols for specific application needs (e.g. NETBLT [Cla87a, Cla87b] for bulk data transfer and VMTP [Cher86] for transactional applications). This approach has a major limitation: the diversity of the applications increases the complexity of the transport by the support of several protocols. Each application would then choose a protocol corresponding to its specific needs. More recent research activities propose the synthesis of the so-called "communication subsystems" tailored to provide the service required by the application, from "building blocks" implementing elementary protocol functions such as flow control, error control, connection management (e.g. [Sch93], [Abb93b]). The synthesis of "fine grain" protocol functions should replace the "coarse grain" protocol choice (e.g. TCP or UDP).

At the presentation level, several research activities were centered around the optimization of the ASN.1 Basic Encoding Rules. The cost of the coding and decoding routines is attributed to the heavy Type-Length-Value oriented coding of ASN.1 BER. This motivated the work on "light weight" or XDR-like transfer syntaxes (LWS) [Hui90, Hui89] based on three design principles: (1) avoid unnecessary information in the encoding, (2) use fixed representation when it is possible, and (3) simplify the mapping of the elements by fixed length structures. In parallel, the optimization of the BER implementation functions was conducted and resulted in a drastic improvement of the speed of coding and decoding routines on high performance RISC workstations. We tested an enhanced version of the MAVROS compiler [Hui91] with improved performance for the generated coding and decoding routines [Dab92]. In table 1, we present the results of a performance comparison test for coding and decoding routines of a complex data type (X.400 message) on several workstations.

	Coding time (μs)		Decoding time (μs)		Size (octets)		Decoding (Mbps)	
	BER	LWS	BER	LWS	BER	LWS	BER	LWS
Sun3	1861	2018	6944	6398	540	1029	0.62	1.29
SS 10/30	112	199	318	225	540	1029	13.58	36.59
Dec 5000/300	208	208	372	212	540	1029	11.6	38.8
HP 9000/715	100	100	339	155	540	1029	12.74	53.1
Dec alpha	74	55	176	92.6	540	1029	24.54	88.9

Table 1: Speed of the presentation routines

These results show that on high performance RISC workstations such as SparcStation 10, coding BER is much more efficient than LWS, which implies that the memory access dominates the processing time. However, BER decoding is still limited by the CPU performance. Both coding and decoding routines for LWS are limited by the memory access (due to the large code size stored and loaded). Similar results are obtained on both Dec 3000 and HP workstations. Note that the coding time is the same for both BER and LWS, which means that both memory access and CPU bottlenecks are balanced. For decoding, the processor speed limitation is more pronounced. The "best" results are obtained on Dec-alpha where the 64-bit bus enhances the memory access performance. The figures are "classical": BER is more costly than LWS, and decoding (BER or LWS) is more costly than coding. This is typically due to CPU speed limitations.

From the above results we can learn the following:

- A drastic improvement of the speed of coding and decoding routines for both BER and LWS routines can be obtained by adequate implementation optimization on high performance workstations

- The LWS is interesting if the processor speed is the bottleneck; however, a more compact syntax is desirable in order to reduce the size of the data to be transmitted on the network and the memory bus.

- Memory access is a limiting factor in most cases on RISC workstations. This confirms that integration techniques should result in increased performance on such workstations.

The optimized version of the encoding and decoding functions should facilitate the implementation of the presentation layer as a filter with "streamlined" *encoding* and *transmission* of application data units.

All the activities cited above in this section were focused on the optimization of specific "layers" independently. We argue that the integration of all the application communication requirements (including functions corresponding to the OSI transport, session and presentation layers) in order to generate a single protocol automaton for the application will result in increased performance gain. In this case, the application selects the options that govern the data exchange, i.e., the policy used for the transmission control. In fact, only the application has sufficient knowledge about its own requirements to optimize the parameters of a data exchange in a convenient way. The application can easily adapt to the network resources changes by appropriate steps: instead of reducing the window size at the transport level in response to a congestion indication, a video-conference application may chose to degrade either the quality[1] or the frequency of the images in order to adapt to the available bandwidth.

This approach is not consistent with the layered OSI model, where clear separation of data transmission control mechanisms (lower layers) and data processing functions (higher layer) is made. The OSI approach discharges the application from the transmission control functions by defining the transport service. The applications are *transport service users* according to the layered reference model. On the contrary, the integrated design and implementation approach put the application inside the "control loop", but it still needs to define how the application level parameters will be mapped onto network parameters and control functions. According to the diversity of the applications two solutions are possible:

- either we define application classes and we select appropriate control functions to be integrated into each class.
- or we use a suitable specification language to express the applications needs and we design a generic tool to derive the complete communication subsystem automatically.

The first solution means an extension of the transport service. There has been considerable work on the definition of high speed transport service(s) [Léo92],[Diot92]. However, this solution seems limited due to variety of application profiles. A more focused solution is desirable.

The second approach is to build specialized communication subsystems based on specific application requirements and on a set of optimized building blocks. This corresponds to a horizontal approach to the layered architecture, i.e., applications select control and data manipulation functions based on the service parameters and on the cost of these functions. These functions should be combined in a way that minimizes the memory access cost, in order to build the complete communication subsystem.

These ideas are in line with the ALF/ILP design concepts proposed by Clark and Tennenhouse [Cla90]. The next section will be dedicated to the presentation of these concepts.

[1] The quality of a H.261 image can be controlled by changing he quantizer value and/or the movement detection threshold. These two parameters are used in the INRIA Videoconference System' control algorithm to adapt the quality of the image to the available bandwidth.

3 New protocol architecture

The requirement for a new protocol architecture is clear. Traditional layered protocol architectures such as the ISO OSI model (see for example [IEEE83]) and the ARPA Internet model (see for example [Lei85]), are reaching the very end of their lifetimes.

The artificial separation of concerns in layers is partly due to the hardware architecture of the 70's. The service/protocol concept derives historically from the model presenting interfaces between different service providers. There would be a link, network, transport and session provider, perhaps all of which would have different potential vendors. This approach has proved one thing: The inefficiency of the communication systems pushed buyers to go to other markets. The market appears to be in 3 layers: end system hardware and operating systems; end systems communications stacks; finally transmission networks.

Even in a single vendor software implementation, one could accuse the layered model (TCP/IP or OSI) of causing inefficiency. In fact, the operations of multiplexing and segmentation both hide vital information that lower layers need to optimize their performance. Although it is important to distinguish between the architecture of a protocol suite and the implementation of a specific end system or a relay node, the layered protocol architecture may unnecessarily reduce the engineering alternatives available to an implementor. In fact, a certain degree of flexibility in the way the functions are organized within the layers is needed. A proposed solution is to integrate all transmission control layers in a single block, in essence discarding the highly modular layered model in exchange for performance.

3.1 ALF/ILP

In their SIGCOMM '90 paper [Cla90], Clark and Tennenhouse have proposed Application Level Framing (ALF) as a key architectural principle for the design of a new generation of protocols. ALF is in fact the natural result of advanced networking experiments, which showed the need for "a rule of three units:"

1. a relatively old result is that efficient transmission can only be achieved if the unit of control is exactly equal to the unit of transmission. Obvious penalties for violating this rule are unnecessary large retransmissions in case of errors and inefficient memory usage.

2. the key idea expressed in [Cla90] is that the unit of transmission should also be the unit of processing. Otherwise, large queues will build up in front of the receiving processes and eventually slow down the application.

3. we found, through experience with multimedia services, that adaptive applications are much easier to develop if the unit of processing is also the unit of control.

According to the ALF principle, applications send their data in autonomous "frames" (or Application Data Units (ADUs)) meaningful to the application. It is also desirable that the presentation and transport layers preserve the frame boundaries as they process the data. In fact, this is in line with the widespread view that multiplexing of application data streams should only be done once in the protocol suite. The sending and receiving application should define what data goes in an ADU so that the ADUs can be processed out of order. The ADU will be considered as the unit of "data manipulation", which will simplify the processing. For example, an image server will send messages corresponding to well identified parts of the picture. When a receiver receives such a frame, it can immediately decompress the data and "paint" the corresponding pixels on the screen, thus minimizing response times.

Integrated Layer Processing (ILP) is an engineering principle that has been suggested for addressing the increased cost of data manipulation functions on modern workstations. In fact, the performance of workstations has increased with the advent of modern RISC architectures but not at the same pace as the network bandwidth during past years. Furthermore, access to primary memory is relatively costly compared to cache and registers, and the discrepancy between the processor and memory performance is expected to get worse. The memory access is expected to represent a bottleneck [Dru93].

Protocol processing can be divided into two parts, control functions and data manipulation functions. Example of data manipulation functions are presentation encoding, checksumming, encryption and compression. In the control part there are functions for header and connection state processing. Jacobson et al. have demonstrated that the control part processing can match gigabit network performance for the most common size of PDUs with appropriate implementations [Cla89].

Data manipulation functions present a bottleneck [Cla90], [Gun91]. They consist of two or three phases. First a read phase where data is loaded from memory to cache or registers, then a manipulation or "processing" phase, followed by a write phase for some functions, e.g., presentation encoding. For very *simple functions*, e.g. checksumming or byte swap, the time to read and write to memory dominates the processing time. For other *processing oriented functions*, like encryption and some presentation encodings, the manipulation time dominates with current processor speeds. However, the situation is expected to change with the increase of processor performance: the memory access will be the major bottleneck rather than data processing.

The data manipulation functions are spread over different layers. In a naive protocol suite implementation, the layers are mapped into distinct software or hardware entities which can be seen as atomic entities. The functions of each layer are carried out completely before the protocol data unit is passed to the next layer. This means that the optimization of each layer has to be done separately. Such ordering constraints are in conflict with efficient implementation of data manipulation functions [Wak92], [Cla90].

The main concept behind ILP is to minimize costly memory read/write operations by combining data manipulation oriented functions within one or two processing loops instead of performing them serially as is most often done today . It is expected that the cost reduction due to this optimization will result in better overall performance as it will reduce time consuming memory access. This optimization may be applied within a single data manipulation function (*intra-function optimization*) or across several functions (*inter-function optimization*) [Dab94a]. The interest of the ILP principle (and similar software pipelining principles such as lazy message evaluation and delayed evaluation) has been discussed in [Cla90], [Gun91], [Par93b], [O'M90], [Peh92] and [Abb93a].

Previous work ([Gun91] [Dab94a] [Par93b]) has demonstrated that there is a performance benefit with ILP. The results show reduced processing time for one PDU, as much as a factor of five when six simple data manipulation functions were integrated. These reported results came from experiments that were isolated from the rest of the protocol stacks and where hand-coded assembler routines were used in order to control register allocation and cache behavior. Abbot and Peterson [Abb93b] use a language approach to integrate functions, but with less speedup. In [Par93b] only two functions are integrated, data copying and checksum calculation, but it is a real operational implementation of UDP. Experience with an implementation of the XTP protocol from the OSI95 project [Dab93] showed improved performance when ILP was used. This implementation is in user address space, which also demonstrates that such implementations can perform as well as kernel tuned implementations. We should go one step further and integrate several of the data manipulation functions in a complete, operational stack in order to understand the architectural implications and the achievable speedup.

ALF and ILP were proposed in 1990 and for the time being there is no complete implementation of a communication subsystem based on both concepts. Several reasons make the design of a global framework for integrated implementation based on ALF/ILP concepts now both feasible and attractive:

- The increase of processor speeds and of the discrepancy between processor and memory performance is pushing toward the integration of data manipulation operations.
- Experience with an implementation of the XTP [Dab93] and TCP [Hogl94, Cas94a] protocols showed that user level software implementations can perform as well as kernel tuned TCP, while still keeping the flexibility of configurable software.
- Performance enhancements to the ASN.1 encoding facilitate the implementation of a presentation filter [Dab92].
- Multiple communication services including group communication and variable error control have been studied for a variety of applications (e.g. multimedia video conferences with shared workspace, mobile applications). It is very desirable to have a mechanism to customize communication subsystems to specific application needs.

4 Impact on protocol design

The basic idea of ALF and ILP is that the protocol should adapt to the requirements of the application. The application is acknowledged to be best prepared to choose the proper strategies for treating lost or out-of-order data. This design of a flexible integrated architecture requires on the one hand that the communication system have access to the application semantics, and on the other hand that the application have the means to affect the relevant control and synchronization aspects of the communication system.

In order to validate this integrated approach, we wanted to analyze and measure what the real impact of ALF and ILP was on both the architecture and the performance of the communication system.

This section presents first the experiments we led with ILP and ALF. The most significant results are shown. Then, we discuss the consequences of ALF and ILP based design on the way communication systems should be designed for more integration and more efficiency.

4.1 Experimental results

- ALF evaluation. Two hand-coded implementations of a JPEG player have been designed to investigate the effect of ALF on communication subsystems design and performance. The No ALF implementation runs over an in-kernel TCP. The ALF implementation runs over its own protocol which takes advantage of ALF. The protocol called TPALF is a user-level protocol that runs over UDP/IP. It was modified from the 4.3BSD TCP/IP implementation. The only changes with regard to TCP were to allow out-of-order processing of incoming data and to handle ADUs as opposed to streams of packets. Flow control is done with a sliding-window scheme using the slow-start algorithm. Error control is achieved through both Cumulative and Selective Negative Acknowledgments.

A first issue was to determine the size of the unit of transmission. As JPEG ADUs are small, several ADUs have been concatenated within one *transmission unit* or NDU (Network Data Unit). But ALF means also the ADUs must be preserved through the whole communication system and NDU segmentation must be avoided. ALF is a strategy that organizes the transmitted NDUs into data meaningful to the application. This allows the receiver to process independently and immediately each packet received. When segmentation occurs within the protocol, the received packets cannot be delivered to the application on arrival, and the benefits of ALF are lost (table 2). Thus the size of the NDU should not exceed the size of the minimum MTU (Maximum Transmission Unit) of the network.

MTU	ALF	No ALF (TCP)
512 bytes	14.7 Kbits/s	6.3 Kbits/s

Table 2: Throughput via Internet, NDU size = 512 octets

MTU	ALF	No ALF (TCP)
1460 bytes	7.35 Mbits/s	7.67 Mbits/s

Table 3: Throughput via local Ethernet, NDU size = 1460 octets

The two hand-coded implementations have been also used to investigate whether applications can benefits from out-of-order processing. The comparison between table 2 and table 3 demonstrates how non-ordering gives the possibility of exploiting the internal parallelism of the application. The experiments through local networks (table 3) show that the No ALF implementation[2] performs slightly better when the underlying network is reliable (almost no out-of-sequence data transmission). Through Internet (table 2), where delays and loss can produce out-of-order data delivery, ALF appears to be more efficient (more than 200 %) because the receiver is able to process the ADUs immediately when they arrive, whether they are in order or not. ALF tends to improve the efficiency of the communication sub-system by handling the ordering at the application level. Further details are given in [Diot95].

- Integrated Layer Processing is an implementation concept "to permit the implementor the option of performing all the manipulation steps in one or two integrated processing loops" [Cla90]. Integration of data manipulations by ILP seems to be a promising strategy to increase the performance of communication systems. During the last years the

[2] In the no ALF case, the communications are done over a reliable TCP stream.

performance of processors has been increasing very fast compared to the performance of memories. The memory bottleneck can be reduced by simply avoiding the access to the memory as much as possible, e.g. by eliminating copy operations from the protocol implementations. If avoiding memory access is not further applicable, another solution is to use caches, which are extremely fast but expensive. Usually the fast caches on the processor chips have limited sizes in the range of a few Kbytes (16 Kbytes data cache and 20 Kbytes instruction cache on a SuperSPARC10 processor). To get benefits from the cache it is necessary to keep as much data as possible within the cache.

The concept of ILP tries to gain from both approaches. Theoretically, an ILP protocol stack implementation reads once from the main memory, keeps the read data within registers or cache memory, and performs all the data manipulations of several protocol layers. Processed data are directly written to the destination memory. In this ideal case, the ILP approach requires only one read and one write access to the main memory for each word. All the other operations should work on registers, and eventually on the cache memory. This integrated processing of data is commonly called the ILP loop.

By applying the ILP technique on simple data manipulations like data copying and TCP checksum calculation, performance gains of 50% have been achieved [Ten89]. A similar but also simple experiment nearly shows the same results. The XDR marshalling routine obtained from a stub compiler for an array of 20 integer values has been combined and executed with the TCP checksum routine. When executing the two routines sequentially, the throughput was 70 Mbps, in contrast with the 100 Mbps observed when integrating both functions into a single loop. The performance comparison shows 30% gain in favor of the ILP implementation. The advantages by integrating more data manipulation functions may lead to even larger benefits.

Other simple experiments showed that ILP is sensitive to data manipulation functions. The integration of the DES algorithm (data encryption standard) can reduce the performance gains significantly [Gun91]. But benefits that can be obtained from ILP not only depend on the data manipulation function complexity, but also on other characteristics of these functions like the number and the size of required memory tables, and the necessary interaction between data manipulations and control functions. This means that experiments performed only on data manipulation functions cannot show the real advantages of an ILP implementation.

Designing an experimental ILP implementation of a three level protocol suite based on a user-level TCP, we discovered that ILP reduces the number of memory accesses by 26%, but the relative amount of cache misses could never be reduced compared with a carefully designed non-ILP implementation [Brau95]. ILP throughput improvements are limited and depend heavily on several issues such as complexity of data manipulations, communication subsystem architecture, and host environment characteristics. In the experiment, these reasons decrease the throughput gain to approximately 10% in contrast to the 40 to 50% gain achieved for simple loop experiments. It is shown that ILP is very sensitive to several issues. That makes its use in existing communication systems and workstations debatable.

ILP has the main limitation that it is only applicable with certain types of protocol functions (non-ordering constraint functions) and protocol architectures (header size must be known before data manipulation processing). Another major drawback of ILP is the reduced flexibility, because the use of macros instead of function calls is required to avoid performance loss. Macros do not allow a protocol implementation to be adapted dynamically to changing application requirements or to varying network characteristics.

These experiments with ALF and ILP show that efficient implementation of distributed multimedia applications requires new architectural considerations and not only "enhanced protocol implementation techniques". However, using advanced protocol features such as fixed size headers, different packet types for control information and data, uniform processing unit sizes for different data manipulation functions could also be advantageous for ILP and ALF.

4.2 Network Conscious Applications

ALF implies that applications control their transmission. This may seem an undesirable feature. In fact, ATM is going to provide us with very high bandwidth, that applications should just aim for the best and ask for the corresponding resources. We argue however, that, on the contrary, applications should be designed with "inter-networking" in mind and they should absolutely include the proper adaptation loops. We discuss in this section, the implication of this rule on applications design. Clearly, networks are getting more and more faster. But as the technology progresses, the networking conditions are going to become more and more variable. We will have T1 and T3 lines, 10baseT and 100baseT networks, 640 Mbps Myrinet graphs, 155 Mbps ATM circuits and very variable capacities mobile networks. It should be quite obvious that the applications which can gracefully adapt to multiple environments will outlive those which have to die when their requirements are not met.

It should also be quite obvious that an application which can automatically characterize its environment will be more robust and easier to deploy than an application which has to exchange signalling information with the network for that purpose. Building the characterization loop and the adaptation code inside the application does indeed not prohibit reservation of resources. It simply decouples it from the application design. Adaptability guarantees that the application will always use all the available resources and make the most efficient usage of these resources. That these resources result from a reservation or simply from a best effort network is entirely irrelevant. On the other hand, communication over a network is not easy to guarantee. Communication protocols have been designed to minimize the unreliability of transmission over a net work. Therefore the application adaptivity is always a desirable feature because:

- Even if resource reservation is done, it will be done with a certain level of reliability, specifying a minimum and a desirable (or average) throughput. The adaptation of the application throughput can help guaranteeing the best quality of transmission within the throughput interval defined by QoS negotiation, and optimizing network resource utilization.

- Priorities can be associated to applications (following the level of service required, or the cost payed for the service). If higher level of priority reservation occurs, resource will be taken from lower priority communications. Guarantee will only be effective for higher level priority applications.

- Routers with guaranteed resource may crash. Re-routing on another router, using a new path, does not guarantee that the same resource will be available.

This concept implies that the same application which usually use high data rates may occasionally be constrained to use a few tens of kilobits. The application should then be designed to pro-actively select the most important data for transmission when the resource is scarce, rather than merely accept to be slowed down or to let the network randomly drop some messages.

We argue that this adaptation concept is very useful even for time-constrained applications (i.e. multimedia, voice, video). The devotees of "QoS models" will oppose this argument; they would rather prefer a transmission to be stopped than to see their Quality of Service not respected. Imagine being at home, watching a TV controlled by a QoS manager. You wouldn't like your TV to be switched off if a temporary degradation of the QoS occurs. Stop a communication on QoS degradation is a nice theoretical behavior; but in most of the multimedia applications, temporary and controlled graceful degradation of quality will be preferred to the connection being released.

We have experimented the "Network Consciousness" concept on IVS [Turl94b], a software system to transmit audio and video data over the Internet. It includes PCM and ADPCM audio codecs, as well as a H.261 video codec. In the initial versions of IVS, the user could control manually the maximal output rate of the coder. The variable video output rate was sent into a buffer which was drained at a constant rate. Then, the amount of data in the buffer was used as a feedback information to adapt the parameters of the coder in order not to exceed the maximal output rate allowed. The output rate was controlled by changing either the video frame rate or the quantizer value and the movement detection threshold.

In order to behave like a "good network citizen", we improved the application by adding a feedback control mechanism that prevents the video application from swamping the resources of the Internet. In this mechanism, the parameters of the coder were adjusted according to the network conditions [Bol94a]. Receivers periodically sent back to the coder the observed loss rate and the coder computed an average loss rate to estimate the network load and selected a maximal output rate value according to it.

However, the previous scheme did not scale up to a very large number of receivers: in this case a large amount of feedback is sent to the coder which would cause feedback implosion problem. To scale up to any number of receivers, we implemented a scalable feedback mechanism based on a novel probing mechanism to solicit feedback information in a scalable manner [Bol94b]. The feedback information also includes a loss rate indication sent back from the receivers to the coder. Given this feedback information, the number of receivers with low video quality reception is estimated and the video coder uses it to decide how to adjust its maximal output rate.

This example shows that is both possible and desirable to implement adaptive applications, as implied by the ALF principle.

4.3 Automatic integration

The traditional empirical methodology for protocol design and implementation, which is mostly "intuition based on experience", will not be able to scale up effectively and to produce highly integrated implementations for the new generation of applications. To be able to design efficient implementations of communication support tailored to application characteristics, we claim that the development process of communication support must largely be automated (in a formal framework). This will allow to "easily" generate a large panoply of communication subsystems tailored to application needs. A formal automated design process also allows the correctness of the communication to be checked both in terms of reliability and security, and even the efficiency can be determined given a set of constraints.

The Protocol Compiler is a step toward the support of a new generation communication model, where a distributed application can specify its own communication requirements to be associated to a dedicated transmission control protocol. The control and synchronization aspects of an application are formally specified using ESTEREL [Berr92, Cas94b]. This specification is then used by the protocol compiler to integrate transmission control facilities to the application, and to generate the client and server stubs. The advantages of this approach are the following:

- Like in the classical communication model, the transmission facilities remain transparent to the application designer.
- The control of the transmission is given to the application. The communication facilities are integrated in the application. This allows the application to adapt dynamically its throughput to the available resource on the network.
- The implementation is designed automatically, starting from the application's formal specification. It guarantees that the implementation matches exactly with the original application specification. It is also easier to prove that the communication system provides the application with minimal and efficient communication support.
- The formal approach allows the systematic use of optimization techniques such as optimizing the protocol control automaton, discovering the most frequently used path, in-lining code, etc. These optimizations might lead to better performance than with hand-coded implementations.

Other research groups are currently working on the design and implementation of communication subsystems tailored to application requirements [Oech94], [Plag94], [Sch93], [Rich94], [Diaz94], [Omal94]. The proposed solutions include developing general purpose protocols that allow flexibility. However, these solutions are not operating system independent because the implementations are either part of the kernel, or a server within a micro-kernel based operating system.

The approach that we propose is different, as it starts from the application specification, and as the transmission control facilities are integrated to the application specification before the automated implementation. Flexibility is allowed by the integration of the transmission control mechanism within the application.

A prototype Protocol Compiler has been designed [Diot95]. The JPEG player, hand-coded for the evaluation of the ALF concept, has also been implemented automatically starting from its ESTEREL specifications, and using the Protocol Compiler. With similar protocols, the automated implementation runs 20% faster than the hand-coded on experiments between INRIA (France) and UTS (Australia) (see table 4).

On the code size aspect, the Protocol Compiler produces a code which has quite the same size as the hand-code written in C language. The executable code of the receiver side is even smaller (4% on 200 Kbytes). Recent optimization work on the ESTEREL's code generation phase show that a sensible reduction of the generated code size is possible without a loss of performance. This is very promising for our automated approach.

	Local Ethernet	Internet
handcoded ALF	7.39 Mbits/s	51.3 Kbits/s
ESTEREL	7.42 Mbits/s	62.1 Kbits/s

Table 4: Throughput of both hand-coded and ESTEREL versions

5 Conclusion

Based on the evaluation of ILP and ALF concepts, this paper shows that a new high performance protocol architecture is desirable for efficient implementation of multimedia applications. We propose to base this efficient protocol architecture on three concepts:

- ALF, which is a design principle allowing efficient processing of the application data units;
- Network conscious applications will be able to adapt to resources available on the network, even in association with guaranteed bandwidth services.
- Automated integration of transmission control facilities tailored to the application requirements. We showed its feasibility in the case of an experiment with a JPEG image player.

These results are being applied with the HIPPARCH project to the design of a new generation Protocol Compiler dedicated to multimedia applications. It has been demonstrated that the automated integration of transmission control functions in an application formal specification is possible in practice. Performance results confirm that, in term of code organization, size, and efficiency, the automated approach is almost as efficient as the hand-coded approach. Using this approach, a completely automated and efficient implementation of distributed applications is possible.

Acknowledgments

The authors would like to thank Isabelle Chrisment who developped the JPEG server for providing the results of the tests.

References

[Abb93a] Mark B. Abbott, Larry L. Peterson. "Increasing Network Throughput by Integrating Protocol Layers ", *IEEE/ACM Transactions on Networking*, Vol 1, No 4, October 1993.

[Abb93b] Mark B. Abbott, Larry L. Peterson. "A Language-Based Approach to Protocol Implementation", *IEEE/ACM Transactions on Networking*, Vol 1, No 1, February 1993.

[Berr92] G. Berry, G. Gonthier. "The Esterel Synchronous Programming Language: Design, Semantics, Implementation". *Journal of Science Of Computer Programming*, Vol. 19, Num. 2, pp. 87-152. 1992.

[Bjö93] M. Björkman and P. Gunningberg. "Locking Effects in Multiprocessor Implementation of Protocols", In *Proceedings ACM SIGCOMM'93.*

[Bol94a] J.-C. Bolot, T. Turletti. "A rate control mechanism for packet video in the internet," *in Proceedings of the Conference on Computer Communications, IEEE Infocom '94*, Toronto, Canada, June 1994.

[Bol94b] J-C. Bolot, T. Turletti, I. Wakeman. "Scalable feedback control for multicast video distribution in the Internet", Proc. ACM SIGCOMM '94, Vol. 24, No 4, October 1994, pp. 58-67.

[Brau92] T. Braun and M. Zitterbart. "Parallel Transport System Design", In *Proceedings of the* 4th *IFIP Conference on High Performance Networking*, Liège, Belgium, December 1992.

[Brau95] T. Braun and C. Diot. "Protocol Implementation using ILP", SIGCOMM '95, Boston, August 1995.

[Cas94a] Claude Castelluccia and Walid Dabbous. "Modular Communication Subsystem Implementation using a Synchronous approach", In *Proceedings of USENIX-94, Symposium on High Speed Networking*, Oakland, CA, August 1994.

[Cas94b] C. Castelluccia, I. Chrisment, W. Dabbous, C. Diot, C. Huitema, E. Siegel. "Tailored Protocol Development Using ESTEREL," INRIA Research report, No 2374, October 1994.

[Cher86] D. R. Cheriton, "VMTP: a transport protocol for the next generation of communication systems", In *Proceedings ACM SIGCOMM '86,* Stowe, Vermont, August 1986, pp. 406-415.

[Ches89] G. Chesson, "XTP/PE Design Considerations", In *Protocols for High-Speed Networks,* H. Rudin, R. Williamson, Eds., Elsevier Science Publishers/North-Holland, May 1989.

[Cla90] David D. Clark and David L. Tennenhouse. "Architectural Considerations for a New Generation of Protocols", In *Proceedings ACM SIGCOMM '90,* September 24-27, 1990, Philadelphia, Pennsylvania, pp. 200-208.

[Cla89] David D. Clark, Van Jacobson, John Romkey, Howard Salwen. "An analysis of TCP processing overhead", *IEEE Communications Magazine*, June 1989, pp. 23-29.

[Cla87a] David Clark, Mark Lambert, Lixia Zhang. NETBLT: A Bulk Data Transfer Protocol. Network Information Center, *RFC-998*, SRI International, March, 1987.

[Cla87b] D. Clark, M. Lambert, L. Zhang. "NETBLT: a high throughput transport protocol", *CCR*, Volume 17, Number 5, 1987, pp. 353-359.

[Col85] R. Colella, R. Aronoff, K. Mills. "Performance Improvements for ISO Transport", *Computer Communication Review,* Vol. 15, No. 5, September 1985.

[Coo90] E. C. Cooper, P. A. Steenkiste, R. A. Sansom, and B. D. Zill. "Protocol Implementation on the Nectar Communication Processor", In *Proceedings ACM SIGCOMM '90*, Philadelphia, PA, September 1990, pp. 135-144.

[Dab94a] Walid S. Dabbous. "High performance presentation and transport mechanisms for integrated communication subsystems", In *Proceedings of the 4th Internatinal IFIP Workshop on Protocols for High Speed Networks*, Vancouver, August 1994.

[Dab93] Walid Dabbous, Christian Huitema. *"XTP implementation under Unix"*, Research Report No 2102, Institut National de Recherche en Informatique et en Automatique, November 1993.

[Dab92] Walid Dabbous et al. *"Applicability of the session and the presentation layers for the support of high speed applications"*, Technical Report No 144, Institut National de Recherche en Informatique et en Automatique, October 1992.

[Dab91] W. Dabbous, *"Etude des protocoles de contrôle de transmission à haut débit pour les applications multimédias"*, PhD Thesis, Université de Paris-Sud, March 1991.

[Diaz94] M. Diaz, C. Chassot, and A. Lozes. "From the Partial Order Connection Concept to Partial Order Multimedia Connections". First HIPPARCH workshop, INRIA Sophia Antipolis, December 15-16, 1994.

[Diot92] Christophe Diot, Patrick Coquet and Didier Stunault. *"Specifications of ETS the Enhanced Transport Service"*. Research Report 907-I, LGI-Institut IMAG, May 1992.

[Diot95] C. Diot, I. Chrisment, A. Richards. "Application Level Framing and Automated Implementation", 6th IFIP International Conference on High Performance networking, Palma (Spain), September 1995.

[Dru93] Peter Druschel, Mark B. Abbott, Michael A. Pagels, and Larry Peterson. "Network Subsystem Design", *IEEE network*, July 1993, pp. 8-17.

[Fel93] D. C. Feldmeier. "A Survey of High Performance Protocol Implementation Techniques", *High Performance Networks - Technology and Protocols*, Ed. Ahmad Tantawy, Kluwer Academic Publishers. Boston, MA, 1993, pp. 29-50.

[Gun91] P. Gunningberg, C. Partridge, T. Sirotkin, B. Victor. "Delayed evaluation of gigabit protocols", In *Proceedings of the* 2nd *MultiG Workshop*, 1991.

[Hui91] Christian Huitema. *"MAVROS, Highlights on an ASN.1 compiler"*, INRIA research note, May 1991.

[Hui90] Christian Huitema. *"Definition of the Flat Tree Light Weight Syntax (FTLWS)"*, Internal Document, INRIA Sophia Antipolis, July 1990.

[Hui89] Christian Huitema, Assem Doghri. "Defining faster transfer syntaxes for the OSI Presentation Protocol", *Computer Communication Review*, Vol 19, No 5, Oct 1989, pp. 44-55.

[Hogl94] Anna Hoglander. *Experimental evaluation of TCP in user space.* INRIA internal report, request from `cdiot@sophia.inria.fr`.

[IEEE83] IEEE Special Issue on Open Systems Interconnection (OSI). Proc. IEEE, vol. 71, no. 12, December 1983, pp 1329-1488.

[Kan88] Hemant Kanakia, David R. Cheriton. "The VMP Network Adapter Board (NAB): High-Performance Network Communication for Multiprocessors", In *Proceedings SIGCOMM '88*, Stanford, CA, 1988, pp. 175-187.

[Lap92] T. F. La Porta and M. Schwartz. "A high-Speed Protocol Parallel Implementation: Design and Analysis", In *Proceedings of the* 4th *IFIP Conference on High Performance Networking*, Liège, Belgium, December 1992.

[Lei85] B.M. Leiner, R.H. Cole, J.B. Postel, D. Mills. "The DARPA Internet Protocol Suite", In *Proceedings INFOCOM'85*, IEEE, March 1985.

[Léo92] L. Léonard. "Enhanced Transport Service Specification". *Deliverable ULg-4*, OSI 95 project, October 1992.

[O'M90] S. W. O'Malley and L. L. Peterson. "A Highly Layered Architecture for High-Speed Networks", In *Proceedings of the IFIP Workshop on Protocols for high speed networks II,* Palo Alto, CA, 1990, pp. 141-156.

[Omal94] S. W. O'Malley, T. Proebsting, and A. B. Montz. "USC : A Universal Stub Compiler". In Proceedings of ACM SIGCOMM'94. Vol. 24, No 4. October 1994.

[Oech94] P. Oechslin, S. Leue. "Enhancing Integrated Layer Processing using Common Case Anticipation and Data Dependence Analysis", In Proceedings of the 1st International Workshop on High Performance Protocol Architectures, December 15-16, 1994, Sophia-Antipolis, France

[Par93b] C. Partridge and S. Pink. A Faster UDP. Submitted to *IEEE Transaction on Networking.*

[Peh92] Bjorn Pehrson, Per Gunningberg and Stephen Pink "Distributed Multimedia Applications on Gigabit Networks", *IEEE Network Magazine*, Vol 6, No 1, January 1992, pp. 26-35.

[Plag94] T. Plagemann, B. Plattner, M. Vogt, T. Walter. "A Model for Dynamic Configuration of Light-Weight Protocols", In *Proceedings of the third workshop on FTDCS*, Tapei. Taiwan. pp. 100-110. April 1992.

[Rich94] A. Richards, A. Seneviratne, M. Fry and V. Witana. "Tailoring the Transport Protocol for Giga Bit Networks". In *Proceedings of the Australian Telecommunication Networks and Applications Conference.* 5-7 December 1994.

[Rüt92] Erich Rütsche and Matthias Kaiserwerth. "TCP/IP on the Parallel Protocol Engine", In *Proceedings of the* 4th *IFIP Conference on High Performance Networking*, Liège, Belgium, December 1992.

[Sch93] D. Schmidt, B. Stiller, T. Suda, A.N. Tantawy, and M. Zitterbart. "Language Support for Flexible Application-Tailored Protocol Configuration", *Proceedings of LCN '93.*

[Shen94] S. Shenker. "Fundamental Design Issues for the Future Internet", preprint submitted to JSAC. 1994.

[Ten89] David L. Tennenhouse. "Layered Multiplexing considered harmful", In *Proceedings of the IFIP Workshop on Protocols for high speed networks,* Zurich, Switzerland, 9-11 May, 1989.

[Turl94a] T. Turletti, C. Huitema. "Packetization of H.261 video streams", Internet Draft, Sept. 1994.

[Turl94b] T. Turletti. "The INRIA Videoconferencing System (IVS)", ConneXions - The Interoperability Report, Vol. 8, No 10, October 1994, pp. 20-24.

[Wak92] Ian Wakeman, Jon Crowcroft, Zheng Wang, and Dejan Sirovica, "Layering considered harmful", *IEEE Network*, January 1992, p. 7-16.

[Wat87] Richard W. Watson, Sandy A. Mamrak. "Gaining efficiency in transport services by appropriate design and implementation choices", *ACM transactions on computer systems,* Vol 5, No. 2, May 1987, pp 97-120.

6

Enabling High Bandwidth Applications by High-Performance Multicast Transfer Protocol Processing

G. Carle, J. Schiller
University of Karlsruhe, Institute of Telematics,
Zirkel 2, 76128 Karlsruhe, Germany
Phone: +49 721 608-[4027,4003], Fax: +49 721 388097
e-mail: [carle,schiller]@telematik.informatik.uni-karlsruhe.de

Abstract

A large range of applications exists with demand for high-performance point-to-point and point-to-multipoint communication. Existing communication subsystems frequently represent a major performance bottleneck. To overcome this bottleneck, a framework for high-performance multicast transfer protocol processing is presented, based on hardware support for multicast error control in transmitters and in dedicated intermediate systems called Group Communication Servers. The design of a protocol processing coprocessor for selective retransmissions by end systems and servers in multicast scenarios is presented. High scalability for a large number of receivers can be ensured by the deployment of a VLSI component for list management of acknowledgement processing. The integration of the VLSI component into a generic coprocessor (the Generic ATM Protocol Processing Unit, GAPPU) is shown. Details of processing delays and implementation costs of the proposed hardware implementation are given, and compared with measurements of a typical software implementation.

Keywords

ATM, group communication, VLSI, coprocessor, error control

1 INTRODUCTION

Emerging applications, mostly, require both high performance as well as support of a wide variety of real-time and non-real-time communication services. For example, audio, video, and message passing of distributed systems may require different services. Networks, (e.g., ATM-based networks) are able to fulfil the basic requirements by providing data rates exceeding a gigabit per second and by supporting different kinds of services. However, current communication subsystems (including higher layer protocols) are not able to deliver the available network performance to the applications.
In the evolution of high speed networking, various multipoint communication services will be of increasing importance. Examples of applications that require point-to-multipoint (Multicast, 1:N) as

well as multipoint-to-multipoint (Multipeer, M:N) communication can be found in the areas of computer-supported cooperative work (CSCW), distributed control, and distributed computing as, e.g., in workstation clusters (Heinrichs, 1993). For a growing number of applications such as multimedia collaboration systems, the provision of a multicast service with a specific quality of service (QoS) in terms of throughput, delay, and reliability is crucial.

If multipoint communication is not supported by the network or by the end-to-end protocols, multiple point-to-point connections must be used for distribution of identical information to the members of a group. The support of multicasting is beneficial in various ways: It saves bandwidth, reduces processing effort for the end systems, reduces the mean delay for the receivers, and simplifies addressing and connection management.

Various issues need to be addressed in order to provide group communication services in high-speed networks (Waters, 1992), (Bubenik, 1992). Intermediate systems need to incorporate a copy function for support of 1:N connections. Communication protocols must be capable of managing multipoint connections, and group management functions need to be provided for administration of members joining and leaving a group. A key problem that must be solved to provide a reliable multipoint service is the recovery from packet losses due to congestion in the network nodes and end systems.

Different approaches (Ito, 1992), (Strayer, 1992), (Feldmaier, 1994), (Sterbenz, 1991), (Braun, 1993a) on implementing high performance communication subsystems have been undertaken during the last few years: software optimisation, parallel processing, hardware support, and dedicated VLSI components. Some of the approaches deal with efficient implementations of standard protocols such as OSI TP4 or TCP. Others developed protocols especially suited for advanced implementation environments.

The use of dedicated VLSI components is mostly limited to very simple communication protocols, only (e.g., (Balraj, 1992), and (Krishnakumar, 1993)). In this paper, we present VLSI support that is especially targeted towards more complex multicast protocols. As an example for the provision of specific support for processing intensive functions, the implementation of a dedicated coprocessor for selective retransmissions in a multicast environment is shown. It is also shown how this coprocessor can be integrated into a protocol processing unit featuring parallel processing and direct ATM access.

This paper is organised as follows: Section 2 gives an overview of multipoint communication in high-speed networks and presents the conceptual framework for integrating VLSI components for multicast support into end systems and Group Communication Servers. Section 3 discusses the functionality of a dedicated coprocessor for managing retransmission, and presents performance results as well as implementation complexity of the discussed component. Section 4 summarises the paper and points out some future directions.

2 MULTIPOINT COMMUNICATION IN HIGH-SPEED NETWORKS

2.1 Error Control

The dominant factor which causes high speed networks to discard packets is buffer overflow due to congestion. The probability for packet loss may vary over a wide range, depending on the applied strategy for congestion control. For multicast connections, the problem of packet losses is even more crucial than for unicast connections. It is more difficult to ensure a low packet loss rate. Losses occur more frequently, and every loss causes costly processing for a multicast transmitter.

For applications that cannot tolerate packet losses of the network, error control mechanisms are required. Error control is a difficult task in networks that offer high bandwidth over long distances, where a large amount of data may be in transit. Two mechanisms are available for error correction: Automatic Repeat ReQuest (ARQ) and Forward Error Correction (FEC).

In contrast to the retransmission schemes, FEC promises a number of advantages (McAuley, 1990). The delay for error recovery is independent of the distance, and large bandwidth-delay products do not lead to high buffer requirements. Therefore, FEC is a promising approach in high-speed networks. In contrast to ARQ mechanisms, FEC is not affected by the number of receivers. However, FEC has three main disadvantages when applied for error correction in high speed networks. It is computationally demanding, leading to complex VLSI components. It requires constantly additional bandwidth, limiting the achievable efficiency and increasing packet loss during periods of congestion. The latter limits the usefulness of FEC in many cases. For an accurate assessment of FEC it must be considered that its best performance is achieved for random errors, while packet losses frequently occur in bursts (Biersack, 1993). The question when to apply FEC for real-time applications in high-speed networks requires extended assessments of various trade-offs. FEC has certain attractive properties in high-speed WANs and for multipoint connections. However, only retransmission schemes are able to provide fully reliable services. In many cases, retransmission schemes are superior to FEC in terms of the achievable throughput, the delay properties, or the implementation costs.

Protocols based on ARQ mechanisms are widely used in current data link and transport protocols. However, for high-performance multicast communication, there are still many open questions concerning acknowledgement and retransmission strategy, achievable performance and implementation. Retransmissions may be performed as go-back-N (e. g, in TCP) or as selective repeat (e. g., offered in XTP (XTP Forum, 1994a) and PATROCLOS (Braun, 1993b)). While go-back-N schemes are appropriate for point-to-point communication with low error rates and moderate path capacities, selective repeat schemes are essential for high-performance multicast communication in wide-area networks that may observe congestion (Carle, 1994). Large groups require that the transmitter stores and manages a large amount of status information of the receivers. The number of retransmissions is growing for larger group sizes, decreasing the achievable performance. Additionally, the transmitter must be capable of processing a large number of control information. If reliable communication to every multicast receiver is required, a substantial part of the transmitter complexity is growing proportionally with the group size. In addition, individual receivers may limit the service quality of the whole group. To overcome these problems, a scheme that provides reliable delivery of messages to K out of N receivers may be applied (K-reliable service, (Santoso, 1992)).

2.2 High-Performance Multicast Services

In order to meet the QoS requirements of many real-time applications, it is a common approach to meet the application reliability requirements without performing error control mechanisms. In situations where it is difficult to provide a network bearer service which meets the reliability requirements of the application directly, the following strategy may be applied: providing high protocol processing capability with a low latency and for ensuring that real-time requirements are met even after one or two retransmissions of a message. This strategy potentially offers a way for a better utilisation of network resources in particular for highly bursty source, as it allows to increase the load of intermediate systems up to a level in which losses relatively frequent.

In the past it was frequently debated whether real-time applications can be based on services with retransmissions. In (Dempsey, 1993), it was shown by simulation that a real-time retransmission scheme is feasible within the end-to-end delay constraints of packet voice transmissions for overall one-way delays with an average of 12 ms and a maximum of 36 ms. While the authors of (Dempsey, 1993) used relatively high network access delays and protocol processing delays in transmitter and receiver in modelling the one-way delay, the propagation delay of 5 ms for a fibre-optic transmission over a distance of 1000 km shows that retransmission schemes for real-time applications may also be applied for relatively large distances.

A conceptual framework was described (Carle, 1994) for the use of error control mechanisms best suited for a specific multipoint communication scenario at locations that allow highest performance. The integration of specialised multicast components into the end systems represents an important step towards a high performance reliable multicast service. Further improvements of performance and efficiency may be achieved by the integration of dedicated servers in the network that provide support for group communication. In many cases of multicasting, the achievable throughput degrades fast for a growing group size. A significant advantage can be achieved if a hierarchical approach for multicast error control is chosen.

The support for protocol processing presented in this paper allows selective retransmissions to multiple receivers. This error control functionality may be part of a transport protocol, such as XTP Revision 4.0 (XTP Forum, 1994b) in combination with a connectionless network layer. This functionality may also be part of a transfer protocols combining layer 3 and layer 4 functions, such as XTP Revision 3.7 and PATROCLOS (Braun, 1993a) with multicast extensions. Such a transfer protocol may be used over a conventional LLC service, or over an adaptation layer service as for example offered by AAL5.

The framework on which this paper is based applies to protocols that use gaps of packet sequence numbers for positive and negative acknowledgements. Typical protocols use either sequence numbers identifying the first byte of the payload (as for example TCP, XTP and PATROCLOS), or they use packet sequence numbers (such as TP4, SNR (Sabnani, 1990), and SSCOP (ITU, 1994)). While byte sequence numbers usually have a length of 32 bit (or even 64 bit in XTP 4.0), packet sequence numbers with a length of 24 bits are sufficient even for high-speed WANs.

Figure 1 presents a network scenario with multicast mechanisms in the transport component of end systems and in dedicated servers. The term Group Communication Server describes an intermediate system with multicast error control capability which may be attached to conventional subnetworks or to an ATM network. It may be combined with routing functionality of, e.g., an XTP router.

The Group Communication Server (GCS) presented in this paper may integrate a number of mechanisms that can be grouped into three main tasks:

- Provision of a high-quality multipoint service with efficient use of network resources;
- Provision of processing support for multicast transmitters;
- Support of heterogeneous hierarchical multicasting.

For the first task, performing error control in the server permits to increase network efficiency and to reduce delays introduced by retransmissions. Allowing retransmissions originating from the server avoids unnecessary retransmissions over common branches of a multicast tree. In order to ensure low delay, the server does not guarantee an in-sequence forwarding of packets. Instead, it will forward every packet to the receivers as soon as possible. In combination with a network node with copy function, it is not required that the server processes a packet before forwarding to the receivers. Instead, copies may be forwarded in parallel to the server and the receivers. This guarantees minimal delay while allowing that the server detects losses prior to the receiver and initiates a retransmission by the sender.

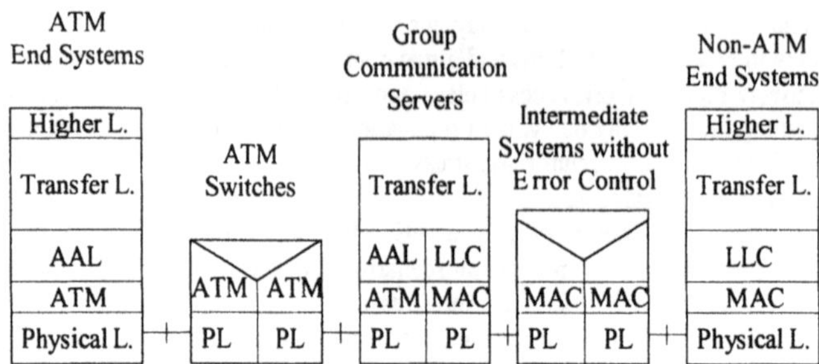

Figure 1 Support for reliable multicasting in servers and end systems.

For the second task, the GCS releases the burden of a transmitter that deals with a large number of receivers, providing scalability. Instead of communicating with all receivers of a group simultaneously, it is possible for a sender to communicate with a small number of GCSs, where each of them provides reliable delivery to a subset of the receivers. Integrating hardware support for reliable high performance multipoint communication in a server allows better use of dedicated resources such as coprocessors. For end systems, it is not required to have VLSI components for multicast error control. It will be sufficient to have access to a local GCS for participation in a high performance multipoint communication over long distances. Then, the error control mechanisms of individual end systems have only negligible influence on the overall performance, as simple error control mechanisms are sufficient for communication with a local GCS.

For the third task, a GCS may use the potential of diversifying outgoing data streams, allowing support of different qualities of service for individual servers or subgroups, may apply filtering functions for specific data streams. It may support different subnetworks, such as FDDI and ATM, where a different set of protocol parameters may be appropriate, and may also support different error control schemes, such as Go-back-N and selective repeat.

3 VLSI FOR RETRANSMISSION SUPPORT

The processing overhead associated with handling selective retransmissions and the required data structures may be extremely high compared to other protocol functions. As an example, the following protocol processing latencies may be observed for an XTP (XTP Forum, 1994a) implementation on Digital Alpha Workstations (150 MHz, 6.66 ns cycle time). The function to insert a new gap in a list needs 824 4-byte commands in the best case and takes, therefore, approximately 5.4 µs. The best case occurs if the new entry can be inserted at the beginning of the list. If the new entry has to be inserted after the first 10 entries, it needs 4054 commands or approximately 27.03 µs due to the search operations in the list. These calculations assume that the processor is not interrupted during execution of this function and all data is stored in the fast processor cache. In this example, XTP was implemented using C without special inline assembly code. Data sent at a rate of 1 Gbit/s results in more than 122,000 1024-byte packets per second. If the retransmission of each packet has to be controlled, this results in a new entry in the list in less than 10 µs.

Clearly, retransmission support is a time-critical task especially in a multicast environment. Therefore, we propose dedicated VLSI support for this task. The retransmission support presented in the following section can handle negative selective, positive selective, and positive cumulative acknowledgements and can be used for gaps managed by the receiver to support the acknowledgement function, or for gaps managed by the transmitter to support the retransmission mechanism. The ALU has a set of commands to set, delete, insert, and read gaps for unicast or multicast connections. It can be used in Group Communication Servers as well as in other high performance end systems (Braun, 1994).

3.1 Logical Representation of Data

A dynamic linked list stores gaps of transmitted data in the following representation: *[seq_no_1, seq_no_2]* with *seq_no_1* and *seq_no_2* representing the beginning and end of a gap. These gaps are connected via linked lists (cf. Figure 2). For every multicast connection (MC), the pointers to the receivers participating in that connection are stored. For every receiver, the ALU stores a pointer to the appropriate list of gaps. Additionally, the ALU manages special lists for every multicast connection.

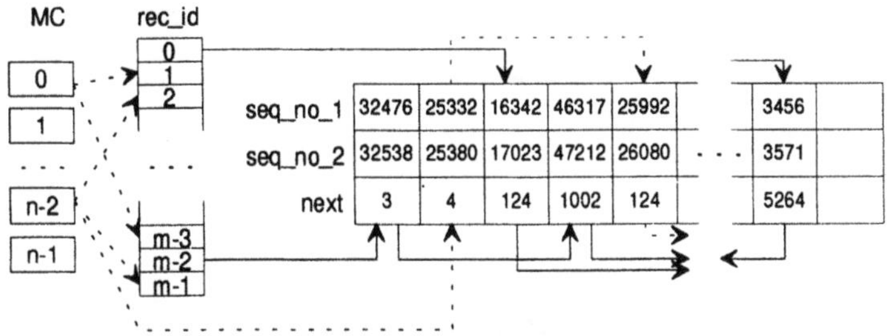

Figure 2 Logical structure of linked lists for retransmission support.

Depending on the implemented protocol, retransmission of data can be performed by multicast to all receivers, or by individual retransmissions to the appropriate receivers.

3.2 Operations of the Retransmission ALU

The component performs not only the insert and delete operations for the list, but also joins two adjoining gaps and updates the group list for a given reliability. Appendix A gives some examples of implemented operations of the retransmission ALU. Every operation sets the error flag if it failed due to memory overflow or violation of several conditions, such as $high\_ack \leq seq\_no \leq high\_seq$ and other range checking.

3.3 Implementation Architecture for Retransmission Support

Figure 3 shows an overview of the internal structure of the ALU. The retransmission ALU consists of a data memory that stores sequence numbers representing gaps, pointers of linked lists, and state information, such as connection and multicast identification, register number of an anchor element, and other flags indicating the state of a connection. The I/O-bus connects the input/output-port (32 bit) of the retransmission ALU with the 5 register banks (Ai through Ei, $0 \leq i \leq 3$). From these registers data can be transferred to the memory via the move unit.

Two simple ALUs (*ALU A*, 32 bit and *ALU D*, 8 bit) perform the operations OR, XOR, AND, ADD, SUB, and NEG. Two specialised 32 bit modulo 2^{32} comparators (*COMP 1* and *COMP 2*) perform fast comparisons needed for list operations. The ALUs and the comparators can work concurrently if no data dependencies exist.

For the command `set_gap_2`, for example, first of all the command itself and the connection identification are read from the I/O-bus into the registers E and D, respectively. In the next two cycles the central control unit reads the sequence numbers (`seq_no_1`, `seq_no_2`) into the registers A and B, respectively. After reading the complete command and performing several range checking operations, the loop for searching the right position to insert the new entry in the list starts. Therefore, the first entry of the appropriate list is loaded into the registers A and B, respectively, and compared with the new entry. The loop terminates if the new entry fits. Otherwise, the next entry is loaded and compared.

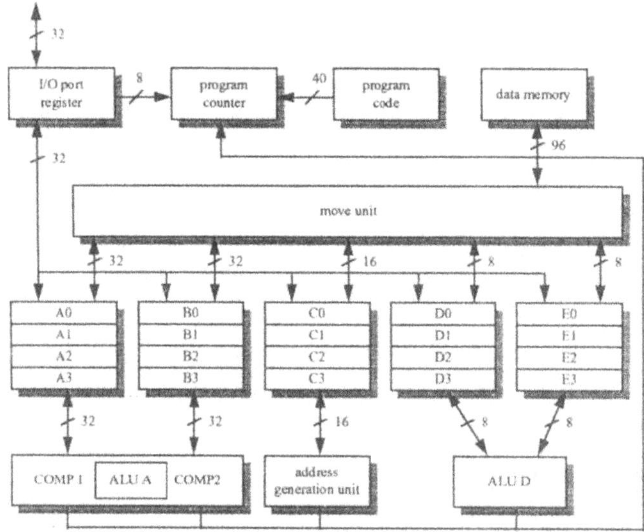

Figure 3 Functional architecture of the retransmission ALU.

3.4 Memory Management

The retransmission processor stores all list data in a single address space. Therefore, it is necessary to design a special memory management unit. Memory is divided into 8 separated areas (cf. Figure 4), each with a dedicated LIFO to manage the pointers to the memory.

The *connection list* stores information related to a certain unicast connection, such as the lowest and highest sequence numbers in use, a pointer to the appropriate unicast list, and additional status information (number of gaps, empty etc.). The gaps of an unicast connection are stored in the *unicast list* with starting and ending sequence numbers and a pointer to the next entry. Maximum memory for unicast applications can be calculated as follows:

$$memory\_depth = NR + NG*NR$$
$$memory\_size = memory\_depth * 96\ Bit$$

with *NR* denoting the maximum number of receivers and *NG* the maximum number of gaps stored per receiver.

65535	multicast auxiliary list 2
	multicast auxiliary list 1
	multicast list
	multicast start list
	group list
	group start list
	unicast list
0	connection list

Figure 4 Memory structure for the retransmission processor.

Additional memory areas are needed to support multicast. The *group start list* stores besides status information a pointer to the connection lists of its first member and a pointer to the appropriate group list. The *group list* itself stores the pointers to the other members of a group. This is done to provide a maximum of flexibility, weather there are 500 multicast connections with up to 10 receivers, 10 multicast connections with up to 300 receivers, or other combinations. The *multicast start list* and the *multicast list* store similar to the connection list and the unicast list the lowest and highest sequence number in use inside a multicast group and all gap information. The *multicast auxiliary list 1* and *2* are used to support list operations. Maximum memory for multicast application is calculated as follows:

$$memory\_depth = NR + NG*NR + MC + MC*(RM-1) + (RM-1) + RM*NG + NG + NG$$

$$memory\_size = memory\_depth * 96\ Bit$$

with *MC* denoting the maximum number of multicast connections, and *RM* denoting the maximum number of receivers per multicast connection. A multicast scenario with, e.g., NR = 1024 receivers with a maximum of NG = 30 gaps per receiver and 32 multicast connections with up to RM = 256 receivers each, results in a maximum memory need of 47931 * 96 Bit = 562 kByte.

One essential feature of this implementation is its inherent flexibility. Protocols like SSCOP (ITU, 1994) or RMC-AAL (Carle, 1994) use sequence numbers of 24 bits instead of instead of 32 bit sequence numbers of XTP. The ALU can be easily adapted to smaller sequence numbers. An ATM WAN with 5000 km maximum distance and a link capacity of 600 Mbit/s has a round trip capacity of approximately 70000 cells. The component will be dimensioned as follows. If the cell loss rate is known to be less than 10^{-4} for the time interval of one RTT, at most 7 cells will be lost during one RTT. If independent cell losses are assumed, at most 7 packets are corrupted during this time interval. Therefore, assuming an average of 8 gaps will be sufficient. With a maximum of 1024 connections in parallel, approximately 1024 + 8*1024 = 9216 entries will be needed in the list for unicast connections only. Now the memory width is only 24+24+16+8+8 = 80 bit. For this scenario an additional 9216*80 = 90kbyte RAM is needed. For the multicast scenario shown above, memory size can be calculated as follows.

$$memory\_size = [1024 + 8*1024 + 32 + 32*255 + 255 + 256*8 + 8 + 8]*80\ bit = 2.4kbyte$$

The complete VLSI coprocessor for this scenario is described in section 3.6.

3.5 Microcode Examples of the Retransmission ALU

To provide a maximum of flexibility all functions of the *Retransmission ALU* are translated into a sequence of microcode operations. These operations are adapted to the implementation architecture shown in Figure 3. The *program counter* controls the microprogram via a special micro sequencer.

Essential microcode operations of the *Retransmission ALU* are listed in the following table. The operations of the ALUs, the central control unit and the comparators are always executed in parallel in one clock cycle. A representative example for the use of the microcode operations can be found in Appendix B, where range-checking for the operation `delete_gap` is listed.

Table 1 Microcode examples of the retransmission ALU

operations	*comment*
`RMOVE S, D`	move a complete row of entries from the registers or RAM into the registers or RAM. S, D ∈ {Ri, RAM; $0 \leq i \leq 3$}, Rn = (An, Bn, Cn, Dn, En), $S \neq D$
`ANOP`	no operation, ALU A
`MOVE I/O, D`	move data from the I/O-bus into the register D; D ∈ {Ai, Bi; $0 \leq i \leq 3$}
`TBBC Ri.n, ra`	test bit n of register Ri and branch to relative address ra if clear; R ∈ {Di, Ei; $0 \leq i \leq 3$}, $0 \leq n \leq 7$
`CBMOD Ri, Rj, Rk, Rl, ra1, ra2, ra3`	compare $Ri \lesssim Rj \leq Rk$ and $Rj \leq Rk \leq Rl$ modulo 2^{32} and branch to: result = 00 then PC := PC + 1; result = 01 then PC := PC + ra1; result = 10 then PC := PC + ra2; result = 11 then PC := PC + ra3; PC: program counter; Ri, Rj, Rk, Rl ∈ {Ai, Bi; $0 \leq i \leq 3$}
`AADD S, D`	S + D -> D; S, D ∈ {Ai, Bi; $0 \leq i \leq 3$}

3.6 Implementation Results

The retransmission processor was designed, simulated, and synthesised using the hardware description language VHDL (IEEE, 1987) in combination with commercial design tools. The control logic of the processor needs 28800 gates, the critical path is 45 ns using a 0.7 µm CMOS standard cell library. The two comparators COMP 1 and COMP 2 together with the ALU A need 8000 gates. The comparators are implemented as four 8 bit carry-select adders forming a ripple carry adder. The address generation unit consists of 2000 gates plus 7300 gates for memory management. 1600 gates are used in the 8 bit ALU D. The data registers Ai through Ei need 8400 altogether. For the program counter including its micro-sequencer 1500 gates are needed. The move unit is distributed over the registers and, therefore, included in their gate count. It needs 10 ns to decode a microcode operation, 7 ns to fetch the appropriate data from a register, a maximum of 23 ns to execute the operation, and 5 ns to store the results in the registers. Performing only one half of the CBMOD (c.f. Table 1) operation on an Alpha processor needs 11 operations which results in a duration of more than 72 ns (6.6 ns cycle time, incl. load/store). The coprocessor needs only 45 ns for the complete CBMOD operation and, therefore, this is the point of further optimisations. This implementation also allows to perform up to four operations in parallel. The die size of this chip for the control logic is 49 mm² if the chip is adapted to XTP (c.f. Appendix C). Assuming RMC-AAL as protocol, not only the data paths can be smaller, but also operations like CBMOD will be faster. The reason for this is the trade-of for speed and chip-area. For the comparators now only three 8 bit carry select adders will be needed and, therefore, one does not have to wait for the forth carry select adder until it gets the carry-bit from the third unit and finally present the result at the output. For further increase in speed the whole adder could be built as carry select adder or other fast implementation variants. Table 2 shows the influence of four different protocols on the chip size and speed if this component is used to support list processing. The differences are mainly due to different sizes of the sequence numbers.

Table 2 Influence of different protocols on synthesis results

	XTP 3.6	*XTP 4.0*	*SSCOP, RMC-AAL*
number of gates			
COMP1, COMP2, ALU A	8 000	16 000	6 000
address generation	2 000	2 000	2 000
memory management	7 300	7 300	7 300
ALU D	1 600	1 600	1 600
registers A through E	8 400	16 000	7 000
program counter, µsequencer	1 500	1 500	1 500
Σ	28 800	44 400	25 400
critical path (ns)			
decoding	10	10	10
fetch data	7	7	7
execution	23	35	20
store data	5	5	5
Σ	45	57	42

3.7 System integration

Distribution of protocol processing tasks onto a number of units operating in parallel plays a key role in the provision of high performance services (Zitterbart, 1993). The functionality of transfer protocols to be executed in end systems and in the GCS can be distributed onto several general purpose processors. Additional performance improvements can be achieved by additional support with dedicated coprocessors. Advances in VLSI technology allow to integrate multiple RISC processors, memory, a switching unit, and additional components for special processing and I/O onto a single chip.

Figure 5 shows an integration of the list processing unit into a generic coprocessor with direct ATM interfaces that provides multiple processing units coupled by a switching unit. This architecture is similar to the TMS320C80 (Texas Instruments, 1994). The unit called GAPPU (Generic ATM Protocol Processing Unit) is unique in combining parallel processing with direct ATM access and components for retransmission support. In Figure 5, the 6 microprocessors provided by GAPPU are used for a software implementation of basic GCS modules (Receiver Processor, Transmit Processor, Send Manger, Ack Manager, Frame Manager Receive, Frame Manager Transmit).

Special coprocessors provide more complex time-critical functions. One example is the list coprocessor discussed above. UTOPIA is used as interface to ATM, a DMA unit performs all data transfer operations between host and network interface. For use in interworking units further network interfaces could be added. All components can communicate via a simple crossbar switch which supports two priority levels. To guarantee certain quality of service a fair round-robin scheduling strategy is used with the crossbar. Up to now, the list coprocessor, the timer unit and the FEC unit together with the crossbar and memory have been implemented using VHDL and powerful synthesis tools.

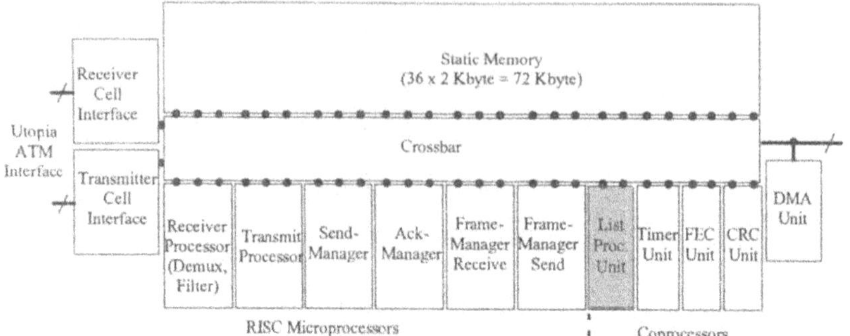

Figure 5 Architecture of GAPPU, the Generic ATM Protocol Processing Unit.

4 SUMMARY AND FUTURE WORK

Within this paper, a framework for the provision of high performance multicast services has been presented which has the potential to fulfil the requirements of upcoming distributed applications. It is based on VLSI components dedicated to specific processing tasks that are to be integrated into end systems and Group Communication Servers. Details of retransmission support have been discussed. The architecture of the generic protocol processing unit shows how multiple microprocessors and specialised VLSI components can be combined on a single chip.

However, not only high protocol processing performance and efficient use of network resources, but specifically service integration are required for forthcoming communication subsystems. Components may be parametrized based on the requested application service, providing a high degree of flexibility. Currently, the implementation of additional components for CRC and memory management are under development..

Acknowledgement

The support by the Graduiertenkolleg „Controllability of Complex Systems“ (DFG Vo287/5-2) is gratefully acknowledged.

5 REFERENCES

Balraj, T.; Yemini, Y. (1992) *Putting the Transport Layer on VLSI - the PROMPT Protocol Chip;* in: Pehrson, B.; Gunningberg, P.; Pink, S. (eds.): Protocols for High-Speed Networks, III, North-Holland, pp. 19-34

Biersack, E. W. (1993) *Performance Evaluation of Forward Error Correction in an ATM Environment;* IEEE Journal on Selected Areas in Communication, Volume 11, Number 4, pp. 631-640, May 1993

Braun, T.; Zitterbart, M. (1993a) *Parallel Transport System Design;* in: Danthine, A.; Spaniol, O. (eds.): High Performance Networking, IV, IFIP, North-Holland, pp. 397-412

Braun, T. (1993b) *A Parallel Transport Subsystem for Cell-Based High-Speed Networks;* Ph.D. Thesis (in German), University of Karlsruhe, Germany, VDI-Verlag, Düsseldorf, Germany

Braun, T.; Schiller, J.; Zitterbart, M. (1994) *A Highly Modular VLSI Implementation Architecture for Parallel Transport Protocols;* IFIP 4th International Workshop on Protocols for High-Speed Networks, Vancouver, Canada, August 10-12, 1994

Bubenik, R.; Gaddis, M.; DeHart, J. (1992) *Communicating with virtual paths and virtual channels*; Proceedings of the Eleventh Annual Joint Conference of the IEEE Computer and Communications Societies INFOCOM'92, pp. 1035 - 1042, Florence, Italy, May 1992

Carle, G. (1994) *Adaptation Layer and Group Communication Server for Reliable Multipoint Services in ATM Networks;* in: Steinmetz, R. (ed.): Multimedia: Advanced Teleservices and High-Speed Communication Architectures, Springer, 1994, pp. 124-138

Dempsey, B.; Liebherr, J.; Weaver, A. (1993) *A New Error Control Scheme for Packetized Voice over High-Speed Local Area Networks*, Proceedings of 18th Conference on Local Computer Networks, September 19-22, 1993, Minneapolis, Minnesota, U.S.A, pp. 91-100

Feldmeier, D.C. (1994) *An Overview of the TP++ Transport Protocol; in:* Tantawy A.N. (ed.): High Performance Communication, Kluwer Academic Publishers

Heinrichs, B.; Jakobs, K.; Carone, A. (1993) *High performance transfer services to support multimedia group communications*; Computer Communications, Volume 16, Number 9, September 1993

IEEE (1987) *Standard VHDL Language Reference Manual;* IEEE Std 1076-1987

Ito, M.; Takeuchi, L.; Neufeld, G. (1992) *Evaluation of a Multiprocessing Approach for OSI Protocol Processing*; Proceedings of the First International Conference on Computer Communications and Networks, San Diego, CA, USA, June 8-10, 1992

ITU-T (1994) Draft Recommendation Q.2110: *B-ISDN Adaptation Layer - Service Specific Connection Oriented Protocol (SSCOP)*, Geneva

Krishnakumar, A.S.; Kneuer, J.G.; Shaw, A.J. (1993) *HIPOD: An Architecture for High-Speed Protocol Implementations*; in: Danthine, A.; Spaniol, O. (eds.): High Performance Networking, IV, IFIP, North-Holland, pp. 383-396

McAuley, A. (1990) *Reliable Broadband Communication Using a Burst Erasure Correcting Code*; Presented at ACM SIGCOMM '90, Philadelphia, PA, U.S.A., September 1990

Sabnani, K.; Netravali, A.; Roome, R. (1990) *Design and Implementation of a High Speed Transport Protocol*, IEEE Transactions on Communications, Vol. 38, No. 11, pp. 2010-2024, November 1990

Santoso, H.; Fdida, S. (1993) *Transport Layer Multicast: An Enhancement for XTP Bucket Error Control*, in: Danthine, A.; Spaniol, O. (eds.): High Performance Networking, IV, IFIP, North-Holland

Sterbenz, J.P.G.; Parulkar, G.M. (1991) *AXON Host-Network Interface Architecture for Gigabit Communications;* in: Johnson, M. J. (ed.): Protocols for High-Speed Networks, II, North-Holland, pp. 211-236

Strayer, W.T.; Dempsey, B.J.; Weaver, A.C. (1992) *XTP: The Xpress Transfer Protocol;* Addison-Wesley Publishing Company

Texas Instruments (1994) *TMS320C80 Multimedia Video Processor (MVP): Technical Brief*, Texas Instruments, Houston, Texas

The XTP Forum (1994a) *XTP Protocol Definition Proposed Revision 3.7*

The XTP Forum (1994b) *XTP Protocol Definition Proposed Revision 4.0*

Waters, A. G. (1992) *Multicast Provision for High Speed Networks*; 4th IFIP Conference on High Performance Networking HPN'92, Liège, Belgium, December 1992

Zitterbart, M.; Tantawy, A.N.; Stiller, B.; Braun, T. (1993) *On Transport Systems For ATM Networks.* Proceedings of IEEE Tricomm, Raleigh, North Carolina, U.S.A., April 1993

Appendix A: Example Operations of Retransmission ALU

operation	*input parameters*	*output parameters*	*comment*
init_list	rec_id, seq_no		initializes a new list for the connection rec_id with the initial sequence number seq_no, sets the error flag if rec_id is already in use
close_list	rec_id		closes the list for connection rec_id, sets the error flag if the list does not exist
init_mcg	mc_con_id, rel		initializes a new multicast group with the identification mc_con_id and the reliability rel (rel ≥ number of connections denotes full reliability)
close_mcg	mc_con_id		closes a mulitcast group and deletes all linked lists
add_mcg	mc_con_id, rec_id		adds a new connection rec_id to an existing multicast group mc_con_id
set_rel	mc_con_id, k		sets the value k for the reliability of the multicast group mc_con_id
set_high_ack	rec_id, seq_no		sets the high_ack register to the value of seq_no; sequence numbers less than high_ack have been already acknowledged
shift_high_seq	rec_id, length		shifts the high_seq register to high_seq + length
set_gap_1	rec_id, seq_no, length		inserts new entry (seq_no, seq_no + length); overlapping entries are automatically joined or deleted, respectively
del_gap_1	rec_id, seq_no, length		deletes an existing entry, a part of an existing entry, or several existing entries, the deleted part is of the form (seq_no, seq_no + length); if necessary an entry is automatically divided into two new entries
read_reg	rec_id, reg_id	cont	reads the contents cont of the register reg_id (e.g. high_ack, high_seq, number_of_gaps)
read_mc_reg	mc_con_id, reg_id	cont	reads the contents cont of the multicast register reg_id (e.g. number_of_gaps)
get_gap_1	rec_id, ptr	seq_no, length, next	reads the entry ptr points to; if ptr = 0, the first gap is read out, if next = 0 the entry represented by (seq_no, length) is the last one, otherwise next point always to the next entry of the list
get_mc_gap_1	mc_con_id, ptr	seq_no, length, next	analogous to get_gap_1, but now the entries of the multicast group mc_is arc read out

Dimensioning of the component: *rec_id, k* ∈ [0, 255]; *mc_con_id* ∈ [0, 63]; *seq_no, seq_no_1, seq_no_2, length, cont* ∈ $[0, 2^{32}-1]$; *reg_id* ∈ [0, 15]; *ptr, next* ∈ $[0, 2^{16}-1]$

Appendix B: Microcode

```
-- DEL_GAP_1
-- delete an existing entry, a part of an existing entry, or several existing
-- entries, the deleted part is of the format (seq_no_start, seq_no_end)
-- if necessary an entry is automatically divided into two new entries
-- start address: 002C (preprocessing for subroutine delete gap)

002C: MOVE I/O, C0        -- load connection ID
      RMOVE[L] C0,R1; MOVE I/O,A0   -- load context and seq_no_start
                MOVE I/O,B0 -- load seq_no_end
      TBBC D1.7,D1_ERR       -- exists list?
      INC A1,A2              -- high_ack+1
      INC B1,B2              -- high_seq+1
      MOVE A7,A3             -- load first gap L0
      CBMOD A3,A0,A2,B2,5,4,4 -- compare: high_ack-L0, seq_no_start, high_ack+1
                             -- high_ack+1, seq_no_start, high_seq+1
      SETBIT ACK             -- error
      SETBIT BUSY
      JMP WAIT
      MOVE A2,A0             -- seq_no_start:=high_ack+1
      CBMOD A3,B0,A2,B2,2,3,3 -- compare: high_ack-L0, seq_no_end, high_ack+1
                             -- high_ack+1, seq_no_end, high_seq+1
      MOVE B0,B1             -- high_seq:=seq_no_end
      JMP DEL_GAP            -- delete gap subroutine
      SETBIT BUSY
      SETBIT ACK
      JMP WAIT
D1_ERR:     SETBIT ACK
      SETBIT BUSY
      JMP WAIT
```

Appendix C: Chip Layout

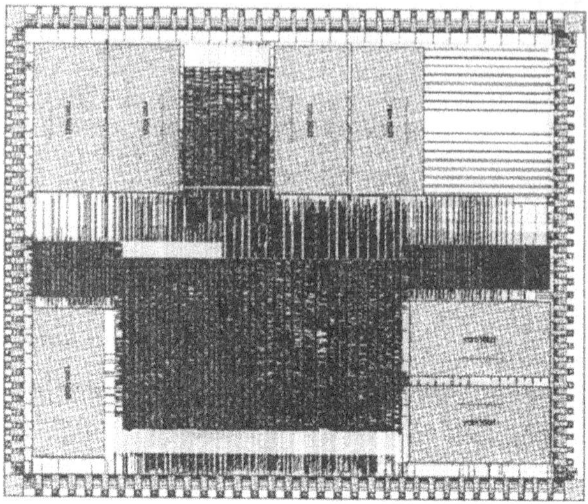

Biographies

Georg Carle

Since Sept. 1992, Georg Carle is preparing a PhD in Computer Science at the University of Karlsruhe, Institute of Telematics. He obtained a degree in Electrical Engineering from the University of Stuttgart, where he performed his diploma project at the Institute for Communication Switching and Data Techniques. In 1990, he performed a seven month research project at the Département Communications, Télécom Paris. In 1989, he obtained the degree Master of Science in Digital Systems at Brunel University, London, U.K. His main areas of interest are protocol engineering for ATM networks, high performance protocol implementation, and performance evaluation.

Jochen Schiller

In 1993 Jochen Schiller received his diploma degree in Computer Science from the University of Karlsruhe, Germany. Currently he finishes his PhD thesis at the Institute of Telematics. He participates in the fellowship program 'Controllability of Complex Systems' of the faculty of Computer Science. Key aspects of his work are the efficient implementation of protocol functions in hardware, hardware/software codesign and the support of automatic synthesis of high-performance communication systems from formal specifications. He is member of the IEEE Computer and Communication Society since 1993.

PART THREE

Switching

7

Performance of an ATM LAN switch with back-pressure function

Hiroyuki Ohsaki, Naoki Wakamiya, Masayuki Murata and Hideo Miyahara
Department of Information and Computer Sciences
Faculty of Engineering Science, Osaka University
1-3 Machikaneyama, Toyonaka, Osaka 560, Japan
(Phone) +81-6-850-6588
(Fax) +81-6-850-6589
(E-mail) oosaki@ics.es.osaka-u.ac.jp

Abstract

Traffic control schemes for ATM networks can be classified into two categories: reactive congestion control and preventive congestion control. Reactive congestion controlcan be effective in ATM local area networks as well as preventive congestion control. A possible scheme to realize efficient reactive congestion control is a switch architecture, which possesses buffers on both sides of input and output ports with a back-pressure function. Especially, when this switch is applied to ATM LANs for data transfer services, its performance should be evaluated by taking into account the bursty traffic, which is a main purpose of the current paper. In this paper, we show the maximum throughput of such an ATM switch with back-pressure function under bursty traffic through an analytic method. In addition to a balanced traffic condition, unbalanced traffic and a mixture of bursty and stream traffic are also considered. Through numerical examples, we show the effects of the average packet length and the output buffer size on the perfomance of the switch quantitively.

Keywords

ATM LAN, input and output buffer switch, back-pressure function, bursty traffic

1 INTRODUCTION

An ATM (Asynchronous Transfer Mode) technology realizes B-ISDN (Broadband Integrated Services Digital Network) by asynchronously treating various multimedia information such as data, voice and video. The benefit of the ATM technique is enjoyed by a statistical multiplexing of multimedia traffic by dividing it into fixed size packets (called cells). Much efforts of researches, developments and standardizations have been extensively devoted to public wide area ATM networks. In addition, the ATM technology also seems to be promising for realization of new high speed local area networks (LANs) to cope with a rapid advance of high-speed and multimedia-oriented computers.

Traffic control is an important issue for an efficient utilization of network resources in an ATM based network including wide and local area networks. Traffic control schemes can be classified into two categories; reactive congestion control and preventive congestion control. The reactive congestion control is the way to resolve network congestion after its occurrence. The preventive congestion control is, on the contrary, to prevent a network from its falling into congestion. The latter is now widely recognized as an effective way in wide area networks since the propagation delay is not negligible and QOS (Quality Of Service) requirements should be preserved in a strict manner. In ATM LANs, however, the propagation delay is small and is used in a private environment. Hence, preventive congestion control becomes meaningful because of its easier implementation.

To implement preventive congestion control in ATM LANs, Fan et al. recently proposes a switch architecture which possesses buffers on both sides of input and output ports with a back-pressure function [Ili92a]. The back-pressure function is provided to avoid a temporary congestion by prohibiting transmission of cells from input buffer to output buffer when the number of cells in output buffer exceeds a some threshold value. The performance of this kind of the switch has been analyzed by Iliadis in [Ili92a, Ili90, Ili92b]. However, he assumed that interarrival times of cells at each input port follow a geometric distribution. Especially when the above switch is applied to ATM LANs for supporting data transfer service, its performance should be evaluated by taking into account the bursty nature of arriving traffic, i.e., packets coming from the upper protocol layers. In this paper, we show the performance of an ATM LAN switch with back-pressure function against bursty traffic, that is, the continuously arriving cells (forming a packet) which are destined for the same output port are treated for the analysis.

This paper is organized as follows. In Section 2, an analytic model of the ATM switch we will evaluate is described. In Section 3, the steady state probability of our model is derived. In Section 4, the maximum throughput is derived based on the results of Section 3. Our subjects of investigation are extended to an imbalance traffic at input and output ports, and a mixture with stream traffic as well. Finally, in Section 5, we conclude our paper with some remarks.

2 ANALYTIC MODEL

In this section, we describe an ATM LAN switch with back-pressure function followed by an introduction of our analytic model. The number of input ports (and output ports) is assumed to be N. Our ATM switch is equipped with buffers at both sides of input and output ports (see Figure 1), and the buffer sizes are defined as N_I and N_O, respectively. The switching speed of cells from input buffer to output buffer is N times faster than the link speed, that is, at most N cells may be transferred from input buffer to output buffer in a time slot. It is called the back-pressure function to prohibit transmission of cells from input buffer to output buffer by signaling back from output buffer to input buffer when the number of cells in output buffer exceeds a some predefined threshold value [RFM94]. By this control, a cell overflow at output buffer can be avoided (see Figure 2). However, it introduces HOL (Head Of Line) blocking of cells at input buffer, which results in the limitation of the switch performance.

We assume that a stream of successively arriving cells forms a packet, and the number of cells in the packet follows a geometric distribution with mean $\overline{BL}$. Let p denote the probability that

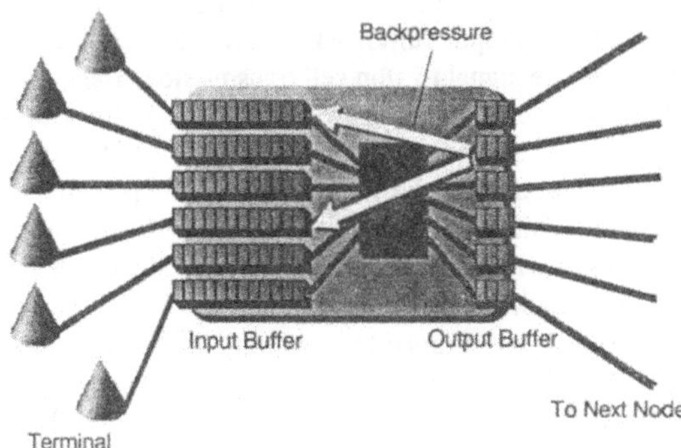

Figure 1 The ATM switch with back-pressure function.

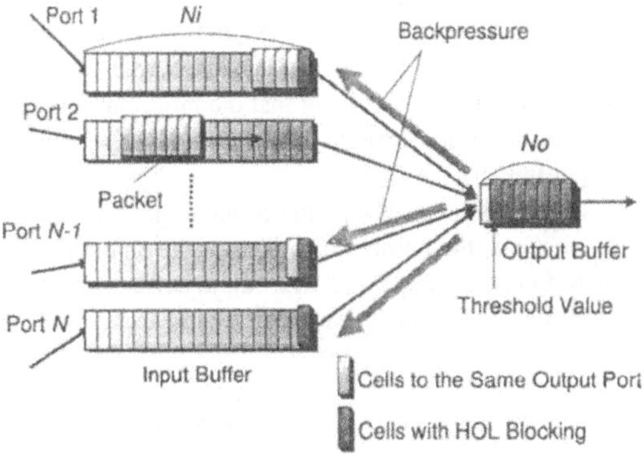

Figure 2 The analytic model.

a newly arriving cells belong to the same packet which is arriving at same input port. Thus, we have a relation;

$$\overline{BL} = \sum_{i=1}^{\infty} i(1-p)p^{i-1} = \frac{1}{1-p}. \tag{1}$$

We will assume that all cells are stored under first-in-and-first-out (FIFO) discipline at input buffer.

The practical threshold value at output buffer would be $N_O - N$ as proposed in [RFM94]. However, as an ideal case, we assume that the HOL cells are transferred from input buffer to output buffer at random until the output buffer is filled up. In other words, when the output buffer is fully occupied with cells, input buffers which have HOL cells destined for that output buffer receives the back-pressure signal to stop cell transmission. Then, all HOL cells are awaited at the head of each input buffer. As soon as the cell in output buffer are transmitted onto the output link, one of HOL cells is selected at random and transmitted to the output buffer. Therefore, it is considered that HOL cells destined for the same output port form a virtual queue which we will call a HOL queue. While HOL cells are actually stored at the HOL queue, it can be regarded that HOL packets form the HOL queue in our modeling. Therefore, in what follows, we will use "HOL cell" and "HOL packet" without discrimination.

The switch size N will be assumed to be infinity in the following analysis. By introducing the assumption of the infinite switch size, we can focus on one single output port and its associated HOL queue. The infinite switch size gives the performance limitation as shown in [Ili90, KHM86]. For example, when compared with the finite case, the maximum throughput with the infinite case gives an upper limit. Further, it is known that the close value are obtained when N reaches at 16 or 32 when the cell interarrivals follow a geometric distribution [Ili90]. In this paper, we will examine this fact in the case of bursty traffic in Section 4.

In this paper, we will first assume the capacity of the input buffer N_I to be infinity in obtaining the maximum throughput and the packet delay distribution. This assumption is realistic because the memory speed of the output buffer should be N times faster than the link speed. Thus, the capacity of the output buffer should be limited. On the other hand, the input buffer can be operated at the same speed as the link, which results in that the large capacity can be equipped. In what follows, we consider a discrete time system in which its slot time equals to a cell transmission time on the input/output link.

Under assumptions described in the above, the system state is represented by two random variables Q_k and H_k, where Q_k is the number of cells at some output buffer and H_k is the number of HOL cells at input buffers associated with that output buffer, respectively. In the next section, the steady state probability of the doublet of two random variables (Q_k, H_k) is derived.

3 DERIVATION OF STEADY STATE PROBABILITY

In the following sections, we focus on a single output port and its associated HOL queue without loss of generality. Let H_k and Q_k denote the random variables for the number of HOL cells and the number of cells in the output buffer at k-th slot, respectively. We further introduce A_k for a random variable to represent the number of HOL packets newly arriving at the HOL queue at the beginning of k-th slot. By defining a symbol $(x)^+ = \max(0, x)$, we have the following equations.

1. $H_{k-1} + A_k \leq N_O - (Q_{k-1} - 1)^+$, that is, all HOL cells can be transferred to the output port: At first, we have

$$Q_k = (Q_{k-1} - 1)^+ + H_{k-1} + A_k. \tag{2}$$

Let B_k be the number of the HOL packets which further generate HOL cells at the current k-th slot. When there exist i HOL packets in HOL queue, the probability that B_k becomes j

is:

$$b_{i,j} = \binom{i}{j} p^j (1-p)^{i-j}, \tag{3}$$

and we have

$$H_k = B_k. \tag{4}$$

2. $H_{k-1} + A_k > N_O - (Q_{k-1} - 1)^+$, that is, some HOL cells cannot be transferred to the output port at k-th slot:
   $N_O - (Q_{k-1} - 1)^+$ HOL cells are transferred to the output buffer, and C_k cells out of them further generate HOL cells in the current k-th slot. Therefore, $H_{k-1} + A_k - (N_O - (Q_{k-1} - 1)^+)$ cells are kept waiting at the HOL queue. Hence,

$$Q_k = N_O, \tag{5}$$

$$H_k = H_{k-1} + A_k - (N_O - (Q_{k-1} - 1)^+) + C_k. \tag{6}$$

Since the switch size N is assumed to be infinity, arrivals of packets at input ports in time slot are assumed to follow a Poisson distribution [KHM86]. Therefore,

$$a_j \equiv P[A = j] = P[A_k = j] = \frac{\lambda_p^j e^{-\lambda_p}}{j!}. \tag{7}$$

In the above equation, λ_p is the mean arrival rate of packets at each input port. By defining λ_c as the mean arrival rate of cells at input ports, we have

$$\lambda_c = \lambda_p \overline{BL}. \tag{8}$$

Now, we consider $s_{n,m,n',m'}$, the transition probability from state $[Q_{k-1} = n, H_{k-1} = m]$ to state $[Q_k = n', H_k = m']$. The transition probability $s_{n,m,n',m'}$ is obtained as follows.

1. When $n' < N_O$, that is, when the back-pressure function in not utilized:
   From Equation (2), we have

$$A_k = Q_k - (Q_{k-1} - 1)^+ - H_{k-1}. \tag{9}$$

   When m' packets of $Q_k - (Q_{k-1} - 1)^+$ HOL packets further generate cells at the next slot, we have a relation

$$s_{n,m,n',m'} = a_{n'-(n-1)^+-m} b_{n'-(n-1)^+,m'}. \tag{10}$$

2. When $n' = N_O$, that is, when the back-pressure function is used:
   From Equation (6), we have

$$A_k = N_O - (Q_{k-1} - 1)^+ - H_{k-1} + (H_k - C_k). \tag{11}$$

Since C_k packets of $N_O - (Q_{k-1} - 1)^+$ HOL packets further generate cells at the next slot, we have a relation

$$s_{n,m,n',m'} = \sum_{i=0}^{m'} a_{n'-(n-1)^+-m+i} b_{n'-(n-1)^+,m'-i}. \tag{12}$$

Once we have $s_{n,m,n',m'}$, the steady state probability $r_{n,m}$,

$$r_{n,m} = \lim_{k\to\infty} P[Q_k = n, H_k = m] = P[Q = n, H = m] \tag{13}$$

is obtained from Equations (10) and (12).

1. When the state is $[Q = 0, H = 0]$, the output port becomes idle, i.e., we have

$$r_{0,0} = 1 - \rho, \tag{14}$$

where ρ is defined as the maximum throughput normalized by the link capacity. By our assumption that the size of the input buffer is infinity, the maximum throughput ρ is equivalent to the cell arrival rate λ_c in steady state if it exists.
2. By considering all states that may be changed to state $[Q = n - 1, H = 0]$, we have (see Figure 3)

$$r_{n,0} = \frac{1}{s_{n,0,n-1,0}} \left\{ r_{n-1,0} - \sum_{i=0}^{n-1} \sum_{j=0}^{i} s_{i,j,n-1,0} r_{i,j} \right\} \quad (0 < n \le N_O). \tag{15}$$

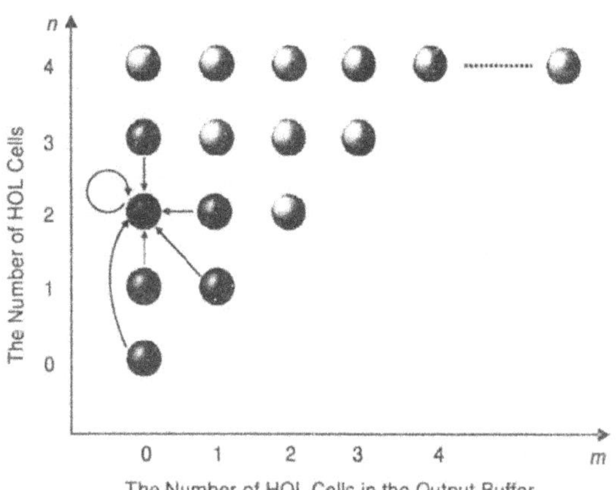

Figure 3 State transition diagram in the case of $m = 0$ and $0 < n \le N_O$.

3. By considering all states that may be changed to state $[Q = n, H = m]$, we have (see Figure 4)

$$r_{n,m} = \frac{1}{1 - s_{n,m,n,m}}\left\{\sum_{i=0}^{n-1}\sum_{j=0}^{i} s_{i,j,n,m} r_{i,j} + \sum_{k=0}^{m-1} s_{n,k,n,m} r_{n,k}\right\} \quad (0 < m, n < N_O). \tag{16}$$

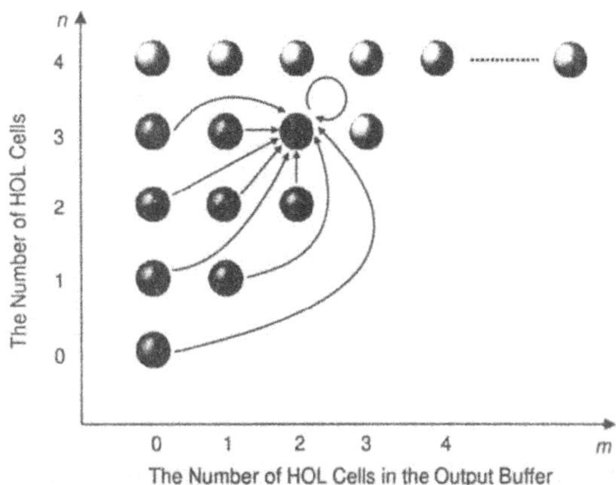

Figure 4 State transition diagram in the case of $0 < m$ and $n < N_O$.

4. By considering all states that may be changed to the state $[Q = N_O, H = m - 1]$, we have (see Figure 5)

$$r_{N_O,m} = \frac{1}{s_{N_O,m,N_O,m-1}}\left\{r_{N_O,m-1} - \sum_{i=0}^{N_O-1}\sum_{j=0}^{i} s_{i,j,N_O,m-1} r_{i,j} - \sum_{k=0}^{m-1} r_{N_O,k}\right\} \quad (0 < m). \tag{17}$$

4 MAXIMUM THROUGHPUT ANALYSIS

In this section, we obtain the maximum throughput using the steady state probability derived in Section 3, under balanced traffic condition in Subsection 4.1, under output unbalanced traffic condition in Subsection 4.2, and under input unbalanced traffic condition in Subsection 4.3. We will further consider the case of a mixture of bursty and stream traffic in Section 4.4.

4.1 Case of Balanced Traffic Condition

In this subsection, a balanced traffic condition is assumed, that is, a mean packet arrival rate at every input ports is identical and each packet determines its output port with an equal probability $1/N$.

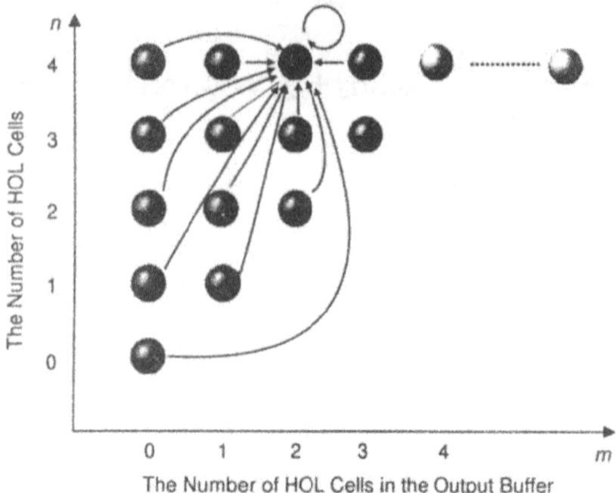

Figure 5 State transition diagram in the case of $0 < m$ and $n = N_O$.

In order to obtain the maximum throughput of our ATM switch, we consider the case where all input ports are saturated so that packets are always waiting in HOL queues. In this case, we have

$$\sum_{i=1}^{N} A^i = N - \sum_{i=1}^{N} H^i, \tag{18}$$

where A^i is the random variable which represents the number of arriving packets destined for output port i in a slot and H^i is the random variable for the number of HOL cells destined for output port i. By dividing the above equation by N and letting N to be infinity, we have

$$\lambda_p = 1 - \overline{H}, \tag{19}$$

where $\overline{H}$ is the average number of HOL cells, and can be expressed with $r_{n,m}$ derived in Section 3 as

$$\overline{H} = \sum_{n=0}^{N_O} \sum_{m=1}^{\infty} m\, r_{n,m}. \tag{20}$$

From Equations (8) and (19), we have

$$\lambda_c = (1 - \overline{H})\overline{BL}. \tag{21}$$

The maximum throughput ρ can be obtained by substituting λ_c in the above equation with ρ and solving it for ρ. Since $\overline{H}$ depends on ρ, ρ is solved iteratively by virtue of a standard iteration technique such as a bisection method [WHPV88].

In Figures 6 and 7, the maximum throughput ρ is plotted against the average packet length $\overline{BL}$ and the output buffer size N_O, respectively. These figures show that the packet length drastically degrades the maximum throughput. Further, we may observe that the size of output buffers must be larger than the average packet length to gain a sufficient throughput.

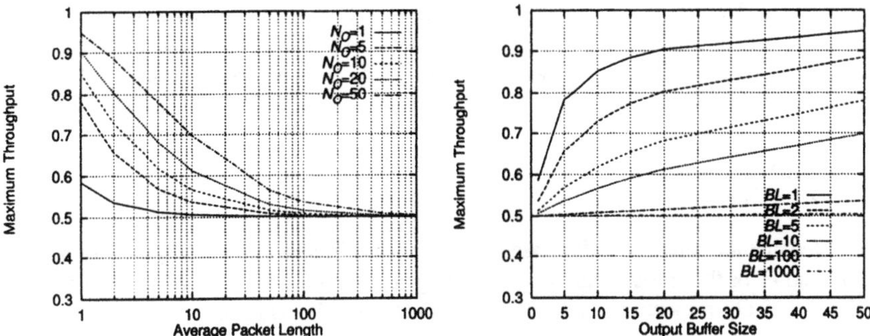

Figure 6 Maximum throughput vs. the average packet length.

Figure 7 Maximum throughput vs. the output buffer size.

Figure 8 shows the simulation results in the case where the switch size is finite for $N_O = 1$ and $N_O = 50$. The results of the analysis become slightly smaller than those of the simulation. Here, we note that the maximum throughput for $N_O = 1$ is exactly same as a well known value of input queueing, 0.585 [KHM86].

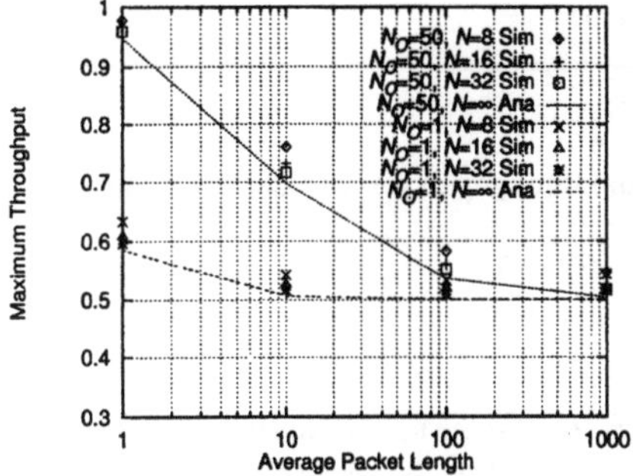

Figure 8 Comparison with simulation results.

4.2 Case of Unbalanced Traffic at Output Ports

In this section, output unbalanced traffic is treated following the approach presented in [Ili92a]. Output buffers are divided into two groups called O_1 and O_2. Let q_O be a ratio of the number of output ports belonging to the group O_1 as

$$q_O \equiv \frac{|O_1|}{N}. \tag{22}$$

The packet arrival rate at each input port is identical. However, each packet arriving at the input port selects one of output ports in group O_1 with probability P_{G1} or one of output ports in group O_2 with probability P_{G2}. By assuming $P_{G1} \geq P_{G2}$ without loss of generality, the relative probability r_O is denoted as

$$r_O \equiv \frac{P_{G1}}{P_{G1} + P_{G2}} \geq 0.5. \tag{23}$$

It is noted that the balanced traffic case is a special case by setting $q_O = 0$, $q_O = 1$ or $r_O = 0.5$. Let P_1 and P_2 be the probabilities that an arriving packet is destined to an output port belonging to the O_1 and O_2, respectively, we have from Equations (22) and (23),

$$P_1 = \frac{q_O r_O}{1 - q_O - r_O + 2q_O r_O} \tag{24}$$

$$P_2 = \frac{1 - q_O - r_O + q_O r_O}{1 - q_O - r_O + 2q_O r_O}, \tag{25}$$

where λ_p is defined as the packet arrival rate at input ports, and λ_{p1} and λ_{p2} are the packet arrival rates at output ports belonging to the group O_1 and O_2, respectively. We then obtain

$$\lambda_{p1} = \frac{r_O \lambda_p}{1 - q_O - r_O + 2q_O r_O} \tag{26}$$

$$\lambda_{p2} = \frac{(1 - r_O)\lambda_p}{1 - q_O - r_O + 2q_O r_O}. \tag{27}$$

For deriving the maximum throughput, we consider a relation:

$$\sum_{i=1}^{N} A^i = N - \left(\sum_{i=1}^{|O_1|} H_1^i + \sum_{i=1}^{|O_2|} H_2^i \right), \tag{28}$$

where random variables H_1^i (H_2^i) is the number of HOL cells destined for the output port belonging to the group O_1 (O_2). By dividing the above equation by N and letting N to be infinity, we have

$$\lambda_p = 1 - \{q_O \overline{H}_1 + (1 - q_O)\overline{H}_2\}, \tag{29}$$

where $\overline{H}_1$ and $\overline{H}_2$ are the average number of HOL cells destined for the group O_1 and O_2, respectively. From Equation (8), we have

$$\lambda_c = \left[1 - \{q_O \overline{H}_1 + (1 - q_O)\overline{H}_2\}\right] \overline{BL}. \tag{30}$$

The maximum throughput ρ can be obtained by substituting λ_c in the above equation with ρ and solving for ρ in the same manner presented in Section 4.1.

In Figures 9 and 10, the relations between q_O and the maximum throughput are plotted for $\overline{BL} = 1$ and $\overline{BL} = 10$, respectively. These figures show that an unbalanced traffic and a larger packet size cause degradation of the maximum throughput.

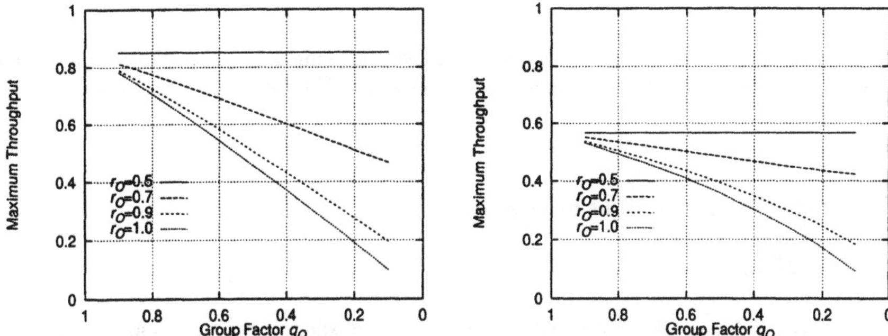

Figure 9 Unbalanced traffic at output ports ($N_O = 10, \overline{BL} = 1$).

Figure 10 Unbalanced traffic at output ports ($N_O = 10, \overline{BL} = 10$).

4.3 Case of Unbalanced Traffic at Input Ports

In this subsection, we evaluate the performance of the switch under the unbalanced traffic condition at the input ports. Similarly to the previous subsection, input ports are divided into two groups I_1 and I_2. Let q_I be a ratio of the number of input ports belonging to the group I_1 defined as

$$q_I \equiv \frac{|I_1|}{N}. \tag{31}$$

λ_{p1} and λ_{p2} are mean packet arrival rates at the groups I_1 and I_2, respectively. Assuming that $\lambda_{p1} \geq \lambda_{p2}$ is assumed without loss of generality, we introduce r_I as

$$r_I \equiv \frac{\lambda_{p1}}{\lambda_{p1} + \lambda_{p2}} \geq 0.5. \tag{32}$$

It is noted that the balanced traffic case is the special case by setting $q_I = 0, q_I = 1$ or $r_I = 0.5$. We assume that each packet arriving at the input port chooses the output port with a same probability $1/N$. By letting λ_p denote the packet arrival rate at each output port, λ_{p1} and λ_{p2} are given as

$$\lambda_{p1} = \frac{\lambda_p r_I}{1 - q_I - r_I + 2q_I r_I} \tag{33}$$

$$\lambda_{p2} = \frac{\lambda_p(1 - r_I)}{1 - q_I - r_I + 2q_I r_I}. \tag{34}$$

To obtain the maximum throughput, we consider the case where input ports are saturated. Recalling that we assume $\lambda_{p1} \geq \lambda_{p2}$, the input buffers belonging to the group I_1 is saturated first. Thus, we have a relation

$$\sum_{i=1}^{|O_1|} A_1^i = |O_1| - \sum_{i=1}^{|O_1|} H^i, \tag{35}$$

where the random variable A_1^i is the number of packets arriving at input port i belonging to the group I_1. By dividing the above equation by N and letting N to be infinity, we have

$$\lambda_{p1} = 1 - \overline{H}. \tag{36}$$

From Equation (8), the following relation holds:

$$\lambda_{c1} = (1 - \overline{H})\overline{BL}, \tag{37}$$

where λ_{c1} is the mean packet arrival rate at each input port belonging to the group I_1. The maximum throughput ρ can be obtained by substituting λ_{c1} in the above equation with ρ and solving for ρ as in the same manner presented in Subsection 4.1.

Figures 11 and 12 show the maximum throughput dependent on q_I for $\overline{BL} = 1$ and $\overline{BL} = 10$, respectively. These figures show that an unbalanced traffic condition and a larger packet size degrade the maximum throughput. The result for $\overline{BL} = 1$ is almost same as that for output unbalanced traffic (see Figure 9). On the other hand, the result for $\overline{BL} = 10$ show higher performance than that of output unbalanced traffic (see Figure 10). This is because unbalanced traffic at input ports causes less HOL blocking than at output ports.

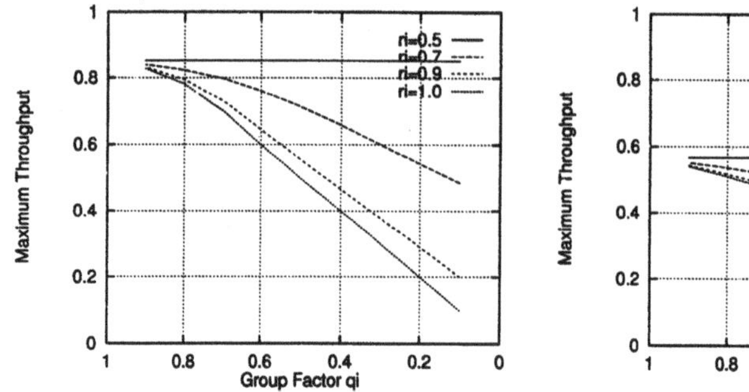

Figure 11 Unbalanced traffic at input ports ($N_O = 10, \overline{BL} = 1$).

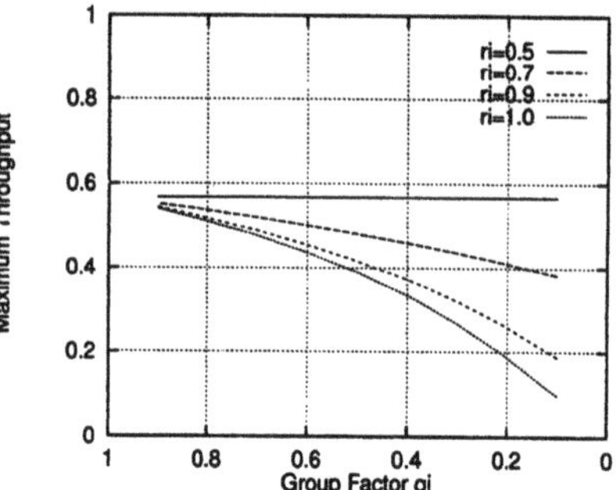

Figure 12 Unbalanced traffic at input ports ($N_O = 10, \overline{BL} = 10$).

4.4 Case of Mixture with Stream Traffic

Finally, we derive the maximum throughput in the case where the bursty traffic and the stream traffic coexist. Here, we assume that the stream traffic occupies some portion of the link with constant peak rate. For example, this class of traffic can support an uncompressed video transfer service.

Let R denote the peak rate of stream traffic normalized by the link capacity. The switch can simultaneously accept $m(\leq \lfloor 1/R \rfloor)$ calls of stream traffic. We assume that call arrivals of the stream traffic follow a Poisson distribution with mean λ_{CBR}, and its service time (or call holding time) does an exponential distribution with mean $1/\mu_{CBR}$. While both bursty and stream traffic share a link, cells from the stream traffic are given a higher priority. Namely, cells from stream traffic arriving at the input port are transferred to its destined output port prior to cells from bursty traffic [RFM94]. By this control mechanism, it can be considered that bursty traffic can utilize $1 - nR$ of the link capacity when n calls of stream traffic are accepted. We note here that if compressed video transfer service is accommodated as stream traffic, a more capacity can be utilized by the bursty traffic. Thus, the maximum throughput derived in the below should be regarded as the "minimum" guaranteed throughput for the bursty traffic.

Since the stream traffic is given a high priority, it can be modeled by a M/M/m/m queueing system. By letting π_n be the probability that n calls of stream traffic are accepted in steady state, we have (e.g., [BG87])

$$\pi_n = \left[\sum_{n=0}^{m} \left(\frac{\lambda_{CBR}}{\mu_{CBR}} \right)^n \frac{1}{n!} \right]^{-1} \left(\frac{\lambda_{CBR}}{\mu_{CBR}} \right)^n \frac{1}{n!}. \tag{38}$$

Since the service time of steam traffic can be assumed to be much longer than cell or the packet transmission time of bursty traffic, an available link capacity to bursty traffic is regarded to be constant when the number of accepted calls of stream traffic is fixed. By letting ρ_n be the maximum throughput for bursty traffic when n calls of the stream traffic are accepted, we have [RFM94]

$$\rho_n = (1 - nR)\rho, \tag{39}$$

where ρ is defined as the maximum throughput of bursty traffic when all link capacity is allocated to bursty traffic, and has been already derived in Subsection 4.1. Consequently, the "averaged" maximum throughput ρ' is obtained as

$$\rho' = \sum_{n=0}^{m} \pi_n \rho_n. \tag{40}$$

Figure 13 shows the maximum throughput of bursty traffic and stream dependent on an offered traffic load for stream traffic for $N_O = 50$, $\mu_{CBR} = 0.1$, $R = 0.2$ and $m = 5$. From this figure, we can observe the natural idea that the larger the average packet length is, the smaller the maximum allowable throughput of bursty traffic is. Therefore, the available bandwidth allocated to the stream traffic should be limited in some way to avoid a degradation of bursty traffic efficiency to some degree. One possible approach is to decrease m, the maximum number of calls of stream traffic that the switch can accept. In Figure 14, the maximum throughput of both bursty traffic and stream traffic dependent on the offered traffic load for stream traffic for $\overline{BL} = 1$ and several values of m. It shows that the performance degradation of bursty traffic can be avoided to some extent by limiting m.

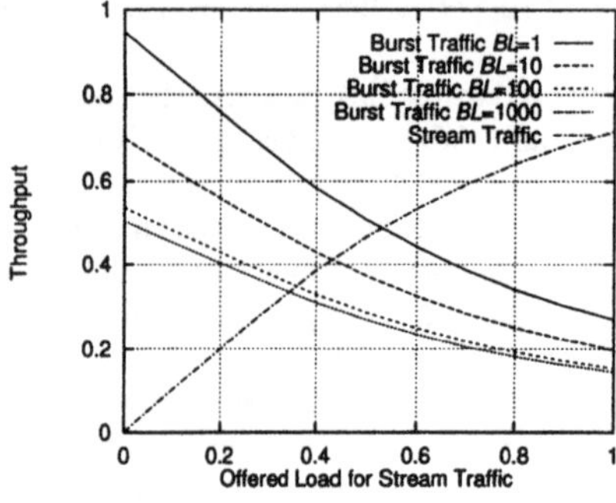

Figure 13 Throughput vs. the offered load of stream traffic.

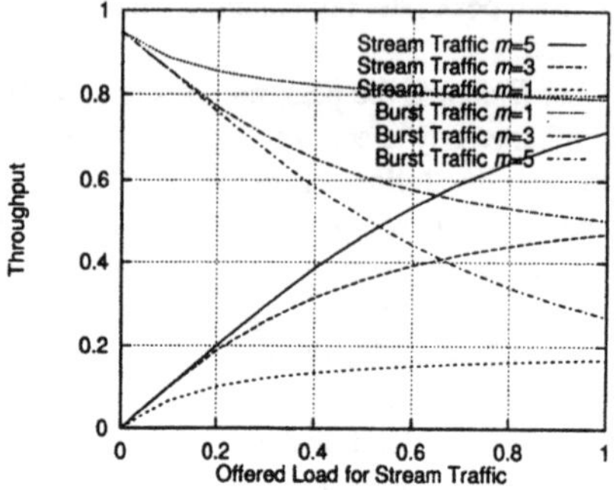

Figure 14 Effect of an available link capacity limitation on stream traffic.

5 CONCLUSION

In this paper, an ATM switch with input and output buffer equipped with back-pressure function was treated. We have analyzed its performance under bursty traffic condition for applying it to ATM LANs. We have derived the maximum throughput under the assumption that the switch size is infinite. Consequently, we have shown that larger packet lengths drastically degrade the performance of the switch. However, it is possible to sustain such a degradation to some extent by larger output buffers. At least, the output buffer size comparable to the average packet length is necessary to gain a sufficient performance. We have already analyzed the packet delay distribution and the approximate packet loss probability of the switch. These results can be found in [OWMM95].

Last, we note that our analytic approach described in the current paper can be applied to other cases, e.g., the case where the switching speed is L $(1 \leq L \leq N)$ times faster than the link speed (see, e.g., [YMKH92]), or the case where, when $L'(> L)$ cells are simultaneously destined for the same output buffer, $(L' - L)$ cells are lost or kept awaiting at the input buffer.

For further works, we should evaluate the performance of the network in which two or more ATM switches are interconnected. In such a network, even when a long term congestion introduces large queue length at the input buffer, cell losses may be avoided to send a back-pressure signal to the upper adjacent switches.

ACKNOWLEDGMENT

We would like to thank Dr. Hiroshi Suzuki and Dr. Ruixue Fan with NEC Corporation, C&C System Laboratories, for their invaluable suggestions.

REFERENCES

[BG87] Dimitri Bertsekas and Robert Gallager. *Data Networks*. Prentice-Hall, Englewood Cliffs, New Jersey, 1987.

[Ili90] Ilias Iliadis. Head of the line arbitration of packet switches with input and output queueing. In *Fourth International Conference on Data Communication Systems and their Performance*, pages 85–98, Barcelona, Spain, June 1990.

[Ili92a] Ilias Iliadis. Performance of a packet switch with input and output queueing under unbalanced traffic. In *Proceedings of IEEE INFOCOM '92*, volume 2, pages 743–752 (5D.4), Florence, Italy, May 1992.

[Ili92b] Ilias Iliadis. Synchronous versus asynchronous operation of a packet switch with combined input and output queueing. *Performance Evaluation*, (16):241–250, 1992.

[KHM86] Mark J. Karol, Michael G. Hluchyj, and Samuel P. Morgan. Input vs. output queueing on a space-division packet switch. In *Proceedings of IEEE GLOBECOM '86*, pages 659–665, Houston, Texas, December 1986.

[OWMM95] Hiroyuki Ohsaki, Naoki Wakamiya, Masayuki Murata, and Hideo Miyahara. Performance of an input/output buffered type ATM LAN switch with back-pressure function. *submitted to IEEE Transactions on Networking*, 1995.

[RFM94] Kenji Yamada Ruixue Fan, Hiroshi Suzuki and Noritaka Matsuura. Expandable ATOM switch architecture (XATOM) for ATM lans. *ICC '94*, 5 1994.

[WHPV88] Saul A. Teukolsky William H. Press, Brian P. Flannery and William T. Vetterling. *Numerical Recipes in C.* Cambridge University Press, 1988.

[YMKH92] Yuji Oie, Masayuki Murata, Koji Kubota, and Hideo Miyahara. Performance analysis of nonblocking packet switches with input / output buffers. *IEEE Transactions on Communications*, 40(8):1294–1297, August 1992.

8

A Study of Switch Models for the Scalable Coherent Interface

B. Wu[1, 2], A. Bogaerts[2] and B. Skaali[1]

1) Department of Physics, University of Oslo, 0136 Oslo, Norway
Tel: +47 22 856428, Fax: +47 22 856422, Email: bin.wu@fys.uio.no

2) ECP-DS, CERN, 1211 Geneva 23, Switzerland

Abstract

The Scalable Coherent Interface (SCI) specifies a topology-independent communication protocol with the possibility of connecting up to 64 K nodes. SCI switches are the key components in building large SCI systems effectively. This paper presents four alternative architectures for a possible SCI switch implementation. Of these four models, two could be seen as crossbar-based switches. The main difference between the two crossbar switch models is the different structure of queues: either a parallel structure, called SwitchLink or serial, called CrossSwitch. The other two models are a ring-based switch model, which is actually connecting four ports into an internal SCI ring and a bus-based switch model which uses an internal bidirectional bus. We will describe each model in detail and compare them by simulation.

Keywords

Scalable Coherent Interface, SCI switch, switch model, simulation, performance

1 THE SCALABLE COHERENT INTERFACE

Connecting several processors, memories and devices is often performed by a single bus for protocol simplicity and low cost. The drawback of the bus structure is bandwidth limitation because of its one-at-a-time feature. Some other multiprocessor systems do provide high performance, but have a very specific design and are often very expensive. With a distributed cache coherence protocol, 1 Gbyte/s per link speed (16 bits data parallel running at 500 MHz), The Scalable Coherence Interface (SCI), IEEE Standard 1596-1992, provides an effective solution to the problem. It uses point-to-point unidirectional links to connect up to 64 K nodes (IEEE Std. 1596 (1992) and Gustavson (1992)).

The basic building block of an SCI network is a ring of two or more nodes connected by unidirectional links. An SCI node would have the basic elements as shown in Figure 1.a. A two-node SCI ring is also shown in the Figure 1.b.

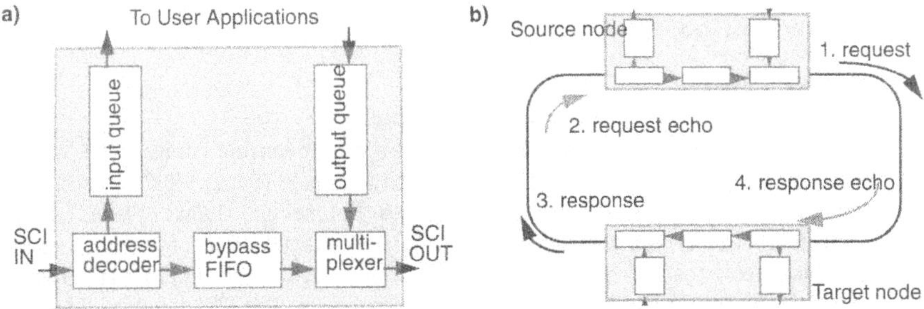

Figure 1 **a)** A block diagram of an SCI node; **b)** A two-node SCI ring.

SCI uses a packet-transmission protocol. Most SCI transactions such as read and write are split in two subactions, a request and a response packet with request echo and response echo respectively. The exception is the move transaction that is without a response subaction. A packet contains target address (16 bits of node address that results in a maximum of 64 K nodes in an SCI system, plus 48 bits of internal address), command, source address (16 bits) and control bits followed by 0, 16, 64 or 256 bytes of data and a CRC checksum.

An SCI node is able to transmit packets while concurrently accepting other packets addressed to itself and passing packets addressed to other nodes. Because an input packet might arrive while the node is transmitting an internally generated packet, queue storage (in FIFO order) is provided to hold the packets received while the packet is being sent. New send packets are inserted by a sending node from its output queue, subject to the bandwidth allocation protocol in SCI. The packet is saved until a confirming echo packet is received. The addressed target node strips the send packet and creates an echo packet, which is returned to tell the source node whether the send packet was accepted by target's input queue, or rejected due to queue overflow or other errors. The source either discards the corresponding send packet or retransmits it if the sending was unsuccessful. The sequence of a successful sending is shown in Figure 1.b. Since a node transmits only when its bypass FIFO is empty, the minimum bypass FIFO size is determined by the longest packet that originates from the node. Idle symbols received between packets provide an opportunity to empty the bypass FIFO in preparation for the next transmission.

Input and output queues are needed in order to match node processing rates to the higher link-transfer rate. Since nodes are normally capable of both sending and receiving, request and response subactions are processed through separate queues, which are not drawn in Figure 1, to avoid system deadlocks on these full-duplex nodes. The depth of queues is important, as the deeper the queue, the more packets the node can send and receive concurrently. For the output queues, more outstanding requests could be issued while waiting for the "good" echo of the first send packet. From the input queue side, more packets could be saved for processing if the back-end logic is not fast enough to keep up with SCI speed.

Near the incoming link of the node interface, there is an address decoder that checks the target address of the incoming packets. A decision is then made to accept the packets into the node's input queue or to send it through the bypass FIFO to downstream nodes.

2 SCI SWITCHES

2.1 Why SCI switch

The simplest SCI system is a ring, but an SCI ring structure is sensitive to hardware failures and limited by its peak load and large SCI rings do not scale (Scott, 1992). So different topologies are investigated to interconnect many nodes (Bothner and Hulaas (1991), Johnson and Goodman (1992) and Wu, et al., (1994)). An SCI switch that consists of two or more SCI node-like interfaces (called SCI ports) to different rings, with appropriate routing mechanisms, is a key component in building these topologies. The SCI standard, however, does not directly specify an SCI switch structure. A wide variety of mechanisms are possible.

2.2 SCI switch taxonomies

Generic switch architectures have several classifications based on different attributes (Feng (1981) and Newman (1992)). One classification that also applies to SCI is based on the internal structure of the switch fabric, i.e. time- or space-division methodology. In the first, the use of a physical resource, say conducting medium or memory, is multiplexed among several input-output connections, based on discrete time slots. A bus is an example of a physical conducting medium that can accommodate time-division multiplexing. In a space-division methodology, the switch fabric can support multiple connections at a given moment. The

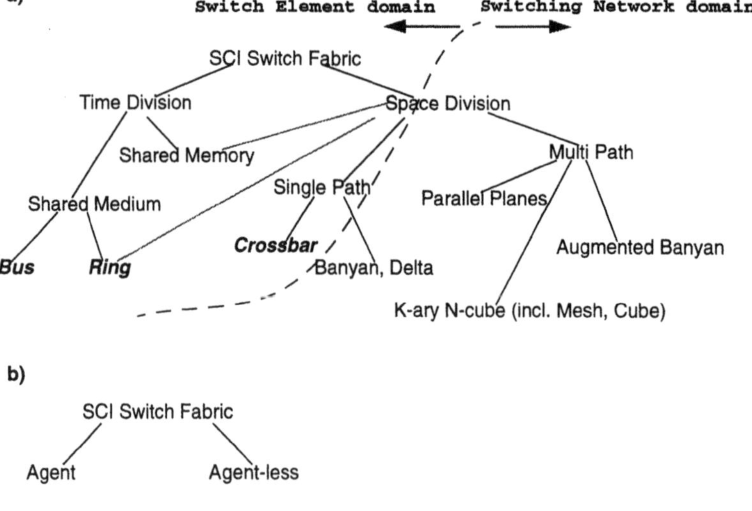

Figure 2 a, b) Two taxonomies of SCI switch fabric.

connections are based on the availability of nonconflicting physical paths within the fabric. A crossbar is an example. Figure 2.a illustrates such a classification adapted for SCI-based network. Due to the complexity of SCI switches, switches with hundreds of input and output links are not realistic to build with the existing technology. The left part of the dashed line implies what is possible to implement on a single chip (including MCM) while the right reflects the possible implementations on a PCB, in a cluster system or a LAN system.

Time- and space-division multiplexing can be combined. For example, a space-division switching network may be interconnected by several time-division switch elements in a hierarchical fashion.

Another classification of SCI switches in two types is agent and agent-less types. An agent would be a switch that takes control of the operation, returning an echo to the originator and taking charge of forwarding the packet. Agent-less would be a non-intelligent switch, which just forwards packets and echoes which come to it.

This paper will focus on agent type operation and switch elements with a small number of ports. We use 4-switch as an example. The reason is that for a frequency of 500 MHz, the switch should be designed on a single chip. Only so could timing skew be tightly specified because components are inherently well matched in a single chip design.

2.3 General switch model

An SCI 4-switch is by definition a switch implementing SCI protocols on its four ports as shown in Figure 3. The interconnections between the four ports could be a bus, a ring of the time-division class, or a crossbar of the space-division switches.

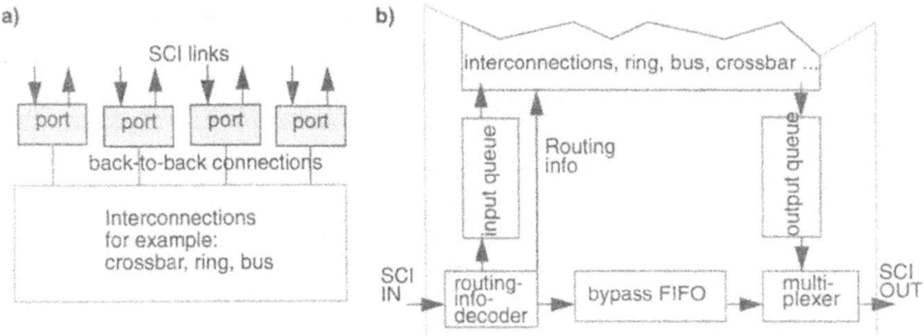

Figure 3 **a)** A block diagram of SCI 4-switch; **b)** A block diagram of an SCI switch port.

An SCI port should have the same functionalities as an SCI node (Figure 1.b), except that the port is simpler in not having the cache coherency logic, but more complicated in back-side connection and routing logic designs (Figure 3.b).

2.4 CrossSwitch model

One of the SCI switch models is named CrossSwitch. The term "Cross" indicates the capability of connecting any input to any output. The block diagram of a CrossSwitch is the same as in Figure 3.a with crossbar interconnect. Not shown in Figure 4 are the bypass FIFOs.

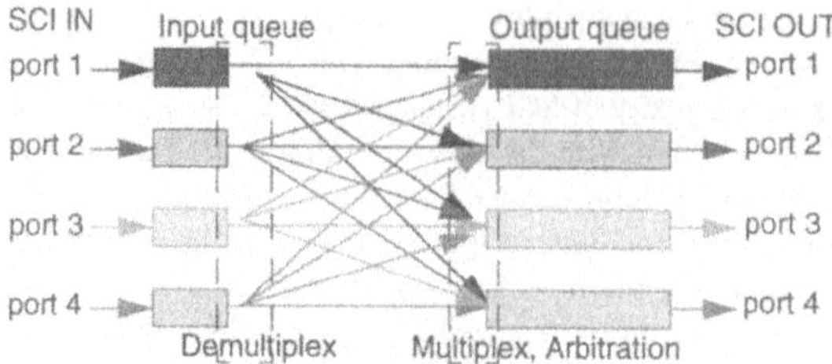

Figure 4 CrossSwitch model data flow with crossbar structure.

It is quite obvious that there will be competition between any two packets that want to go to the same output queue simultaneously, even if the output queue has enough space for both packets. Thus, the loser in the competition will be stored in the input queue. To avoid overflow (which has the consequence that many retries will happen) a deep input queue is recommended.

2.5 SwitchLink model

The SwitchLink model can also be classified as a crossbar-interconnected switch. It solves the contention problem of the CrossSwitch by its internal parallelism. Figure 5 shows the detail of the parallel queue structure of the SwitchLink model.

With sixteen queues in each 4-switch, one can achieve complete parallelism. Packets from different input links go through its private path to its corresponding queue. No packet will be blocked for getting into an output queue even if two packets come from different ports, for example port1 and port3, are routed simultaneously to port2.

The drawback is that parallel queue structure means potential waste of resources if the traffic is not random which unfortunately it is very often the case. On the other hand, the input queue could be omitted.

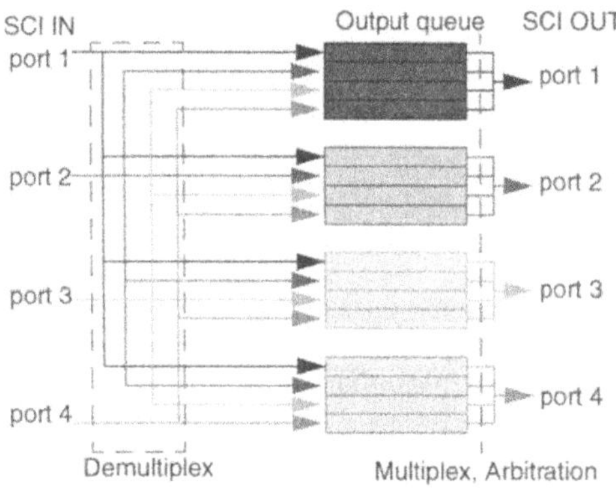

Figure 5 SwitchLink data flow with parallel output queues structure.

2.6 Bus-based switch model

Putting everything on a single chip with so many interconnections makes the hardware implementation for SwitchLink or CrossSwitch very difficult. The internal-bus scheme is investigated because it is simple and cheap.

A bidirectional bus with an arbitration facility is used to connect SCI ports together (Figure 6). The transfer of SCI packets between different ports is similar to a burst write operation on most buses. In principle, the bus could connect many SCI switch ports and other devices, but with the same argument of using SCI instead of a bus, the "one-at-a-time" feature of the bus will be the great obstacle for achieving high performance. One of the solutions is to increase the bus speed. In Figure 10 we will see simulation results showing the effect of bus speed on the switch performance.

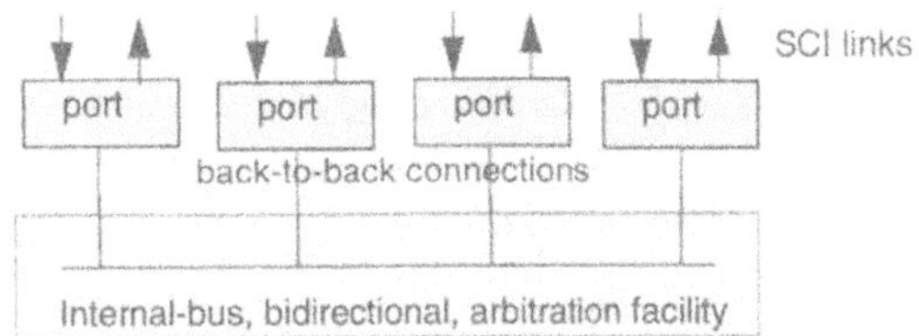

Figure 6 Internal-bus based switch model.

2.7 Ring-based switch model

Neither the crossbar models nor the bus-based model uses point-to-point technique for the interconnections, which implies that it is difficult (costly) to implement hardware, as logic running at 500 MHz has already pushed technology close to its limit. Point-to-point unidirectional communication could greatly reduce the non-ideal-transmission-line problems.

A switch model based on point-to-point connection could be like the one shown in Figure 7. The simplified switch port (s-port) may have the same CSR structure (IEEE Std. 1212, 1991) as the corresponding switch port and it can be implemented without queues. The back-to-back connections are point-to-point transmit channels which could be very fast.

Though SCI rings do not scale indefinitely, they provide better performance than a bus for random traffic pattern (Bogaerts, et al., 1992), as we will see in the simulation results.

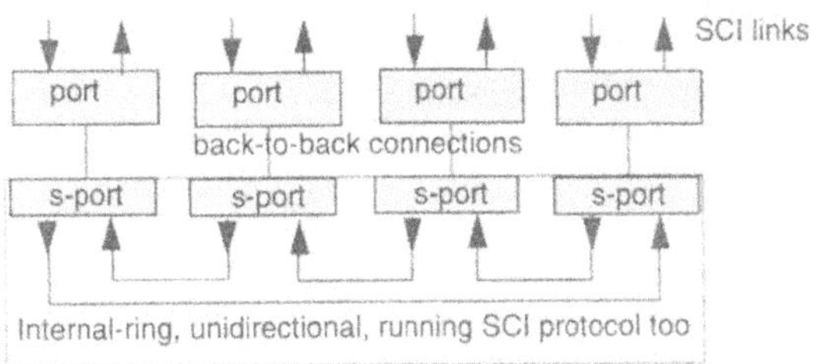

Figure 7 A 4-switch with internal ring structure.

3 MODELLING AND SIMULATION

3.1 Simulation tools and simulator

SCILab (Wu and Bogaerts, 1995) is a simulation environment for studying the behavior of SCI based networks. It is aimed at studying the influence of linkspeed, interface design and SCI protocols under various loading conditions for different topologies. Effects due to the SCI bandwidth allocation, queue acceptance and cache coherency protocols are taken into account. Large SCI systems consisting of rings interconnected by switches can be investigated. Examples of SCI nodes whose behavior can be specified by user definable parameters are also included in the model.

The simulations produce statistics on the SCI link traffic, bandwidth available to SCI nodes such as processors and memories, and the latency of SCI packets traveling through a network from source to target.

We ran each of our simulations for 500 000 ns and each SCI link at the speed of 500 MHz with 16-bit-wide data path. The bypass FIFO and cable delay is 15 ns. Switch routing takes 15 ns and internal transmission of packets from input to output uses 10 ns. Packets are routed in virtual cut-through technique.

3.2 Simulation of a 4-port switch

The performance of a switch design could be tested with the setup shown in Figure 8, with a switch in the middle and four SCI rings connected to it. Each ring has only one processor/ memory node that sends and receives packets through the switch. The system is heavily loaded with random traffic.

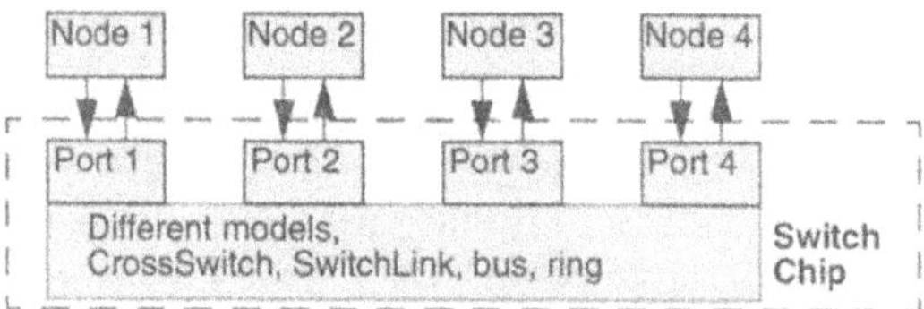

Figure 8 Switch performance test setup.

When we were simulating such a system, we did not allow a node to send packets to itself since a packet sent to a local node will only go through the bypass FIFO of the switch port. Our purpose is to saturate the switch and see the performance. Figure 9 gives a good indication of the performance (Throughput, Latency) of a single switch chip with the architecture of SwitchLink(SL), CrossSwitch(CS), Bus-based and Ring-based switch, respectively.

Simulations were performed for both one-packet-deep and three-packet-deep queues. It should be pointed out that when we say one-packet-deep queue, we mean one-packet-deep output queue and one-packet-deep input queue, except SwitchLink which uses four one-packet-deep output queues (without input queues, and thus use double the queue size in total).

Considering the effect of queue size on system performance, it is not so surprising to see that SwitchLink exhibits the highest throughput and the lowest latency. A Bus-based switch is limited by the bus speed of 1 Gbyte/s no matter how deep the queues are. The performance of

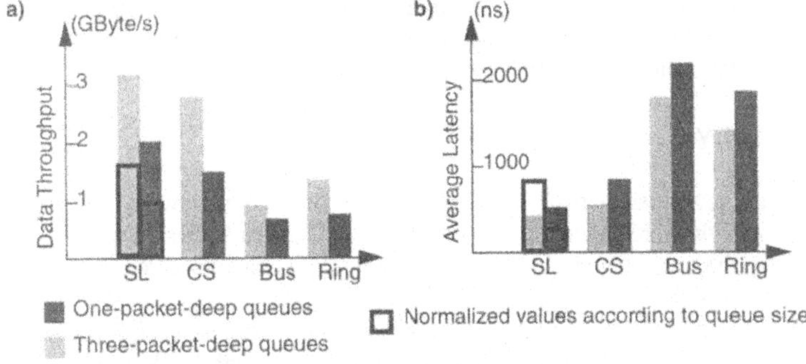

Figure 9 Performance evaluation for SL, CS, Bus- and Ring-based switches.

a Ring-based switch is not good, also due to the limited performance of a single ring. We should emphasize that, among these four models, CrossSwitch is the model that could provide relatively high performance with relatively low cost in the sense of queue size, if we compare with the normalized SwitchLink performance in Figure 9.

The performance of the single switch system can to some extent reflect the performance of a large SCI system, and this is proved by simulations (Wu and Bogaerts, 1993).

3.3 Bus speed

Because a bus is simple and easy to build in hardware but limited by its "one-at-a-time" feature, it is interesting to see whether increased bus speed could help to solve the bottleneck. Figure 10 presents the simulation results for different bus speeds.

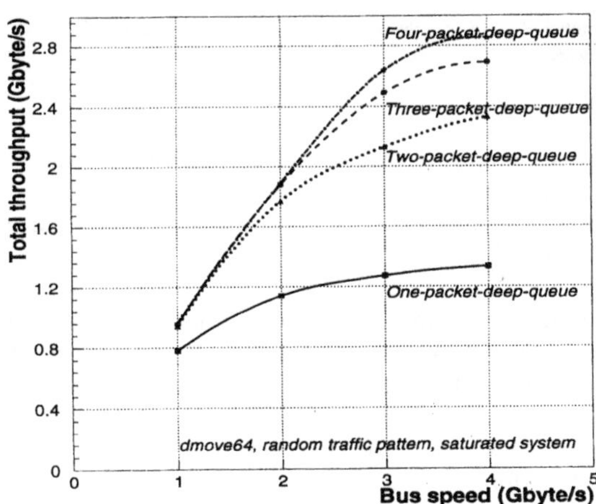

Figure 10 System's total throughput versus internal-bus speed.

It is observed that increasing bus speed does not provide linear increase in system throughput. This is due to the transmission overhead, as well as the obligatory use of store-and-forward switching technique when the bus speed is faster than SCI speed.

4 SUMMARY AND CONCLUSIONS

While SCI does have characteristics that facilitate interconnection between SCI rings, internetworking and switch architecture is not part of the SCI standard. This leaves a high degree of flexibility to the implementor. In this paper we presented and discussed four SCI switch models.

The SCI switch models have been strongly influenced by features of the SCI standard and SCI node chips to achieve high bandwidth and low latency with modest complexity. For comparison, we list some of the important aspects in Table 1.

Table 1 Evaluation of SwitchLink, CrossSwitch, Bus-based and Ring-based switch

Switch Model	SL	CS	Bus[a]	Ring
Complexity	o[b]	o	+	o
Throughput	+	+	-	-
Latency	+	+	o	-
Suitability for burst traffic	-	o	+	o
Suitability for random traffic	+	+	o	o
Performance Scalability vs. # of ports	+	+	-	-
# queues for a 4-switch (n queue)[c]	4x4xn	4x2xn	4x2xn	4x2xn[d]

a. Internal-bus runs at 1 Gbyte/s

b. +: good; o: average; -: poor.

c. kxmxn where k is the number of ports, m is the number of queues, n is the depth of queues in unit of SCI packets.

d. s-port does not have queues.

Though back-plane buses have severe limitations, we still find them quite attractive because of their simplicity and low cost when the number of connected devices is not large. Higher bus speed could also help in some sense to increase performance. Between cost and performance, there are trade-offs to make.

Though still at its early stage, a wide range of applications are possible for SCI now, such as SCI/Sbus, SCI/VME, SCI/PCI and SCI/ATM interface, etc. SCI rings of several nodes have been up and running for almost two years. SCI switches are also catching up, like the UNISYS 2x2 switch (Buggert, et al., 1994), the Dolphin CMOS switch and the TOPSCI 4-switch (Wu, 1994).

Switch designs should capitalize on several of the most important features of SCI, namely scalability, simplicity and high performance. Scalability allows cascading of elementary switches into a larger composite switch as well as the growth of a system which needs more SCI ports. Simplicity eases the design, implementation, and operation of systems requiring high throughput and low latency.

Even being quite familiar with the requirements for SCI switches and their behaviors, it is still a challenge to design switches operating at 1 Gbyte per second per link in GaAs technology, or 200 Mbyte per second in CMOS technology.

5 ACKNOWLEDGMENT

We are grateful to colleagues in RD24 group at CERN and EURIKA TOPSCI project members, for the ideas that assisted this work. SwitchLink idea was initiated by Roberto Divia at CERN and later improved during brain-storm discussions within RD24 group at CERN.

Bin Wu is supported by the Norwegian Research Council. This work is partly supported by Thomson TCS, France.

6 REFERENCES

Bogaerts, A., Divia, R., Muller, H. and Renardy, J. (1992) SCI based Data Acquisition Architectures, *IEEE Transactions on Nuclear Science,* Vol. 39, No. 2, Apr. 1992

Bothner, J. and Hulaas T. (1991) Various interconnects for SCI-based systems, *Proceedings of Open Bus Systems'91,* Paris, 1991

Buggert, J., Desai, V., Herzberg, L. and Kibria, K. (1994) The Unisys Datapump and Switch, *Proceedings the 1st International Workshop on SCI-based High Performance Low-Cost Computing,* Santa Clara, CA., Aug. 1994

Feng, T. (1981) A Survey of Interconnection Networks, *IEEE Computer,* December 1981, pp 12-27

IEEE Std 1212 (1991) *Control and Status Register (CSR) Architecture for Microcomputer Buses,* IEEE Service Center

IEEE Std. 1596 (1992) *The Scalable Coherent Interface,* IEEE Service Center

Johnson, R. and Goodman, J. (1992) Synthesizing General Topologies from Rings, *Proceedings of ICPP,* Aug. 1992

Gustavson, D.B. (1992) The Scalable Coherent Interface and related Standards Projects, *IEEE Micro,* Feb. 1992, pp. 10-22

Newman, P. (1992) ATM Technology for Corporate Networks, *IEEE Comm. Magazine,* Vol. 30, N. 4, Apr. 1992, pp. 90-101

Scott, S., Goodman, J., Vernon, M. (1992) Performance of the SCI Ring, *Proceedings IEEE ISCA 92,* Queensland, May 1992

Wu, B. and Bogaerts, A. (1993) Several Details of SCI Switch Models, *RD24 note,* CERN, Nov. 1993

Wu, B., Bogaerts, A., Kristiansen, E., Muller, H., Perea, E. and Skaali, B. (1994) Applications of the Scalable Coherent Interface in Multistage Networks, *Proceedings IEEE TENCON'94, Frontiers of Computer Technology,* Singapore, Aug. 22-26, 1994

Wu, B. (1994) SCI Switches, *Proceedings Int'l Data Acquisition Conference on Event Building and Event Data Readout in Medium and High Energy Physics Experiments,* Batavia, IL. Oct 26-28, 1994

Wu, B. and Bogaerts, A. (1995) SCILab-A Simulation Environment for the Scalable Coherent Interface, *Proceedings IEEE MASCOTS'95,* Durham, NC., Jan. 16-18 1995

7 BIOGRAPHY

Bin Wu is a Ph.D candidate at the University of Oslo. He received a research fellowship from the Norwegian Research Council in 1993 and he is a Scientific Associate at CERN, Geneva, Switzerland. His research interests include high-speed networks, switch technologies, VLSI design, network modelling and simulation. Wu received a BSc and a MSc degree from Department of Physics, University of Oslo, both in Electronic Engineering, in 1991 and 1992 respectively.

Andre Bogaerts graduated in 1970 in Physics and Mathematics at the University of Utrecht, the Netherlands. From 1970-1974 he was employed by the Dutch Organization for Fundamental Research (FOM) where he joined an international team for a High Energy Physics experiment at the Intersecting Storage Rings (ISR) at CERN, Geneva, Switzerland. Since 1974 he has been employed by CERN. His main interest is in Data Acquisition Systems for High Energy Physics Experiments. He has been involved in the standardization of the Scalable Coherent Interface (SCI) and, since 1992, he has been leading a team to investigate the use of SCI for future experiments at CERN's Large Hadron Collider (LHC).

Bernhard Skaali graduated as Master of Science in Physics in 1966 and Doctor of Philosophy in Nuclear physics in 1976, from the University of Oslo. Research assistant 1966-72, Senior lecturer 1973-82, and Professor from 1983 at the Department of Physics, University of Oslo. Current main research areas: instrumentation and DAQ for high energy physics. Experiments: LEP-DELPHI-CERN from 1985, LHC-ALICE from 1994. Stays abroad: Tandem Acc. Laboratory, Niels Bohr Institute, Copenhagen, 1969-70 (17 months), Scientific Associate, CERN SPS div, 1976-77 (12 months), CERN DD div 1980-81 (6 months), CERN PPE div 1988-89 (17 months).

9

Optimization Of Logically Rearrangeable Multihop Lightwave Networks With Genetic Algorithms

C. Gazen
Bogazici University, Computer Engineering Department
80815 Bebek, Istanbul, Turkey
fax: (90-212)2631540/1896

C. Ersoy
Bogazici University, Computer Engineering Department
80815 Bebek, Istanbul, Turkey
e-mail: ersoy@boun.edu.tr
fax: (90-212)2631540/1896
tel: (90-212)2631500/1861

Abstract

Multihop lightwave networks are a way of utilizing the large bandwidth of optical fibers. In these networks, each node has a fixed number of transmitters and receivers connected to a common optical medium. A multihop topology is implemented logically by assigning different wavelengths to pairs of transmitters and receivers. By using tunable lasers or receivers, it is possible to modify the topology dynamically, when node failures occur or traffic loads change.

The reconfigurability of logical multihop lightwave networks requires that optimal topologies and flow assignments be found. In this paper, optimization by genetic algorithms is investigated. The genetic algorithm takes topologies as individuals of its population, and tries to find optimal ones by mating, mutating and eliminating them. During the evolution of solutions, minimum hop routing with flow deviation is used to assign flows, and evaluate the fitness of topologies.

The algorithm is tested with different sets of parameters and types of traffic matrices. The solutions found by the genetic algorithm are comparable to the solutions found by the existing heuristic algorithms.

Keywords
multihop architecture, genetic algorithms, flow deviation, logically rearrangeable networks

1 INTRODUCTION

Although the optical fiber has been used very efficiently in long distance communications, its use as the medium of transmission in multi-access applications is still a problem. There are two factors limiting the use of optical fibers in local area networks (Henry, 1989). The first is the power problem. In long distance communications, the optical fiber is used between two points, the sender and the receiver, whereas in local area networks there are many users sharing a single line, and the energy associated with a signal is divided among the users. Since the successful reception of a signal requires a certain amount of energy, the number of users sharing a line is limited. The second problem is the electronic-bottleneck. Although lightwave technology can support many Tb/s of throughput, the electronic devices that carry the signals from one fiber to another, such as in a ring architecture, are at best capable of speeds of a few Gb/s. Among the various approaches for avoiding both of these problems is wavelength division multiplexing (Henry, 1989)(Goodman, 1989).

Wavelength division multiplexing is the lightwave counterpart of frequency division multiplexing for electronic transmission (Goodman, 1989). The available bandwidth on an optical fiber is divided into channels. When a sender and a receiver want to communicate, they first decide on a channel and then use that channel for transmission. Although various architectures are possible with this approach, most require quickly-tunable lasers and receivers, and this is a disadvantage.

The multihop architecture overcomes the disadvantages of relying on quickly-tunable lasers, by assigning fixed channels to sets of transmitters and receivers. In this architecture, each node has a number of receivers and transmitters that are connected to a common multi-wavelength optical network. The multihop topology is implemented by assigning channels to transmitters and receivers that are logically connected. A message may be transmitted at different wavelengths through many nodes before reaching its destination, if the sender and receiver are not assigned to a common channel.

In the original multihop architecture, nodes are connected in a perfect-shuffle topology (Acampora, 1988). This topology decreases the average number of nodes a packet has to pass through before reaching its destination to $\log n$ from $n/2$, the average for a ring topology. As a result, the total traffic capacity of the network is increased. Although the perfect shuffle topology is very efficient for uniform traffic patterns, it does not perform so well with unbalanced traffic loads (Eisenberg, 1988). Other logical connection patterns and design methods that can handle unbalanced traffic loads have also been proposed in (Sivarajan, 1991), (Bannister, 1990), and (Labourdette, 1991).

The logical multihop architecture has a number of advantages. Since the topology is logical, it can be modified to respond to node failures and traffic load changes, as long as the optical network functions properly. Compared with a simple wavelength division multiplexing architecture, this topology requires either tunable lasers or tunable receivers, neither of which need to be quickly-tunable.

To use the logically rearrangeable multi-hop lightwave networks as efficiently as possible, it is necessary to find some procedure that will find an optimum topology and flow assignment for a given traffic load. The problem has been formulated as an optimization problem and solved with an heuristic algorithm in (Labourdette, 1991). In this paper, the use of genetic algorithms on optimizing the logically rearrangeable multi-hop lightwave networks will be studied.

2 PROBLEM DEFINITION

The problem, as formulated in (Labourdette, 1991), is to find a topology, and an assignment of flows, given the traffic matrix, such that the maximum ratio of the flow to the capacity of the link it is assigned to, is minimized. Such a solution allows the traffic matrix to be scaled up optimally without exceeding the capacity of the links. However, the original problem is slightly modified so that representation and manipulation of solutions by the genetic algorithm is easier. The original problem allows two nodes to be connected by more than one link as long as the maximum number of transmitters and receivers is not exceeded. In this paper, either two nodes are connected by one link or they are not connected. The resulting formulation is:

$$\text{minimize} \quad \max_{i,j} \left\{ f_{ij} / C_{ij} \right\} \tag{1}$$

$$\text{subject to}: \quad \sum_j f_{ij}^{st} - \sum_j f_{ji}^{st} = \begin{cases} \lambda_{st}, & \text{if } s = i; \\ -\lambda_{st}, & \text{if } t = i; \; \forall i,s,t \\ 0, & \text{otherwise} \end{cases} \tag{2}$$

$$f_{ij} = \sum_{s,t} f_{ij}^{st} \quad \forall i,j \tag{3}$$

$$\sum_j Z_{ij} = T \quad \forall i \tag{4}$$

$$\sum_i Z_{ij} = T \quad \forall j \tag{5}$$

$$Z_{ij} = 0,1 \text{ integer} \quad \forall i,j \tag{6}$$

$$f_{ij}^{st} \geq 0 \quad \forall i,j \tag{7}$$

where f_{ij} = total flow on link from node i to node j (8)

f_{ij}^{st} = traffic flowing from node s to node t on the link from node i to node j (9)

λ_{st} = traffic due from node s to node t (10)

C_{ij} = capacity of link from node s to node t (11)

Z_{ij} = number of directed links from node i to node j (12)

T = number of transmitters / receivers per node (13)

Equations (2) and (3) are flow conservation constraints, and equations (4) and (5) are the degree constraints that set the number of transmitters and receivers per node to T. The problem minimizes the maximum ratio of flows to capacities of links, but because the topology is logical, the capacities can be assumed to be constant and C_{ij} can be removed from the objective function (1).

3 GENETIC ALGORITHMS

Genetic algorithms are a method of searching solutions in the solution space by imitating the natural selection process (Androulakis, 1991)(Holland, 1991). In genetic algorithms, a population of solutions is created initially. Then by using genetic operators, such as mutation and crossover, a new generation is evolved. The fitness of each individual determines whether it will survive or not. After a number of iterations, or some other criterion is met, it is hoped that a near optimal solution is found.

In simple genetic algorithms, solutions are represented by strings. Each string consists of the same number of characters from some alphabet. Initially, the population is a random collection of such strings. At each iteration, individuals from the population are selected for breeding with a probability proportional to their fitness. Individuals are mated in pairs by the crossover operator to generate offsprings. The crossover operator divides each parent string at the same random position and then combines the left substring from the first parent with the right substring from the second parent to produce one of the offspring. The remaining substrings are concatenated to create the second offspring. After the population is replaced with the new generation, the mutation operator is applied. If an individual is to go under mutation, then one of the characters of its string representation is selected at random, and changed to some other character. This process is repeated until the termination criterion is satisfied.

As the above description shows, the simple genetic algorithm assumes that the crossover and mutation operators produce a high proportion of feasible solutions. However, in many problems, simply concatenating two substrings of feasible solutions, or modifying single characters do not produce feasible solutions. In such cases, there are two alternatives. If the operators produce sufficient numbers of feasible solutions, it is possible to let the genetic algorithm destroy the unfeasible ones by assigning them low fitness values. Otherwise, it becomes necessary to modify the simple operators so that only feasible individuals result from their application.

4 OPTIMIZATION USING GENETIC ALGORITHMS

4.1 Strategy

The optimization of logically rearrangeable multihop lightwave networks is a difficult problem, because the physical network and therefore the mathematical formulation imposes only one restriction on the topology, that the number of incoming and outgoing arcs is fixed. Not considering the assignment of flows, the size of the solution space grows exponentially as the number of nodes in the system increases. The problem is made even more difficult by the fact that given a fixed topology, finding an optimum flow assignment is itself a complex problem. The only constraint on flow assignment is the flow conservation criterion. Therefore, attacking the problem as a whole by trying to solve both the topology and assignment problems at the same time, is almost hopeless.

A more promising approach is to divide the problem into two independent problems: the connectivity problem and the assignment problem. Solving the connectivity problem yields optimal topologies and solving the assignment problem yields optimal flow assignments on fixed topologies. Since the genetic algorithm is much better suited for discrete optimization problems, it will be used to solve the connectivity problem. The flow assignment problem will be solved on fixed topologies with minimum-hop routing and flow-deviation. The overall algorithm is:

create initial population of topologies
while stopping criterion is not met
evaluate each topology using minimum-hop routing
mate individuals to create new generation
mutate generation

The division of the problem reduces the size of the solution space that each algorithm must work on. The genetic algorithm works only on topologies without considering the optimization of flows. Therefore the individuals of the population need to satisfy conditions (4), (5), and (6) only.

The individuals of the genetic algorithm's population are graph topologies. The second problem, assignment of flows, is solved on these topologies generated by the genetic algorithm. The minimum hop routing is applied to each individual to assign flows and evaluate its fitness for survival. Since the topologies that the routing algorithm works on satisfy conditions (4), (5), and (6) already, only conditions (2), (3), and (7) are considered during routing.

4.2 Representing Graphs as Genes

Manipulation of topologies with the genetic algorithm requires that they are represented in some suitable format. Although more compact alternatives are possible, the characteristic matrix of a graph is quite an adequate representation. A graph is represented by $n*n$ bits arranged in an n by n matrix, where n is the number of nodes of the graph. A '1' in row i and column j of the matrix stands for an arc from node i to j and a '0' represents that node i and node j are not connected. The most important advantage of this representation is that it is very easy to check if a graph satisfies the degree constraints. If the number of '1's in every row and every column of the matrix equals the fixed value, then the graph satisfies the constraint. To accelerate the convergence of the genetic algorithm, the genetic operators will be defined so as to generate only feasible individuals.

4.3 The Genetic Operators

Generating Random Graphs

Generating random graphs for the initial population is not very easy because of the degree constraints. To generate a random graph, it is initialized to contain no links. At each iteration, an available bit in the matrix is chosen at random which is then changed to a '1'. In other words, a new link is added to the graph. Addition of a graph can make other bits unavailable future links, because the node might have reached its capacity for incoming or outgoing links. The unavailability of links, in turn, can necessitate that some other links be created, so that the fixed-number of links criterion is met (Figure 1).

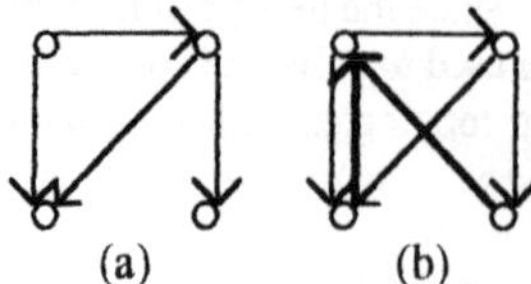

Figure 1 The links that *should* be added after random addition of some links.

The Mutation Operator

This operator takes a single graph and then modifies it by removing one of its links and adding a different link. The result is a different graph still satisfying the constraints. The algorithm is:

> *let (u,v) be some link chosen at random*
> *find (x,y) randomly such that (u,y) and (x,v) do not exist*
> *remove (u,v) and (x,y)*
> *add (u,y) and (x,v)*

The algorithm works well with the representation because the graph matrix can easily be searched to find the links and then be modified.

The Crossover Operator

This operator takes two graphs and produces two offsprings. First, it superimposes the matrices of the two graphs to find the different entries. The graphs can differ at a position in two ways. Either the first has a link and the other does not, or the first does not have a link and the second does. The positions of the differing entries are marked as '01' or '10' on a separate matrix. On this graph, a closed path with corners at the differing bits is found at random. The path should also have the property that, along the path consecutive corners are of different types: A '10' corner should be followed by a '01' corner and a '01' corner by a '10' corner. Flipping the original graphs' bits that are at the same positions as the corners of the path, create two offsprings that still satisfy the constraints. An example will clarify the procedure:

$$g = \begin{matrix} 0 & 1 & 1 & 0 & 0 \\ 0 & 0 & 1 & 1 & 0 \\ 0 & 0 & 0 & 1 & 1 \\ 1 & 0 & 0 & 0 & 1 \\ 1 & 1 & 0 & 0 & 0 \end{matrix}, \quad h = \begin{matrix} 0 & 0 & 1 & 1 & 0 \\ 1 & 0 & 1 & 0 & 0 \\ 1 & 0 & 0 & 1 & 0 \\ 0 & 1 & 0 & 0 & 1 \\ 0 & 1 & 0 & 0 & 1 \end{matrix}$$

To mate graphs g and h, a temporary graph matrix is created to show the positions of the differing bits:

$$
t = \begin{matrix} 0 & 10 & 0 & 01 & 0 \\ 01 & 0 & 0 & 10 & 0 \\ 01 & 0 & 0 & 0 & 10 \\ 10 & 01 & 0 & 0 & 0 \\ 10 & 0 & 0 & 0 & 01 \end{matrix}
$$

On this graph, a closed path, such as (1,2)-(1,4)-(2,4)-(2,1)-(4,1)-(4,2)-(1,2), is randomly found. Flipping the bits at these positions in the original matrices produces the two offsprings:

$$
\mathit{offg} = \begin{matrix} 0 & 0 & 1 & 1 & 0 \\ 1 & 0 & 1 & 0 & 0 \\ 0 & 0 & 0 & 1 & 1 \\ 0 & 1 & 0 & 0 & 1 \\ 1 & 1 & 0 & 0 & 0 \end{matrix}, \quad \mathit{offh} = \begin{matrix} 0 & 1 & 1 & 0 & 0 \\ 0 & 0 & 1 & 1 & 0 \\ 1 & 0 & 0 & 1 & 0 \\ 1 & 0 & 0 & 0 & 1 \\ 0 & 1 & 0 & 0 & 1 \end{matrix}
$$

4.4 Evaluating the Graphs

At each iteration, the individuals need to be evaluated to determine their fitness. Since the problem is to minimize the maximum flow on any link, evaluation requires that flows be assigned to the links. The flow assignment problem is solved with the minimum-hop routing and the flow deviation method. During minimum-hop routing, the flow matrix is built as the shortest paths are found. This allows the algorithm to choose the least used path, when alternate paths exist. Still, minimum-hop routing is not adequate because it does not allow flows to be split between alternate paths. To overcome this deficiency, flow deviation is used.

The flow deviation method works with a flow assignment algorithm, in this case minimum-hop routing. Given a flow assignment, the flow deviation method removes the most heavily used link and then the assignment problem is solved on the resulting graph. The two solutions are then combined linearly so as to improve the result. The best linear combination constant is found by fibonacci search (Zangwill, 1969). The algorithm is:

while there is significant improvement
    given graph g and flow assignment matrix f, find the most heavily used link (i,j)
    remove link (i,j) from g
    find the flow assignment matrix fr on reduced graph g
    *find, by fibonacci search, the constant k such that k*f+(1-k)*fr is optimal*
    *assign f = k*f+(1-k)*fr*
    add link (i,j) back to g

This algorithm, however, is not very effective on graphs with a small number of links, because at the end of the first iteration, the flow assignment matrix contains two very close entries, and removal of any of them in the second iteration causes the other to increase, making improvement impossible. Therefore, the flow deviation method is modified so that the most heavily used link and any other links with close amounts of flow are removed. The flow deviation continues until the removal of the highly loaded links causes the graph to be disconnected.

Although the flow deviation method is successful in decreasing the maximum flow, it has the disadvantage of being a number of times slower than simple minimum hop routing. Each evaluation by flow deviation requires a number of calls to the minimum-hop routing algorithm, and optimization by Fibonacci search between these calls. Instead of simply using flow deviation to evaluate every individual in the population at each generation, it is also possible to improve the flow assignment of the fittest solution without affecting its survival probability.

One problem with assigning flows for evaluating graphs is the case of disconnected graphs. If the traffic matrix is not disconnected, then a disconnected graph cannot carry all the traffic. In such a case, one possibility is to make the survival probability of the disconnected topology equal to zero, and therefore eliminate it in the next generation. However, it may be the case that mating the disconnected graph generates a proper offspring, so it is better to assign a minimal selection probability to disconnected graphs.

4.5 Stopping Criteria

A disadvantage of optimization with genetic algorithms is the difficulty of deciding when to stop. Although statistical variables, such as average fitness of a generation and best fitness value in a generation, are available, their values change almost erratically as generations evolve. Stopping after a certain number of iterations with no improvement or when the change in average fitness is small may cause the algorithm to stop too early or too late, and yet introduces even more parameters that depend on problem size. The stopping criterion used in this paper uses an exponential average of the average fitness values. At each generation, the current value of the exponential average and the average fitness value are combined with the equation $exp\_avg = \alpha * exp\_avg + (1-\alpha) * avg\_fit$, $0 < \alpha < 1$. Assigning small values (0.05) to α, makes it possible to trace the general behavior of the average fitness value. A good stopping criterion is then to stop when the change in the exponential average is small.

5 RESULTS AND DISCUSSION

5.1 Counting the Graphs

Before running the genetic algorithm, it is a good idea to study the number of distinct individuals that can be generated. A recursive function that yields the distinct number of graphs that satisfy equations (4), (5), and (6) for $T=2$ will be defined here. The function has three parameters: the number of source nodes still to be connected, the number of destination nodes still requiring two links, and the number of destination nodes still requiring one link. Adding two outgoing links to a node has three outcomes. The two links may both be connected to nodes with no incoming links yet. This decreases the second parameter by two, and increases the third parameter by two. Another possibility is that, one of the links may be connected to a node with no incoming links and the other to a node with one incoming link. This decreases the second parameter by one, and does not affect the third parameter. Finally, both links may be connected to nodes with one incoming link. This decreases the third parameter by two, and does not affect the second. In all cases, the first parameter is decreased by one. To account for selecting different combinations of destination nodes, each of the three possibilities is multiplied by the number of combinations possible:

$$f(nodes, twos, ones) = \binom{twos}{2} f(nodes-1, twos-2, ones+2) +$$

$$\binom{twos}{1}\binom{ones}{1} f(nodes-1, twos-1, ones) + \binom{ones}{2} f(nodes-1, twos, ones-2) \quad (14)$$

$$f(1, twos, ones) = \begin{cases} 1 & \text{if } twos = 0 \text{ and } ones = 2 \\ 0 & \text{otherwise} \end{cases} \quad (15)$$

$$\binom{n}{r} = \begin{cases} \dfrac{n!}{(n-r)!r!} & \text{if } n \geq r \geq 0 \\ 0 & \text{otherwise} \end{cases} \quad (16)$$

The values of the function for n=2-8 are shown in Table 1. Even for the small network of eight nodes, the number of different topologies is very large. Therefore, it is not possible to try all the possibilities exhaustively.

Table 1 Number of distinct graphs for different graph sizes

Number of nodes	*2*	*3*	*4*	*5*	*6*	*7*	*8*
Number of graphs	1	6	90	2040	67950	3110940	187530840

5.2 Experimenting with the Genetic Algorithm

The traffic data used in testing the genetic algorithm was taken from (Labourdette, 1991). The networks contained eight nodes, each with two receivers and two transmitters, and the traffic data consisted of four different specific types of non-uniform traffic data. To test the algorithm's performance with different graph sizes, the matrices were extended without modifying their underlying structures.

Parameters and Performance

The performance of the genetic algorithm is greatly influenced by a number of parameters, namely the number of individuals in the population, crossover rate and mutation rate. To achieve the best of the genetic algorithm, it is necessary to experiment with these parameters and determine an optimum set. Figure 2 shows the effect of mutation rate on the population. Normally the genetic algorithm converges by settling on some best-fit individuals, but when the mutation rate is too low, the genetic algorithm will converge too soon, before the global optimum solution is found. As the mutation rate is increased, the standard deviation of the fitnesses of the individuals remains at a high level even after some number of generations. Again, too high a rate may be a problem if the number of mutations does not permit the population to become stable.

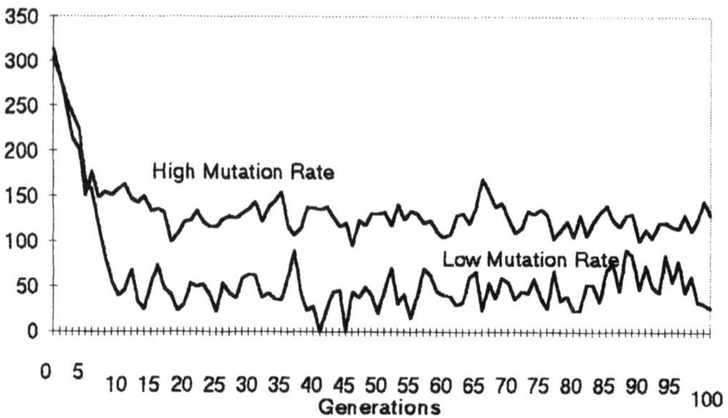

Figure 2 Graph showing effects of mutation rate.

Crossover rate is a parameter closely related to the mutation rate and has similar effects on the evolution of the population. With high crossover rates, the probability that the fittest individual of a generation will mate with some other individual increases, and the best fitness values are not very stable. On the other hand, low crossover rates have the disadvantage of being slow in exploring the solution space.

Using flow deviation method for evaluating each individual in every generation improves the quality of results in early generations but also results in a drastic slow-down, approximately by a factor of ten. Running the algorithm with minimum hop routing with more generations, and improving the final result with flow deviation provides a solution as good as the solution produced with pure flow deviation method.

Time Complexity

The genetic algorithm is almost a random search and its time performance can only be studied experimentally. One important factor, affecting the execution time, is the number of nodes in the input graph. Experiments run with different sized graphs, show that the execution time per generation increases approximately as fast as n^3 (Figure 3). However, this does not necessarily mean that the running time of the algorithm increases as n^3, since the number of generation required to find an optimal solution also varies.

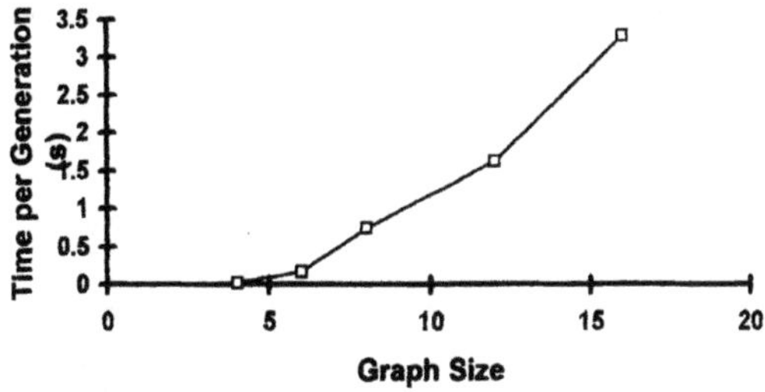

Figure 3 Time behavior of the genetic algorithm with respect to graph size.

In the basic genetic algorithm crossover over rate and mutation rate have negligible effects on running time. However, it is better to modify the genetic algorithm so that only new individuals that created by mutations and crossovers are evaluated at each generation. With this modification increases in crossover and mutation rates also increase the running time slightly.

5.3 Comparison of the Genetic and Heuristic Algorithms

To compare the genetic algorithm with the heuristic algorithm in (Labourdette, 1991), in optimizing logically rearrangeable networks, the genetic algorithm was tested with the traffic data in (Labourdette, 1991). The genetic algorithm was run with a population of 100, 60% crossover rate and 10% mutation rate on a 486DX2-66 PC. The values were collected in 20 runs with each type of traffic.

Table 2 Comparison of optimization algorithms

	Quasi-uniform traffic	*Ring*	*Centralized*	*Disconnected*
Average	6.42	15.11	34.14	32.28
Best solution	6.22	13.63	33.50	29.32
Worst solution	6.55	16.58	34.80	35.67
Running time (s)	38.64	34.63	30.64	31.28
Solution by heuristic algorithm	6.42	13.17	33.50	31.22

The genetic algorithm almost always performs well with quasi-uniform traffic. With other types of traffic, the algorithm produces good results also, but not in every run. This results both from the fact that most topologies behave well with quasi-uniform traffic therefore allowing the genetic algorithm to investigate the solution space freely and also from the fact that the stopping criteria cannot really decide if an optimal solution is reached.

The genetic algorithm's performance in quality of the results was comparable with those of the heuristic algorithm (Table 2). Although, most of the runs did not result in a more optimal solution, the solutions found by the genetic algorithm were within close proximity

improvements of about 30% compared with the perfect shuffle topology, the genetic algorithm outperformed the heuristic method (Figure 4).

0	10	11	9	0.9	0.8	1	1
9	0	11	9	1	0.8	1	0.9
10	12	0	8	0.9	0.9	1.1	1
8	9	10	0	1.1	1	0.8	0.7
0.7	0.8	1.1	1	0	10	11	8
1.2	0.8	0.9	0.9	9	0	9	8
0.8	1.1	1	1.1	10	11	0	11
0.9	1.1	1	1	11	8	9	0

(a)

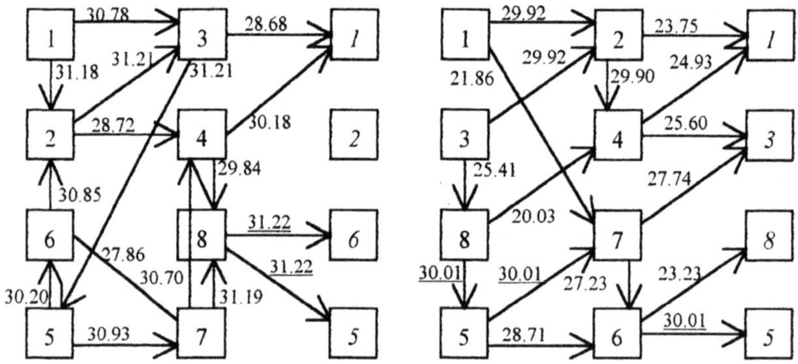

(b) (c)

Figure 4 Comparison of the algorithms. (a) Traffic matrix (b) Flows assigned by heuristic algorithm. (c) Flows assigned by genetic algorithm.

6 CONCLUSION

The optical fiber will be playing even a more important role in communications during the next years. Its use as the medium of transmission in local area networks is made possible with multi-wavelength multi-hop architectures. To use this architecture efficiently, it is necessary to find efficient algorithms that solve the topology and flow assignment problems specific to multi-hop lightwave networks.

In this paper, genetic algorithms are used to find optimal topologies that are further improved with the flow deviation method. To apply the genetic algorithm to the problem, a suitable representation and a set of operators are defined. The operators generate only feasible solutions and keep the individuals within the solution space. Although the algorithm operates quite randomly, it produces consistent results that are comparable to the existing heuristic algorithm's results in quality. The availability of alternate configurations is advantageous in cases where node failures and traffic load changes are frequent.

Improvements to the genetic algorithm are still possible. A more compact representation, faster evaluation algorithms and very finely tuned set of parameters will increase the time performance of the genetic algorithm as well as the quality of results. In time, the genetic algorithm will prove to be even more successful in competing with other optimization algorithms.

7 REFERENCES

Acampora, A.S., Karol, M.J. and Hluchyj, M.G. (1988) Multihop Lightwave Networks: A New Approach to Achieve Terabit Capabilities. *Proc. of IEEE Int. Conf. on Comm.*, pp. 1478-1484, 1988.

Androulakis, I.P. and Venkatasubramanian, V. (1991) A Genetic Algorithmic Framework for Process Design and Optimization. *Computers and Chemical Engineering*, Vol. 15, No.4.

Bannister, J.A., Fratta, L. and Gerla, M. (1990) Topological Design of the Wavelength-Division Optical Networks. *Proc. of the IEEE INFOCOM*, pp. 1005-1013.

Eisenberg, M. and Mehravari, N. (1988) Performance of the Multichannel Multihop Lightwave Network under Nonuniform Traffic. *IEEE JSAC*, pp. 1063-1077, Aug. 1988.

Goodman, M.S. (1989) Multiwavelength Networks and New Approaches to Packet Switching. *IEEE Communications Magazine*, Oct. 1989.

Henry, P.S. (1989) High-Capacity Lightwave Local Area Networks. *IEEE Communications Magazine*, Oct. 1989.

Holland, J.H. (1991) Genetic Algorithms. *Scientific American*, Vol. 267, Jul. 1991.

Labourdette, J.F.P. and Acampora, A.S. (1991) Logically Rearrangeable Multihop Lightwave Networks. *IEEE Transactions on Communications*, Vol. 39, Aug.1991.

Sivarajan, K. and Ramaswani, R. (1991) Multihop Lightwave Networks based on De Brujin Graphs. *Proc. of the IEEE INFOCOM*, pp. 1001-1011.

Zangwill, W.I. (1969) *Nonlinear Programming, A Unified Approach.* Prentice Hall, Englewood Cliffs, NJ.

8 BIOGRAPHY

C. Ersoy is an assistant professor at the Department of Computer Science in Bogazici University. After receiving the B.S. and M.S. degrees in electrical engineering from Bogazici University in 1984 and 1986, respectively, he earned the PhD degree in electrical engineering from Polytechnic University, Brooklyn, NY in 1992. During his M.S. studies between 1984-1986, he worked as an R&D engineer in Northern Telecom, Istanbul. His research interests include optimization and networks.

C. Gazen was born in Istanbul, Turkey in 1972. He received the B.S. degree in computer science from Bogazici University in 1995. He will start studying for the PhD degree in computer science in the University of Southern California in Fall 1995, where he will be employed as a graduate research assistant, working on molecular robotics.

PART FOUR

Queueing Models

10

Closed Form Performance Distributions of a Discrete Time $GI^G/D/1/N$ Queue with Correlated Traffic

Demetres Kouvatsos, Rod Fretwell
University of Bradford
Computer Systems Modelling Research Group, University of Bradford, Bradford BD7 1DP, U.K.
email: {D.D.Kouvatsos,R.J.Fretwell}@comp.brad.ac.uk

Abstract

A novel approach is applied to the study of a queue with general correlated traffic, in that the only features of the traffic which are taken into account are the usual measures of its correlation: the traffic is modelled as a batch renewal process. The batch renewal process is a precise tool for investigation into effects of correlation because it is the least biased choice of process which is completely determined by the infinite sets of measures of the traffic's correlation (e.g. indices of dispersion, covariances or correlation functions).

The general effect of traffic correlation on waiting time, blocking probability and queue length is well known from simulation studies and numerical analysis of a variety of models. The contribution of this paper is to show that these effects are due to correlation alone (and not to any other features of the traffic or of the models used) and to show explicitly how the magnitudes of blocking, waiting time and queue length distribution are determined by the degree of correlation in the traffic.

The study focuses upon a discrete time $GI^G/D/1/N$ queue with single server, general batch renewal arrivals process, deterministic service time and finite capacity N. Closed form expressions for basic performance distributions, such as queue length and waiting time distributions and blocking probability, are derived when the batch renewal process is of the least biased form which might be expected to result from actual traffic measurements at the interior of a network or of some individual traffic source.

The effect of varying degrees of traffic correlation upon basic performance distributions is investigated and illustrative numerical results are presented. Comments on implications of the results on analysis of general discrete time queueing networks with correlated traffic are included.

Keywords

asynchronous transfer mode (ATM), batch renewal process, correlated traffic, least biased process, discrete time queue, performance distributions

1 INTRODUCTION

ATM traffic is both bursty and correlated. Even for traffic sources described as being (bursty) renewal processes, superposition of several such sources generally yields correlated processes. The indices of dispersion have been proposed as appropriate characterisation of bursty, correlated traffic and Markov modulated processes as models of sources of bursty traffic with correlation. Sriram and Whitt (1986) described superposition of bursty sources (modelled by renewal processes) in terms of the indices of dispersion for intervals (IDI). Heffes and Lucantoni (1986) model the superposition of bursty renewal processes approximately by a 2-phase Markov modulated Poisson process (MMPP) matched on three features of the indices of dispersion for counts (IDC) and mean arrival rate. Gusella (1991) estimated indices of dispersion for measured LAN traffic and modelled the traffic approximately by a 2-phase MMPP matched on three features of the IDC and the SCV of one inter-arrival time. It should be noted that i) a 2-phase MMPP is defined by only four parameters and cannot conform entirely to all the indices of dispersion which may be used to characterise traffic of any particular source, ii) the IDI, J_n, and IDC, I_t, of a 2-phase MMPP tend to the same finite limit, i.e. $I_\infty = J_\infty$, and iii) a batch renewal process may be constructed for an arbitrary set of indices of dispersion (with finite $I_\infty = J_\infty$). If the indices of dispersion are all that is known about certain traffic (as might result from measurements of real traffic) then a batch renewal process may be constructed which incorporates all that information and no other: in that sense, batch renewal processes provide a description of the traffic which is both complete and least biased — which models, such as MMPP, with limited parameterisation do not.

Fowler and Leland (1991) have reported LAN traffic with unbounded IDC (i.e. infinite I_∞). However, it is to be expected that performance of restricted buffer systems with deterministic service (as in ATM switches) would not be affected by the magnitude of IDC for long intervals. Recently, Andrade and Martinez-Pascua (1994) have shown that queue length distribution, etc. is affected by IDC only up to a certain size of interval (determined by the buffer size) and "the value of the IDC at infinity has little importance." So it may be expected that, for practical purposes, the finite limit to IDI and IDC in batch renewal processes would not be a disadvantage in traffic models.

The general effect of traffic correlation on waiting time, blocking probability and queue length is well known from simulation studies and numerical analysis of a variety of models. The contribution of this paper is to show that these effects are due to correlation alone (and not to any other features of the traffic or of the models used) and to show explicitly how the magnitudes of blocking, waiting time and queue length distribution are determined by the degree of correlation in the traffic. The traffic is modelled as a batch renewal process which is a precise tool for investigation into effects of correlation because it is the least biased choice of process which is completely determined by the infinite sets of measures of the traffic's correlation (e.g. indices of dispersion, covariances or correlation functions).

Batch renewal processes are defined and their properties described in Section 2. In Section 3, the relationships between the component distributions of the batch renewal arrivals process and the queue length distribution, waiting time and blocking probability in a finite buffer queue with deterministic service and censored batch renewal arrivals process are presented. Such a queueing system is an appropriate model for an ATM multiplexer or partitioned buffer switch.

The analysis is specialised in Section 4 to batch renewal processes in which the component distributions are shifted generalised geometric (shifted GGeo). This form of process appears to be appropriate to measured traffic, especially where the usable data set be limited by (say) the time for which the actual process may be regarded as being stationary. Closed form expressions for queue length distribution, waiting time and blocking probability are given.

Section 5 presents analysis of the effects, on blocking probability, waiting time and queue length, of varying degrees of correlation and illustrates the results by numerical examples.

Finally, conclusions and proposals for extensions to the work, including those towards approximate analysis of general queueing networks with correlated traffic, are given in Section 6

2 BATCH RENEWAL PROCESSES

Definition A random sequence $\{\xi(t) : t = \ldots, -2, -1, 0, 1, 2, \ldots\}$ is stationary in the wide sense (equivalently, stationary in Khinchin's sense) if

- the random function $\xi(t)$ has constant finite mean $\mathbf{E}[\xi(t)] = \xi$ (which is independent of t) and
- the correlation function $\mathrm{Cov}[\xi(t), \xi(s)] \triangleq \mathbf{E}[(\xi(t) - \xi)(\xi(s) - \xi]$ is finite and depends on the lag $t - s$ only.

Observe that $\mathrm{Cov}[\xi(t), \xi(t+\ell)] = \mathrm{Cov}[\xi(t+\ell), \xi(t)]$, by symmetry of the definition, and that $\mathrm{Cov}[\xi(t+\ell), \xi(t)] = \mathrm{Cov}[\xi(t), \xi(t-\ell)]$, by change of variable t to $t-\ell$. Consequently, $\mathrm{Cov}[\xi(t), \xi(t+\ell)] = \mathrm{Cov}[\xi(t), \xi(t-\ell)]$ — only the magnitude of the lag is significant and it is therefore necessary only to consider positive lags.

Consider an arrivals process which is a two dimensional wide sense stationary sequence $\{\alpha(t), \beta(t) : t = \ldots, -2, -1, 0, 1, 2, \ldots\}$, in which realisations of $\alpha(t)$ and $\beta(t)$ are drawn from the positive integers. The $\beta(t)$ are to be interpreted as the number of arrivals (i.e. batch sizes) and the $\alpha(t)$ as the number of slots in intervals between successive batches of arrivals. From $\{\alpha(t), \beta(t)\}$ may be derived two related sequences of interest

- the numbers of arrivals $\{N(t) : t = \ldots, -2, -1, 0, 1, 2, \ldots\}$ at each epoch,
- the intervals $\{X(t) : t = \ldots, -2, -1, 0, 1, 2, \ldots\}$ between successive arrivals.

To be specific, $X(0)$ may be assigned $\alpha(0)$, $X(\beta(1))$ assigned $\alpha(1)$, $X(\beta(1)+\beta(2))$ assigned $\alpha(2)$, etc. and intermediate values, $X(1)$ through $X(\beta(1)-1)$, etc., assigned 0. Similarly, $N(1)$ may be assigned $\beta(1)$, $N(\alpha(2)+1)$ assigned $\beta(2)$, etc. and intermediate values, $N(2)$ through $N(\alpha(2))$, etc., assigned 0.

It is generally true that

$$\mathbf{E}[N(t)^n] = \frac{\mathbf{E}[\beta(t)^n]}{\mathbf{E}[\alpha(t)]}$$

and, thence, generally true that

$$I_1 \triangleq \frac{\mathrm{Var}[N(t)]}{\mathbf{E}[N(t)]} = bC_b^2 + b - \frac{b}{a}$$

where $a = \mathbf{E}[\alpha]$, $b = \mathbf{E}[\beta]$ and C_b^2 is the square coefficient of variation of β and I_1 is the index of dispersion for counts (IDC) over one slot (i.e. at lag 0). Similarly,

$$J_1 \triangleq \frac{\mathrm{Var}[X(t)]}{\mathbf{E}[X(t)]^2} = bC_a^2 + b - 1$$

where C_a^2 is the square coefficient of variation of α and J_1, the index of dispersion for intervals (IDI) for one interval (i.e. at lag 0). J_1 is the square coefficient of variation of X.

To determine the correlation functions for $\{N(t)\}$ and $\{X(t)\}$ more information about $\{\alpha(t), \beta(t)\}$ is required.

Definition A batch renewal arrival process is a process in which there are batches of simultaneous arrivals such that

- the numbers of arrivals in batches are independent identically distributed random variables,
- the intervals between batches are independent identically distributed random variables.
- the batch sizes are independent of the intervals between batches.

It is shown, below, that a discrete time batch renewal arrival process may be constructed to give any degree of correlation between numbers of arrivals at different epochs and, simultaneously, any degree of correlation between interarrivals times at arbitrary lags. Indeed, there is a *one-to-one* correspondence between an arbitrary set of indices of dispersion (or, equivalently, of correlation functions or covariances) and a batch renewal process. Furthermore, the corresponding batch renewal process is the *least biased choice* given only a set of indices of dispersion or of correlation functions. (To say that a process be the "least biased choice" means that, of all possible processes which satisfy the given conditions (e.g. set of indices of dispersion), is chosen that process which involves least arbitrary additional information. For example, in the case of a 2-dimensional joint probability distribution $\mathbf{P}[X = n_1, Y = n_2]$ given only the marginal distributions $\mathbf{P}[X = n_1]$ and $\mathbf{P}[Y = n_2]$, the least biased choice for the joint distribution is $\mathbf{P}[X = n_1, Y = n_2] = \mathbf{P}[X = n_1]\mathbf{P}[Y = n_2]$. The effect is to treat X and Y as being independent. Any other choice would introduce arbitrary information in the form of the dependence between X and Y.)

2.1 Independence or Dependence at Various Lags

Consider a discrete time batch renewal process in which

- the distribution of batch size is given by the probability mass function (*pmf*) $b(n)$, $n = 1, 2, \ldots$, with mean b, square coefficient of variation (SCV) C_b^2 and probability generating function (*pgf*) $B(z) = \sum_{n=1}^{\infty} b(n)z^n$, and

- the distribution of intervals between batches is given by the *pmf* $a(t)$, $t = 1, 2, \ldots$, with mean a, SCV C_a^2 and *pgf* $A(\omega) = \sum_{t=1}^{\infty} a(t)\omega^t$.

Observe that no loss in generality ensues from the assumption that $a(0) = 0$, $b(0) = 0$.

It is readily seen that the stationary distribution of the number n of arrivals at an epoch is given by the *pmf* $\nu(n)$, $n = 0, 1, \ldots$,

$$\nu(n) = \begin{cases} 1 - \dfrac{1}{a} & n = 0 \\ \dfrac{1}{a}\, b(n) & n = 1, 2, \ldots \end{cases} \tag{1}$$

and the conditional probability $\nu_\ell(n; k)$ that there be n arrivals ($n = 0, 1, \ldots$) at an epoch, given that there had been k arrivals ($k = 0, 1, \ldots$) at the epoch ℓ slots earlier ($\ell = 1, 2, \ldots$), is

$$\nu_\ell(n; k) = \begin{cases} 1 - \dfrac{1 - \phi_\ell}{a - 1} & n = 0,\ k = 0 \\ \dfrac{1 - \phi_\ell}{a - 1}\, b(n) & n = 1, 2, \ldots,\ k = 0 \\ 1 - \phi_\ell & n = 0,\ k = 1, 2, \ldots \\ \phi_\ell\, b(n) & n = 1, 2, \ldots,\ k = 1, 2, \ldots \end{cases} \tag{2}$$

where ϕ_ℓ is the probability that there be a batch at any epoch, given that there had been a batch at the epoch ℓ slots earlier. The number of arrivals at an epoch is either independent of or dependent on the number of arrivals at the epoch ℓ slots earlier according to whether $\phi_\ell = 1/a$ or not. Obviously, ϕ_ℓ satisfies the (convolution) relationship

$$\phi_\ell = \begin{cases} 1 & \ell = 0 \\ \sum_{t=1}^{\ell} \phi_{\ell-t}\, a(t) & \ell = 1, 2, \ldots \end{cases} \tag{3}$$

and ϕ_ℓ is generated from the *pgf*

$$\sum_{\ell=0}^{\infty} \phi_\ell \omega^\ell = \frac{1}{1 - A(\omega)} . \tag{4}$$

Note that ϕ_ℓ is determined by $a(1), \ldots, a(\ell)$ only ($\ell = 1, 2, \ldots$) and so $a(\cdot)$ may be constructed to give independence or dependence arbitrarily at any specified lags.

The correlation functions (covariances) at lag ℓ, $\ell = 1, 2, \ldots$, are derived from equations (1) and (2) as

$$\mathrm{Cov}[N(t), N(t+\ell)] = \mathbf{E}[N(t)N(t+\ell)] - \mathbf{E}[N(t)]^2 = \sum_{k=1}^{\infty}\sum_{n=1}^{\infty} n\, k\, \nu(k)\, \nu_\ell(n; k) - \left(\frac{b}{a}\right)^2$$

$$= \left(\sum_{k=1}^{\infty} k\frac{b(k)}{a}\right)\left(\sum_{n=1}^{\infty} n\,\phi_\ell\, b(n)\right) - \left(\frac{b}{a}\right)^2$$

$$= \frac{b^2}{a}\left(\phi_\ell - \frac{1}{a}\right) \quad (5)$$

Hence, utilising equation (4), the variance and covariances are generated by

$$K(\omega) \triangleq \frac{1}{\lambda}\left(\mathrm{Var}[N] + 2\sum_{\ell=1}^{\infty}\mathrm{Cov}[N(t), N(t+\ell)]\,\omega^\ell\right) = bC_b^2 + b\frac{1+A(\omega)}{1-A(\omega)} - \lambda\frac{1+\omega}{1-\omega} \quad (6)$$

where $\lambda = b/a$. Observe that $\lambda K(z)$ is a *pgf* analog of the spectral density function for the random sequence of the number of arrivals at successive epochs. Further, the relationship between correlation functions (covariances) and indices of dispersion may be described by

$$K(\omega) = (1-\omega)^2 I'(\omega)$$

where $I(\omega)$ is the *pgf* of the indices of dispersion I_t for counts and where the prime ($'$) indicates the derivative.

In an essentially similar way, it may be shown that the batch size distribution $b(\cdot)$ may be constructed to give independence or dependence of intervals (between individual pairs of arrivals) arbitrarily at any specified lags and that the correlation functions are generated from

$$L(z) = \lambda^2\left(\mathrm{Var}[X] + 2\sum_{\ell=1}^{\infty}\mathrm{Cov}[X(t), X(t+\ell)]\,z^\ell\right) = bC_a^2 + b\frac{1+B(z)}{1-B(z)} - \frac{1+z}{1-z} \quad (7)$$

where $L(\omega)/\lambda^2$ is a *pgf* analog of the spectral density function for the random sequence of interarrival times,

$$L(z) = (1-z)^2 J'(z)$$

where $J(z)$ is the *pgf* of the indices of dispersion J_n for intervals and where the prime ($'$) indicates the derivative.

It may be shown that equations (6) and (7) together imply a *one-to-one* relationship between the set of correlation functions or covariances (equivalently, indices of dispersion) and the batch renewal process distributions of batch sizes and intervals between batches.

3 CENSORED GI^G/D/1/N QUEUE UNDER DEPARTURES FIRST POLICY

Consider a GI^G/D/1/N queue in discrete time in which arrivals to a full system are turned away and simply lost (i.e. censored arrivals). Events (arrivals and departures) occur at discrete points in time (epochs) only. The intervals between epochs are called *slots* and, without loss of generality, may be regarded as being of constant duration. At an epoch at which both arrivals and departures occur, the departing customers release the places, which they had been occupying, to be available to arriving customers (*departures first* memory management policy). The service time for a customer is one slot and the first customer arriving to an empty system (after any departures) receives service and departs at the end of the slot in which it arrived (*immediate service* policy). By GI^{G} arrivals process is meant the intervals between batches are independent and of general distribution and the batch size distribution is general (batch renewal process). Consider further two processes embedded at points immediately before and immediately after each batch of arrivals. Each process may be described independently by a Markov chain but the processes are mutually dependent. Let

$p_N(n)$ be the steady state probability that there be $n = 0, 1, \ldots, N$ customers in the system (either queueing or receiving service) during a slot (i.e. $\{p_N(n) : n = 0, \ldots, N\}$ is the random observer's distribution),

$p_N^A(n)$ be the steady state probability that a batch of arrivals 'see' $n = 0, \ldots, N-1$ customers in the system (i.e. $\{p_N^A(n) : n = 0, \ldots, N-1\}$ is the stationary distribution of the Markov chain embedded immediately before batch arrivals),

$p_N^D(n)$ be the steady state probability that there be $n = 1, \ldots, N$ customers in the system immediately after a batch of arrivals to the queue (i.e. $\{p_N^D(n) : n = 1, \ldots, N\}$ is the stationary distribution of the Markov chain embedded immediately after batch arrivals).

When, immediately following a batch of arrivals, the system contains k ($k = 1, \ldots, N$) customers and the interval to the next batch is t slots, then there will be one departure at the end of each of the t slots for which there remain customers in the system. If $t \leq k$ that next batch will 'see' $k - t$ customers. If $t > k$ the system will become empty before the next batch arrives. Similarly, when, immediately prior to admission of a batch of arrivals, the system contains k ($k = 0, \ldots, N-1$) customers and the batch size is r, then the buffer will become full if $r \geq N - k$. Otherwise there will be $k + r$ customers immediately after the batch arrives. Consequently, $p_N^A(\cdot)$ and $p_N^D(\cdot)$ are related by

$$p_N^A(n) = \begin{cases} \sum_{k=1}^{N} p_N^D(k) \sum_{t=k}^{\infty} a(t) & n = 0 \\ \sum_{k=n+1}^{N} p_N^D(k) a(k-n) & n = 1, \ldots, N-1 \end{cases} \tag{8}$$

and

$$p_N^D(n) = \begin{cases} \sum_{k=0}^{n-1} p_N^A(k) b(n-k) & n = 1, \ldots, N-1 \\ \sum_{k=0}^{N-1} p_N^A(k) \sum_{r=N-k}^{\infty} b(r) & n = N \end{cases} \tag{9}$$

The relationship between$p_N^D(\cdot)$ and the random observer's probability $p_N(\cdot)$ results from the following considerations.

If, immediately after the arrival of a batch, there be k customers in the system and the interval between that and the next batch be t slots (figure 1) then, during that interval of t slots,

if $t \leq k$ the system visits the states $k, \ldots, k-t+1$ for one slot each,
if $t > k$ the system visits the states $k, \ldots, 1$ for one slot each and resides in state 0 for $t-k$ slots.

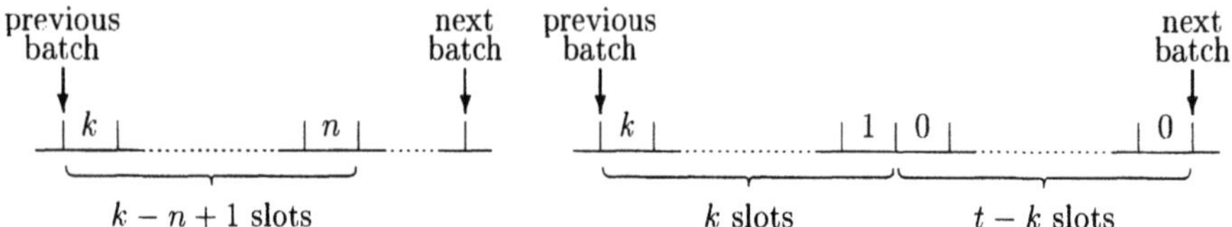

Figure 1 Ways in which state n ($n > 0$) and state 0 may be reached during an interval between batches.

Any arbitrary slot must fall in the interval between two batches. The probability that the interval be of length t is $\frac{1}{a}ta(t)$ and the probability that the slot occupy any particular position within the interval is $1/t$.

Hence,

$$p_N(n) = \begin{cases} \frac{1}{a} \sum_{k=1}^{N} p_N^D(k) \sum_{t=k+1}^{\infty} (t-k)a(t) & n = 0 \\ \frac{1}{a} \sum_{k=n}^{N} p_N^D(k) \sum_{t=k-n+1}^{\infty} a(t) & n = 1, \ldots N \end{cases} \tag{10}$$

3.1 Blocking Probability

If an arriving batch of size $N-k+r$ see k customers in the system, then only the first $N-k$ members of the arriving batch may be admitted and r customers will be blocked. The probability that the arriving batch see k customers is $p_N^A(k)$, the probability that the batch be of size $N-k+r$ is $(N-k+r)\,b(N-k+r)/b$ and the probability of a customer being in one of the r positions, given that the batch be of size $N-k+r$, is $r/(N-k+r)$. Therefore, the blocking probability π_N^B is given by

$$\pi_N^B = \sum_{k=0}^{N-1} p_N^A(k) \sum_{r=1}^{\infty} \frac{r}{b} b(N-k+r) \tag{11}$$

3.2 Waiting Time

The waiting time of a customer is given by its position in the queue at the instant at which it arrive in the queue. Thus, given that there be k customers in the queue (including any in service) at the time of an arriving batch of size r, the customer in position $t-k$ in the batch ($1 \le t-k \le r \le N-k$) will remain in the queue for t slots.

Let $b_k(n)$, $n = 1, \ldots, k$ with mean $\bar{b}_k$ be the effective arrival distribution given that an arriving batch see k, $k = 1, \ldots, N-1$, places available in the buffer.

$$b_k(n) = \begin{cases} b(n) & 1 \le n < k \\ \sum_{r=k}^{\infty} b(r) & n = k \\ 0 & \text{otherwise} \end{cases} \tag{12}$$

Then the mean $\bar{b}_k$ is given by

$$\bar{b}_k = \sum_{n=1}^{k} n\, b_k(n) = \sum_{n=1}^{k} n\, b(n) + k \sum_{n=k+1}^{\infty} b(n) \tag{13}$$

Then waiting time is distributed as $w(t)$, $t = 1, \ldots, N$,

$$w(t) = \sum_{k=0}^{t-1} p_N^A(k) \frac{1}{\bar{b}_{N-k}} \sum_{r=t-k}^{N-k} b_{N-k}(r) = \sum_{k=0}^{t-1} p_N^A(k) \frac{1}{\bar{b}_{N-k}} \sum_{r=t-k}^{\infty} b(r) \tag{14}$$

4 SHIFTED GGEO DISTRIBUTIONS OF BATCH SIZE AND INTERVALS

This section presents particular forms of batch renewal arrivals process which appear to be especially appropriate to models of traffic where there are relatively few measurements from which the correlation functions (covariances) may be estimated. In such cases it is natural to plot the logarithms of covariances against lags and fit a straight line to the plot. Then, if

$$\log \mathrm{Cov}[X(t), X(t+\ell)] \simeq -C - m\ell$$

(for some constants C and m), equation (7) implies that the corresponding batch renewal process has batch size distribution of the form

$$b(n) = \begin{cases} 1-\eta & n = 1 \\ \eta\nu(1-\nu)^{n-2} & n = 2, \ldots \end{cases} \tag{15}$$

in which

$$\eta = \frac{\lambda^2 e^{-C}}{1+\lambda^2 e^{-C}} \left(1 - e^{-m}\right)$$

$$\nu = \frac{1}{1+\lambda^2 e^{-C}} \left(1 - e^{-m}\right)$$

Similarly, if

$$\log \mathrm{Cov}[N(t), N(t+\ell)] \simeq -C - m\ell$$

(for some constants C and m) equation (6) implies that the corresponding batch renewal process has intervals between batches distributed as

$$a(t) = \begin{cases} 1-\sigma & t=1 \\ \sigma\tau(1-\tau)^{t-2} & t=2,\ldots \end{cases} \tag{16}$$

in which

$$\begin{aligned} \sigma &= \frac{e^{-C}}{\lambda^2+e^{-C}}\left(1-e^{-m}\right) \\ \tau &= \frac{\lambda^2}{\lambda^2+e^{-C}}\left(1-e^{-m}\right) \end{aligned}$$

Distributions of form (15) and (16) are known as shifted generalised geometric (shifted GGeo).

This section first discusses the solution of $\mathrm{GI}^{\mathrm{G}}/\mathrm{D}/1/\mathrm{N}$ queues in which the intervals between batches are distributed as a shifted GGeo. (A similar solution method may be applied to $\mathrm{GI}^{\mathrm{G}}/\mathrm{D}/1/\mathrm{N}$ queues in which the batch sizes are distributed as a shifted GGeo.) Then closed form expressions for queue length distribution, waiting time distribution and blocking probability are derived for the interesting case when both batch sizes and intervals are distributed as shifted GGeo.

4.1 $\mathrm{GGeo}^{\mathrm{G}}/\mathrm{D}/1/\mathrm{N}$ Queues

When the distribution of intervals between batches is shifted generalised geometric with parameters σ and τ the correlation functions (covariances) for the numbers of arrivals per slot are

$$\mathrm{Cov}[N(t), N(t+\ell)] = \frac{b^2}{a}\phi_\ell - \left(\frac{b}{a}\right)^2 = \lambda^2\frac{\sigma}{\tau}(1-\sigma-\tau)^\ell = \lambda^2\frac{a-1}{a}{\beta_a}^\ell$$

which shows, for given interval a between batches, the significance of the "correlation factor" $\beta_a \triangleq 1-\sigma-\tau$.

Equation (8) becomes

$$p_N^A(n) = \begin{cases} p_N^D(1) + \sigma\sum\limits_{k=2}^{N} p_N^D(k)(1-\tau)^{k-2} & n=0 \\ (1-\sigma)p_N^D(n+1) + \sigma\tau\sum\limits_{k=n+2}^{N} p_N^D(k)(1-\tau)^{k-n-2} & n=1,\ldots,N-2 \\ (1-\sigma)p_N^D(N) & n=N-1 \end{cases} \tag{17}$$

Consideration of the differences between the $p_N^A(n)$ for successive values of n leads to the difference relations below. For $n = 1, \ldots, N-3$

$$
\begin{aligned}
(1-\tau)p_N^A(n+1) - p_N^A(n) &= (1-\sigma-\tau)p_N^D(n+2) - (1-\sigma)p_N^D(n+1) \\
&= (1-\sigma-\tau)\sum_{k=0}^{n+1} p_N^A(k)b(n-k+2) \\
&\quad -(1-\sigma)\sum_{k=0}^{n} p_N^A(k)b(n-k+1)
\end{aligned}
\tag{18}
$$

and for $n = 0$

$$
\begin{aligned}
(1-\tau)p_N^A(1) - p_N^A(0) &= (1-\sigma-\tau)p_N^D(2) - \tau p_N^D(1) \\
&= (1-\sigma-\tau)\left[p_N^A(0)b(2) + p_N^A(1)b(1)\right] - \tau p_N^A(0)b(1)
\end{aligned}
\tag{19}
$$

The system of linear equations (18) and (19) establish ratios between $p_N^A(n)$ and $p_N^A(0)$ (for $n = 1, \ldots, N-2$) which are independent of N and are the same as in the corresponding unrestricted buffer $\mathrm{GGeo}^{\mathrm{G}}/\mathrm{D}/1$ system. Therefore, writing $p^A(n)$ for the steady state probability that a batch of arrivals to the unrestricted queue 'see' n in the system,

$$p_N^A(n) = \frac{1}{Z}p^A(n) \quad n = 0, \ldots, N-2 \tag{20}$$

for some normalising constant Z, and so, writing

$$P^A(z) = \sum_{n=0}^{\infty} p_N^A(n) z^n$$

for the generating function of $p^A(n)$, gives

$$P^A(z) = \frac{\sigma + \tau - b}{1 - (1-\sigma-\tau)\dfrac{B(z)}{z} - \tau\dfrac{1-B(z)}{1-z}} \tag{21}$$

Given the distribution $b(\cdot)$ explicitly, equation (21) may (in principle) be solved, leading (via equation (20)) to the solution of equation (17). Thence, relations (9), (10), (14) and (11) give queue length distribution, waiting time distribution and blocking probability. This method is shown, in the next subsection, when the batch size is distributed as GGeo.

4.2 $\mathrm{GGeo}^{\mathrm{GGeo}}/\mathrm{D}/1/\mathrm{N}$ Queues

When both the intervals between batches and the batches are distributed as shifted Generalised Geometric equation (21) becomes

$$P^A(z) = \frac{\sigma - \tau\dfrac{\eta}{\nu}}{1 - \tau\left(1 + \dfrac{\eta z}{1-(1-\nu)z}\right) - (1-\sigma-\tau)\left(1 - \dfrac{\eta(1-z)}{1-(1-\nu)z}\right)}$$

$$= \frac{1}{\nu} \frac{(\sigma\nu - \tau\eta)(1-(1-\nu)z)}{(\sigma + (1-\sigma-\tau)\eta) - (\sigma(1-\eta-\nu)+\eta)z} \tag{22}$$

It follows immediately that

$$p^A(n) = \begin{cases} \frac{1}{\nu}(1-x) & n = 0 \\ \frac{\eta}{\nu}\frac{\tau + (1-\sigma-\tau)\nu}{\sigma + (1-\sigma-\tau)\eta}(1-x)x^{n-1} & n = 1, 2, \ldots \end{cases} \tag{23}$$

where the geometric term x is

$$x = \frac{\sigma(1-\eta-\nu)+\eta}{\sigma + (1-\sigma-\tau)\eta} .$$

Then, from equation (17) for $n = N$, equation (9) and the distribution of batch sizes,

$$\begin{aligned} p_N^A(N-1) &= (1-\sigma)p_N^D = (1-\sigma)\sum_{k=0}^{N-1} p_N^A(k) \sum_{r=N-k}^{\infty} b(r) \\ &= \frac{1}{Z}\frac{1-\sigma}{\sigma}\frac{\eta}{\nu}(1-x)x^{N-2} \end{aligned} \tag{24}$$

and so the normalising constant Z is seen to be

$$Z = 1 - \frac{\eta}{\nu}\frac{\tau}{\sigma}x^{N-1} \tag{25}$$

Combining equations (20), (23) and (24) yields

$$p_N^A(n) = \begin{cases} \frac{1}{Z}\frac{1}{\nu}(1-x) & n = 0 \\ \frac{1}{Z}\frac{\eta}{\nu}\frac{\tau + (1-\sigma-\tau)\nu}{\sigma + (1-\sigma-\tau)\eta}(1-x)x^{n-1} & n = 1, \ldots, N-2 \\ \frac{1}{Z}\frac{\eta}{\nu}\frac{1-\sigma}{\sigma}(1-x)x^{N-2} & n = N-1 \end{cases} \tag{26}$$

Applying equation (9) to (26) yields

$$p_N^D(n) = \begin{cases} \frac{1}{Z}\frac{1-\eta}{\nu}(1-x) & n = 1 \\ \frac{1}{Z}\frac{\eta}{\nu}\frac{\tau(1-\eta-\nu)+\nu}{\sigma + (1-\sigma-\tau)\eta}(1-x)x^{n-2} & n = 2, \ldots, N-1 \\ \frac{1}{Z}\frac{\eta}{\nu}\frac{1}{\sigma}(1-x)x^{N-2} & n = N \end{cases} \tag{27}$$

Applying equation (10) to (27) yields

$$p_N(n) = \begin{cases} \frac{1}{Z}\frac{1}{\nu}\frac{1}{\sigma+\tau}(\sigma\nu - \tau\eta) = \frac{1}{Z}(1-\lambda) & n = 0 \\ \frac{1}{Z}\frac{1}{\nu}\frac{\tau}{\sigma+\tau}(1-x) & n = 1 \\ \frac{1}{Z}\frac{\eta}{\nu}\frac{\tau}{\sigma+\tau}\frac{1-(1-\sigma-\tau)(1-\eta-\nu)}{\sigma+(1-\sigma-\tau)\eta}(1-x)x^{n-2} & n = 2,\ldots,N-1 \\ \frac{1}{Z}\frac{\eta}{\nu}\frac{\tau}{\sigma+\tau}\frac{1}{\sigma}(1-x)x^{N-2} & n = N \end{cases} \tag{28}$$

Hence, mean queue length L_N is

$$\begin{aligned} L_N &= \frac{1}{Z}\frac{\tau}{\sigma+\tau}\frac{1}{\nu}\Bigg((\eta+\nu) + \eta\frac{1-(1-\sigma-\tau)(1-\eta-\nu)}{\sigma\nu-\tau\eta} \\ &\quad -\eta\frac{1-(1-\sigma-\tau)(1-\eta-\nu)}{\sigma\nu-\tau\eta}x^{N-1} \\ &\quad + N\frac{\eta}{\sigma}(1-\sigma-\tau)x^{N-1}\Bigg) \end{aligned} \tag{29}$$

From equations (11) and (26) the blocking probability π_N^B is

$$\pi_N^B = \sum_{k=0}^{N-1} p_N^A(k)\sum_{r=1}^{\infty}\frac{r}{b}b(N-k+r) = \frac{1}{Z}\frac{\eta}{\eta+\nu}\frac{\sigma\nu-\tau\eta}{\sigma\nu}x^{N-1} = \frac{1-Z}{Z}\frac{1-\lambda}{\lambda} \tag{30}$$

Hence

$$\frac{\pi_{N+1}^B}{\pi_N^B} \to x = \frac{\sigma(1-\eta-\nu)+\eta}{\sigma+(1-\sigma-\tau)\eta} \quad \text{as} \quad N \to \infty$$

which illustrates the typical log-linear relationship between blocking probability $\pi_N^B$and buffer size N.

From equation (13), the mean effective batch size given k buffer places available to arrivals is given by

$$b_k = (1-\eta) + \eta\nu\sum_{n=2}^{k-1} n(1-\nu)^{n-2} + k\eta\nu\sum_{n=k}^{\infty}(1-\nu)^{n-2} = 1 + \frac{\eta}{\nu}\left(1-(1-\nu)^{k-1}\right) \tag{31}$$

Hence, waiting time is distributed as

$$\begin{aligned} w(t) &= \sum_{k=0}^{t-1} p_N^A(k)\frac{1}{b_{N-k}}\sum_{r=t-k}^{\infty} b(r) \\ &= (t>1)\sum_{k=0}^{t-2} p_N^A(k)\frac{\eta(1-\nu)^{t-k-2}}{1+\frac{\eta}{\nu}\left(1-(1-\nu)^{N-k-1}\right)} + p_N^A(t-1)\frac{1}{1+\frac{\eta}{\nu}\left(1-(1-\nu)^{N-t}\right)} \end{aligned} \tag{32}$$

4.3 Infinite Buffer

In the limit as the buffer size $N \to \infty$, the expressions for mean queue length and waiting time reduce to

$$L = \frac{\tau}{\sigma + \tau} \frac{1}{\nu} \left((\eta + \nu) + \eta \frac{1 - (1 - \sigma - \tau)(1 - \eta - \nu)}{\sigma\nu - \tau\eta} \right) \tag{33}$$

and

$$w(t) = \begin{cases} \dfrac{1}{\eta + \nu}(1 - x) & t = 1 \\ \dfrac{\eta}{\eta + \nu} \dfrac{1 - (1 - \sigma - \tau)(1 - \eta - \nu)}{\sigma + (1 - \sigma - \tau)\eta}(1 - x)x^{t-2} & t = 2, 3, \ldots \end{cases} \tag{34}$$

Hence, the mean waiting time becomes

$$W = 1 + \frac{\eta}{\eta + \nu} \frac{1 - (1 - \sigma - \tau)(1 - \eta - \nu)}{\sigma\nu - \tau\eta} = \frac{L}{\lambda} \tag{35}$$

5 EFFECTS OF CORRELATION

In view of the significance of the following terms in the analysis, it is convenient to introduce symbols for them.

$\beta_a \triangleq (1 - \sigma - \tau)$ as the geometric factor in the correlation function for numbers of arrivals per slot (and in the IDC),

$\beta_b \triangleq (1 - \eta - \nu)$ as the geometric factor in the correlation function for intervals between individual arrivals (and in the IDI),

$x \triangleq \dfrac{\sigma(1 - \eta - \nu) + \eta}{\sigma + (1 - \sigma - \tau)\eta}$ as the geometric factor in the queue length distibution, asymptotic blocking probability, etc.

Further, it is convenient to investigation of queue behaviour when β_a or β_b be close to 1 to define additional symbols $\kappa_a \triangleq (1 - \beta_a)^{-1}$ and $\kappa_b \triangleq (1 - \beta_b)^{-1}$.

5.1 Choice of Reference System

The factors β_a and β_b appear to be good indicators of the type of correlation in the $\text{GGeo}^{\text{GGeo}}$ batch renewal process.

- A β value of 0 implies no correlation (in the *number* or *time* dimension, as appropriate). If $\beta_a = 0$, the process is Batch Bernoulli and there is no correlation between numbers of events at different epochs. If $\beta_b = 0$, the process is renewal and there is no correlation between intervals (i.e. between the interval between one pair of successive events and the interval between another pair of successive events).

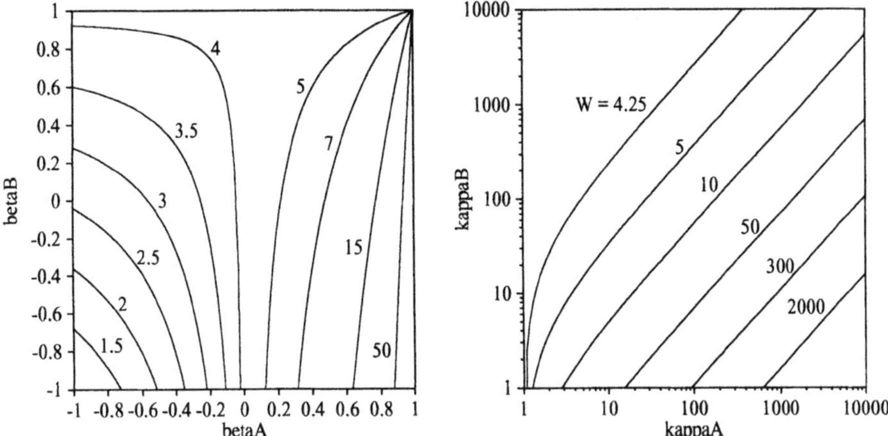

Figure 2 Waiting time (in slots) against correlation factors β_b and $\kappa_b = (1 - \beta_b)^{-1}$ (vertical scale) of batch size and β_a and $\kappa_a = (1 - \beta_a)^{-1}$ (horizontal scale) of intervals between batches for SCV of individual interarrival times $J_1 = 6.25 \times (1 - \lambda)$.

- A positive (negative) value for the β implies positive (negative) correlation in the appropriate (*number* or *time*) dimension. A greater magnitude of the β value implies stronger (positive or negative) correlation in that dimension.
- Only if both $\beta_a = 0$ and $\beta_b = 0$ is the process completely free of correlation.

In order to determine effects on queueing behaviour of correlation arising from a $\mathrm{GGeo}^{\mathrm{GGeo}}$ batch renewal arrivals process, the performance distributions and statistics for the queue must be compared with those of a reference process which is free of correlation but invariant in other significant characteristics. The $\mathrm{GGeo}^{\mathrm{GGeo}}$ process is determined by 4 parameters and, since two degrees of freedom are determined by the choice of the factors β_a and β_b, there remain 2 characteristics to be chosen to be invariant. An obvious requirement is that the intensity λ be invariant.

For the last remaining choice of invariant, it would appear natural, in view of 'traditional' traffic characterisation, to chose J_1, the SCV of intervals between successive arrivals. Figure 2 shows that, for $\beta_a < 0$, mean waiting time and (by Little's Law) mean queue length increase with β_a and with β_b, as would be expected. However, for $\beta_a > 0$, mean waiting time and mean queue length increase with β_a but reduce as β_b increases.

Similar difficulties arise with other obvious choices of other statistics to be invariant: the limiting values of the indices of dispersion $I_\infty = J_\infty$; the mean queue length or mean waiting time in an infinite buffer.

The best choice was found to be when both the mean batch size b and the mean interval a between batches were invariant. This choice is intuitively appealing because the factor β_b (equivalently κ_b) is closely related to the variability in batch sizes and the factor β_a

(equivalently κ_a) is closely related to the variability in both intervals between batches and the individual interarrival times.

$$C_a^2 = \frac{a-1}{a}\left(\frac{2}{1-\beta_a} - 1\right) = \frac{a-1}{a}(2\kappa_a - 1)$$

$$C_b^2 = \frac{b-1}{b}\left(\frac{2}{1-\beta_b} - 1\right) = \frac{b-1}{b}(2\kappa_b - 1)$$

Finally the reference system was chosen to be that with the same mean batch size and same mean interval between batches. Compare figure 3 with figure 2.

5.2 Results

Measures of interest are recast, below, in terms of the parameters a, b, κ_a and κ_b.

The geometric term x in queue length distribution, etc.

$$x = 1 - \frac{a-b}{a\,(b-1)(\kappa_a - 1) + (a-1)\,b\,\kappa_b}$$

The normalising factor Z in queue length distribution, etc. (cf equation (25))

$$Z = 1 - \frac{b-1}{a-1}\,x^{N-1}$$

The blocking probability π_N^B (cf equation (30))

$$\pi_N^B = \frac{1-Z}{Z}\frac{1-\lambda}{\lambda} = \frac{1}{Z}\frac{a-b}{a}\frac{b-1}{a-1}\,x^{N-1}$$

Mean queue length (cf equation (29))

$$L_N = \frac{1}{Z}\left(\frac{b}{a} + \frac{b(b-1)}{a-b}(\kappa_a + \kappa_b - 1)(1 - x^{N-1}) + N\frac{b-1}{a-1}(\kappa_a - 1)x^{N-1}\right)$$

Mean waiting time W in the infinite buffer queue (cf equation (35))

$$W = 1 + \frac{1}{1-x} - \kappa_b = 1 + \frac{a(b-1)}{a-b}(\kappa_a + \kappa_b - 1)$$

Figures 3, 4, 5, 6 and 7 illustrate the effects of varying correlation on mean waiting time in the infinite buffer queue, the factor x which appears as a geometric term in queue length distribution, etc., blocking probability against buffer size and mean queue length in a finite buffer. All the illustrations are for an intensity $\lambda = 0.2$. From the relations given at the begining of this sub-section, in can be appreciated that results for other intensities show similar forms.

Figure 3 shows the impact on waiting time in the infinite buffer queue. The numbers

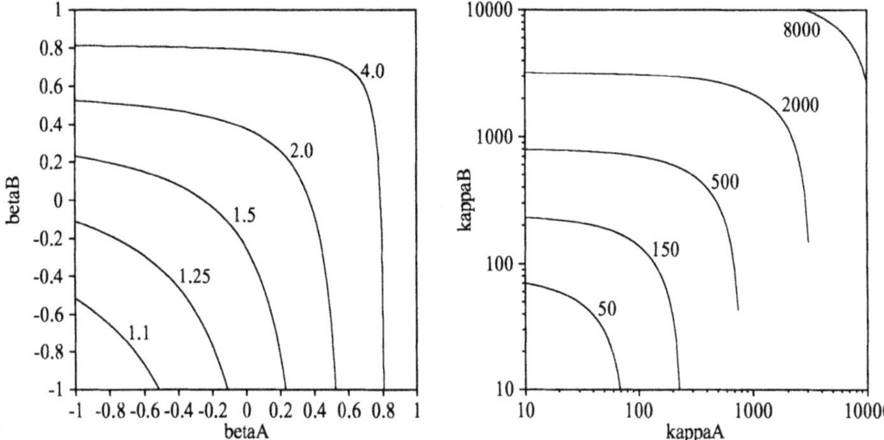

Figure 3 Waiting time (in slots) against correlation factors β_b and $\kappa_b = (1-\beta_b)^{-1}$ (vertical scale) of batch size and β_a and $\kappa_a = (1-\beta_a)^{-1}$ (horizontal scale) of intervals between batches for mean batch size $b = 1.5$, mean interval $a = 7.5$ slots between batches, intensity $\lambda = 0.2$.

on the contours give the waiting time as a number of slots. The right hand of the pair of charts gives an expanded view of the upper right hand corner of the left hand chart ($\beta_a \geq 0.9$, $\beta_b \geq 0.9$). The charts show that waiting time increases increasingly rapidly (and without limit) as either β_a or β_b approach unity.

Figure 4 shows the impact on the geometric term x of various degrees of correlation. The numbers on the contours give the value of x. It is seen that, as either κ_a or κ_b increases (β_a or β_b approaches 1), the value of x increases increasingly rapidly towards unity. For relatively low intensity ($\lambda = 0.2$ in the examples), the effect of variablity in batch size (given by κ_b or β_b) is stronger than that of variability in interval between batches (given by κ_a or β_a). This distinction is more pronounced when the mean batch size b is close to unity, as comparison of the two charts of figure 4 shows.

Blocking probabilty is also markedly effected by correlation in either the *time* dimension or in the *number* dimension. The two charts of figure 5 give blocking probability against buffer size for various values of β_b. The legend on each line is the value of β_b. The charts show that blocking probability increases rapidly with correlation.

The effects of correlation on mean queue length in finite buffers is shown figures 6 and 7. Each figure comprises two charts of mean queue length against buffer size for various values of κ_b, the upper chart for $\kappa_a = 1$ ($\beta_a = 0$, no correlation between interarrival times) and the lower for $\kappa_a = 5$ ($\beta_b = 0.5$, moderate correlation between interarrival times). The effects of positive correlation are marked. However, comparison of figures 6 and 7 shows that the impact of positive correlation in interarrival times is greater when the mean batch size is closer to unity.

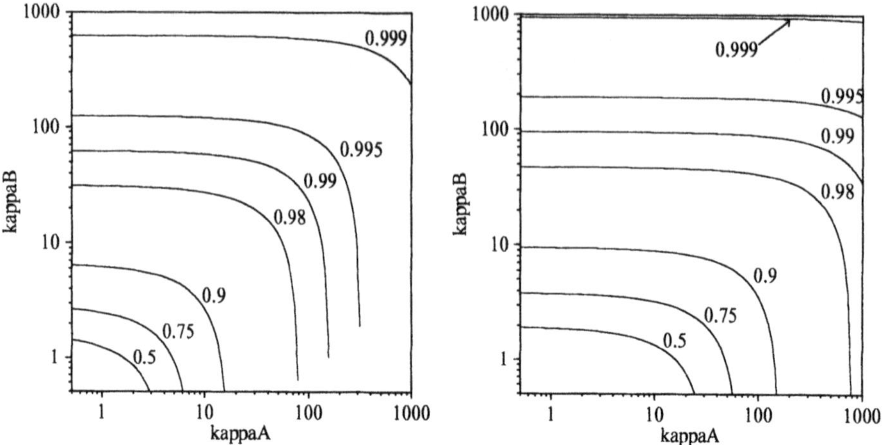

Figure 4 Geometric term x against correlation factors $\kappa_b = (1-\beta_b)^{-1}$ (vertical scale) of batch size and $\kappa_a = (1-\beta_a)^{-1}$ (horizontal scale) of intervals between batches for intensity $\lambda = 0.2$ and, in the left hand chart, mean batch size $b = 1.5$, mean interval $a = 7.5$ slots between batches and, in the right hand chart, mean batch size $b = 1.05$, mean interval $a = 5.25$ slots between batches.

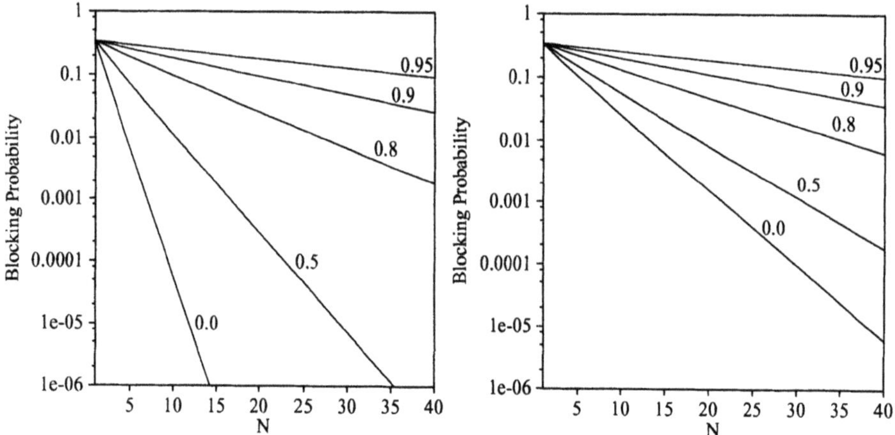

Figure 5 Blocking probability against buffer size for mean batch size $b = 1.5$, mean interval $a = 7.5$ slots between batches, intensity $\lambda = 0.2$ with $\beta_b = 0, 0.5, 0.8, 0.9, 0.95$ ($\kappa_b = 1, 2, 5, 10, 20$) and, in the left hand chart, $\beta_a = 0$ ($\kappa_a = 1$) and, in the right hand chart, $\beta_a = 0.8$ ($\kappa_a = 5$).

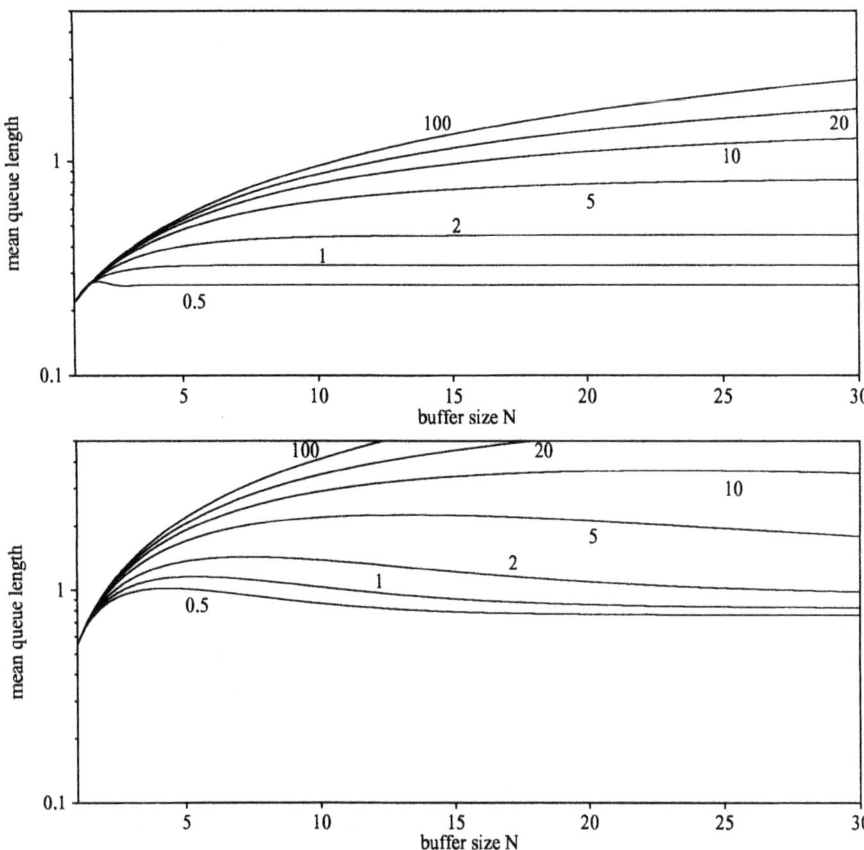

Figure 6 Mean queue length against buffer size N for mean batch size $b = 1.5$, mean interval $a = 7.5$ slots between batches, intensity $\lambda = 0.2$ and various values of κ_b $(1, 2, 5, 10, 20, 100$ and, in the upper chart, $\kappa_a = 1$ and, in the lower chart, $\kappa_a = 5$.

6 CONCLUSIONS AND PROPOSALS FOR FURTHER WORK

A discrete time $GI^G/D/1/N$ queue with single server, general batch renewal arrivals process, deterministic service time and finite capacity N is analysed. Closed form expressions for basic performance distributions, such as queue length and waiting time distributions and blocking probability, are derived when the batch renewal process is of the form which might be expected to result from actual traffic measurements. Those closed form expressions are used to show the effect of varying degrees of traffic correlation upon the basic performance distributions and the results are illustrated by numerical examples.

It is seen that positive correlation has markedly adverse impact on crucial quality of

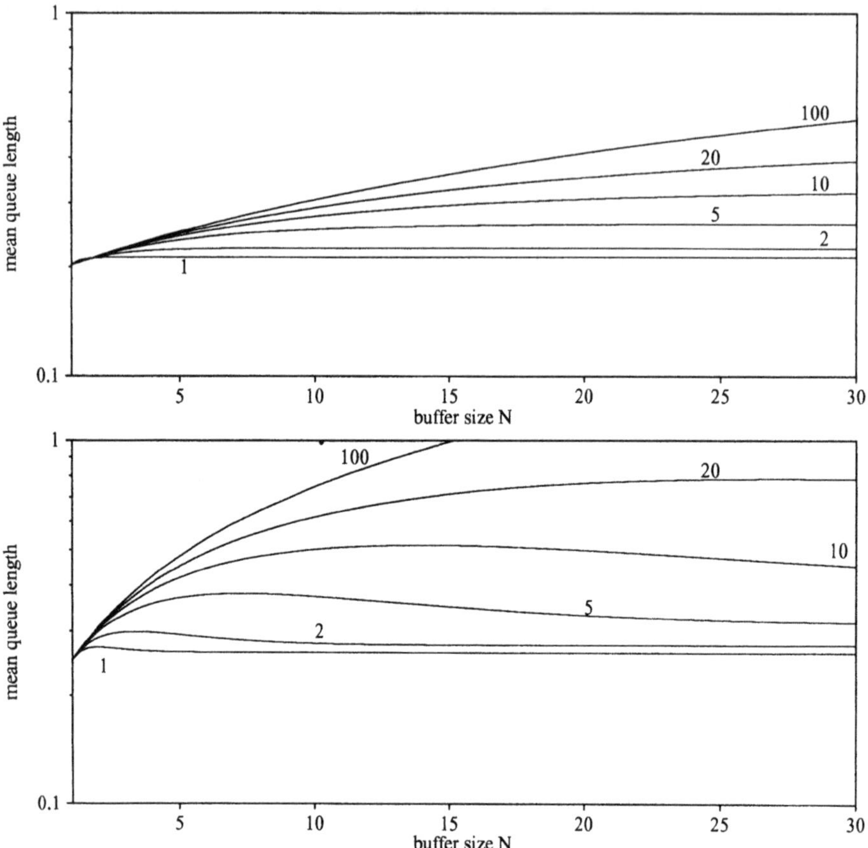

Figure 7 Mean queue length against buffer size N for mean batch size $b = 1.05$, mean interval $a = 5.25$ slots between batches, intensity $\lambda = 0.2$ and various values of κ_b $(1, 2, 5, 10, 20, 100$ and, in the upper chart, $\kappa_a = 1$ and, in the lower chart, $\kappa_a = 5$.

service (QoS) measures such as blocking probability and waiting time. Both correlation of interarrival times and correlation of counts have similar impact.

The importance of the analysis is that it shows explicitly how the magnitudes of blocking, waiting time and queue length distribution are determined by the degree of correlation in the traffic.

Characterisation of the departure process from a $GI^G/D/1/N$ queue is required in order to investigate the transmission of correlation in traffic though a multiplexer or partitioned buffer switch. Further research is required into effects of correlated traffic on the behaviour of queueing networks and, particularly, into propagation of correlation across networks of ATM switches (shared buffer, space division e.g. banyan interconnection networks). These are subjects of current study.

REFERENCES

Julián ANDRADE, M. Jesus MARTINEZ-PASCUA (1994) Use of the IDC to Characterize LAN Traffic, in *Proc. 2nd. Workshop on Performance Modelling and Evaluation of ATM Networks*, (ed. Demetres Kouvatsos), 15/1–15/12

Henry J. FOWLER, Will E. LELAND (1991) Local Area Network Traffic Characteristics, with Implications for Broadband Network Congestion Management, *IEEE JSAC* **9**(7), 1139–1149

Riccardo GUSELLA (1991) Characterizing the Variability of Arrival Processes with Indexes of Dispersion, *IEEE JSAC* **9**(2), 203–211

Harry HEFFES, David M. LUCANTONI (1986) A Markov Modulated Characterization of Packetized Voice and Data Traffic and Related Statistical Multiplexer Performance, *IEEE JSAC* **4**(6), 856–868

Kotikalapudi SRIRAM, Ward WHITT (1986) Characterizing Superposition Arrival Processes in Packet Multiplexers for Voice and Data, *IEEE JSAC* **4**(6), 833–846

APPENDIX 1 LEAST BIASED CHOICE OF PROCESS

In the notation of Section 2, the objective is to find the least biased choice for the wide sense stationary process $\{(\alpha(t), \beta(t)) : t = \ldots, -1, 0, 1, 2, \ldots\}$ given only $\mathbf{E}[N(t)N(t+\ell)]$ and $\mathbf{E}[X(t)X(t+\ell)]$ for all ℓ.

It is shown, by the outline proof below, that the least biased choice is that the $\alpha(t)$ and the $\beta(t)$ each be both stationary (in the strict sense) and independent.

First, introduce additional notation. Let

$a(n,t)$ be $\mathbf{P}[\alpha(t) = n]$ $(n = 1, 2, \ldots)$ with mean $\mathbf{E}[\alpha(t)] = a$,
$b(n,t)$ be $\mathbf{P}[\beta(t) = n]$ with mean $\mathbf{E}[\beta(t)] = b$,
$\phi_\ell(t)$ be $\mathbf{P}[N(t+\ell) \geq 1 | N(t) \geq 1]$ and
$\psi_\ell(t)$ be $\mathbf{P}[X(t+\ell) \geq 1 | X(t) \geq 1]$.

The method is first to show that $\phi_\ell(t)$ be stationary and that the $\beta(\cdot)$ be independent of each other: by similar reasoning, that $\psi_\ell(t)$ be stationary and that the $\alpha(\cdot)$ be independent of each other. Then, using the independence of the $\alpha(\cdot)$ and the stationarity of $\phi_\ell(t)$, it is readily seen that the $\alpha(\cdot)$ be stationary also: by similar reasoning, that the $\beta(\cdot)$ be stationary also.

Now

$$\mathbf{E}[N(t)N(t+\ell)] = \sum_{n=1}^{\infty} \sum_{k=1}^{\infty} n\,k\,\mathbf{P}[N(t) = k, N(t+\ell) = n]$$

Observe that only terms with $n \geq 1$ and $k \geq 1$ contribute to the sums and so it is sufficient to consider

$$\begin{aligned}\mathbf{P}[N(t) = k \geq 1, N(t+\ell) = n \geq 1] &= \mathbf{P}[N(t) = k, N(t+\ell) = n | N(t) \geq 1, N(t+\ell) \geq 1] \\ &\quad \times \mathbf{P}[N(t) \geq 1 | N(t+\ell) \geq 1] \mathbf{P}[N(t) \geq 1] .\end{aligned}$$

Now, $\mathbf{P}[N(t) \geq 1]$ is simply the probability of there being a batch at epoch t, i.e. $\mathbf{P}[N(t) \geq 1] = 1/a$, and $\mathbf{P}[N(t) \geq 1 | N(t+\ell) \geq 1]$ is $\phi_\ell(t)$, by definition, and

$$\mathbf{P}[N(t) = k, N(t+\ell) = n \,|\, N(t) \geq 1, N(t+\ell) \geq 1] = \mathbf{P}[\beta(t_1) = k, \beta(t_2) = n]$$

where t_1 is the index in the sequence $\beta(\cdot)$ which corresponds to the same batch as that indexed by t in the sequence $N(\cdot)$ and where t_2 is the index in the sequence $\beta(\cdot)$ which corresponds to the same batch as that indexed by $t + \ell$ in the sequence $N(\cdot)$.

A well known consequence of the Principle of Maximum Entropy is that, given only the marginal distributions, the least biased choice for the joint distribution is the product of the marginals. Thus, the least biased choice for the distribution $\mathbf{P}[\beta(t_1) = k, \beta(t_2) = n]$ is

$$\mathbf{P}[\beta(t_1) = k, \beta(t_2) = n] = \mathbf{P}[\beta(t_1) = k]\,\mathbf{P}[\beta(t_2 = n] = b(k, t_1)\,b(n, t_2)$$

i.e. that $\beta(t_1)$ and $\beta(t_2)$ are independent. Hence, the least biased choice for process $\{\alpha(t), \beta(t)\}$ requires that $\mathbf{E}[N(t)N(t+\ell)]$ satisfy

$$\mathbf{E}[N(t)N(t+\ell)] = \frac{1}{a}\,\phi_\ell(t) \sum_{n=1}^{\infty}\sum_{k=1}^{\infty} n\,k\,b(k,t_1)\,b(n,t_2) = \frac{1}{a}\,\phi_\ell(t)\,\mathbf{E}[\beta(t)]^2 = \frac{1}{a}\,\phi_\ell(t)\,b^2$$

But, because the process $N(t)$ is wide sense stationary, $\mathbf{E}[N(t)N(t+\ell)]$ must be independent of t. Consequently, $\phi_\ell(t)$ is independent of t: $\phi_\ell(t)$ is stationary and may be written $\phi_\ell(t) = \phi_\ell$.

Similarly, by consideration of $\mathbf{E}[X(t)X(t+\ell)]$, the $\alpha(\cdot)$ are independent and $\psi_\ell(t)$ is stationary.

Thence, using the independence of $\alpha(t)$, $\alpha(t+\ell)$ and the stationarity of $\phi_\ell(t) = \phi_\ell$

$$\phi_1 = \mathbf{P}[\alpha(t) = 1] = a(1,t)$$

so $a(1,t)$ is independent of t,

$$\phi_2 = \mathbf{P}[\alpha(t) = 2] + \mathbf{P}[\alpha(t) = 1, \alpha(t+1) = 1] = a(2,t) + a(1,t)^2$$

so $a(2,t)$ is independent of t, etc.

Thus, $\alpha(t)$ is stationary.

By similar argument on the independence of the $\beta(t)$ and the stationarity of $\psi_\ell(t)$, it may be seen that $\beta(t)$ is stationary.

Finally, because in process $\{(\alpha(t), \beta(t)) : t = \ldots, -1, 0, 1, 2, \ldots\}$ the sequences $\alpha(t)$ and $\beta(t)$ are stationary and mutually independent, the process $\{\alpha(t), \beta(t)\}$ is a batch renewal process by definition.

11

Buffer requirements in ATM-related queueing models with bursty traffic : an alternative approach.

B. Steyaert
SMACS Research Group, Lab. for Commun. Engineering, University of Ghent
Sint-Pietersnieuwstraat 41, B-9000 Gent, Belgium, bart.steyaert@lci.rug.ac.be

Y. Xiong
INRS - Télécommunications
16 Place du Commerce, Verdun Quebec, Canada H3E1H6, xiong@inrs-telecom.uquebec.ca

Abstract

During the past couple of years, a lot of effort has been put into solving all kinds of Markov Modulated discrete-time queueing models, which occur, almost in a natural way, in the performance analysis of slotted systems, such as ATM multiplexers and switching elements. However, in most cases, the practical application of such solutions is limited, due to the large state space that is usually involved. In this paper, we try to set a first step towards obtaining approximate solutions for a discrete-time multiserver queueing model with a general heterogeneous Markov Modulated cell arrival process, that allow accurate predictions concerning the behavior of the buffer occupancy in such a model, and still remains tractable, both from an analytical and a computational point-of-view. We first introduce a solution technique which leads to a closed-form expression for the joint probability generating function of the buffer occupancy and the state of the arrival process, from which an expression for V(z), the probability generating function of the buffer occupancy is easily derived. Based on this result, for the single-server case, we propose an approximation for the boundary probabilities, that reduces all calculations to an absolute minimum. In addition, we show how accurate data for the distribution of the buffer occupancy can be obtained, by using multiple poles of V(z) in the geometric-tail approximation of the distribution.

Keywords

discrete-time queueing, Markov-modulated models, generating functions, tail distribution, buffer dimensioning.

1 INTRODUCTION

As the basic information units to be transferred in ATM (asynchronous transfer mode) based communication networks are fixed-length cells (De Prycker (1991)), buffers in multiplexers and switches can in general be modeled as a discrete-time queueing system where new cells are generated by a superposition of individual traffic sources. The service time of a cell equals its transmission time, which is one slot. Analyzing such a queueing system is essential in the design and evaluation of ATM networks. However, this can be a difficult task, due to the fact that the traffic sources to be connected to the same buffer may have different traffic characteristics (like voice, data and video), and the time-correlated behavior that each of the individual sources might exhibit.

To facilitate the queueing analysis, a traffic source with variable bit rate (VBR) in ATM is usually modeled as a Markov-modulated arrival process (e.g., Markov-modulated Bernoulli process, Markov-modulated Poisson process, etc...). The problem is then reduced to analyzing a queueing system with heterogeneous Markov-modulated arrival streams generating subsequent cell arrivals. Even so, the related queueing analysis is still complicated and requires solving a multi-dimensional Markov chain. In this heterogeneous traffic environment, the matrix-geometric solution technique

(Neuts (1989)), which has been widely used in the performance analysis of various types of related problems, is only suitable for analyzing small systems because of the large state space (Blondia (1992)).

A general solution technique, called the matrix spectral decomposition method, was developed in Li (1991b) (and extended in Zhang (1991)) to analyze the above queueing system. This solution technique is based on a generating–functions approach and uses the properties of Kronecker products to decompose the problem of solving a global system with multiple traffic types into the problem of solving subsystems, each of which consisting of one single traffic type. Thus, calculating the poles of the probability generating function of the buffer occupancy depends only on the traffic source parameters and is independent of the system size and the number of traffic types. The main computational limitation in this method is the memory size required to solve the set of linear equations for the boundary probabilities (Li (1991b)). Furthermore, in this general solution technique, the superposed arrival processes is expressed in a Kronecker products form. The whole derivation is quite complicated and the final results are not easy to apply. Another main solution technique is the fluid–flow approximation (Anick (1982), Stern (1991)), in which a traffic source is described by a Markov–modulated continuous flow process. It equally uses properties of Kronecker products and sums in the decomposition of the overall problem into smaller 'sub– problems'. Consequently, similar comments as above also hold for the fluid–flow method.

The purpose of this paper is two fold : first to present an alternative solution technique, based on a generating–functions approach, for discrete–time queueing analysis in ATM, and secondly to give a good approximation for the tail distribution of the buffer occupancy, an important performance measure that allows an accurate estimate of the required buffer space, crucial for dimensioning purposes in practical engineering. Compared to the matrix spectral decomposition method, the solution technique to be presented below is relatively simple and has following properties: (1) it uses straightforward analysis, again based on a generating–functions approach, instead of Kronecker products, to represent the superposition of arrival processes; (2) no sophisticated computational matrix manipulations are required and the whole derivation is easy to follow; (3) the final results are relatively easy to use. Furthermore, we found, via comparison of a large number of numerical examples, that the tail distribution of the buffer occupancy can be well approximated when only considering a few poles (i.e., the ones with the smallest modulus) of the probability generating function of the buffer occupancy. This paper is an extension of the work presented in Steyaert (1992).

2 TRAFFIC SOURCE DESCRIPTION

Consider a multiplexer model fed by several independent traffic sources, which, according to their traffic characteristics, are grouped together into K distinct classes or types, each class having N_k, $1 \leq k \leq K$, identical and independent sources. A source belonging to class k is modeled as an L_k–state Markov Modulated arrival process, where the states will be labeled by $S_{i,k}$, $1 \leq i \leq L_k$, and where the $L_k x L_k$ probability generating matrix (as in Sohraby (1992))

$$\mathbf{Q}_k(z) = \begin{bmatrix} q_{11,k}(z) & q_{12,k}(z) & \cdots & q_{1L_k,k}(z) \\ q_{21,k}(z) & q_{22,k}(z) & \cdots & q_{2L_k,k}(z) \\ \vdots & & \ddots & \vdots \\ q_{L_k1,k}(z) & q_{L_k2,k}(z) & \cdots & q_{L_kL_k,k}(z) \end{bmatrix} , \tag{1}$$

characterizes the cell arrival process. It is assumed that transitions between states occur at slot boundaries, and let us denote by $p_{ij,k}$, $1 \leq i,j \leq L_k$, the one–step transition probability that a source from the k–th traffic class transits from state $S_{i,k}$ to state $S_{j,k}$ at the end of a slot during which it was in state $S_{i,k}$. Then, the elements $q_{ij,k}(z)$ in the above matrix are given by

$$q_{ij,k}(z) \triangleq G_{ij,k}(z)\, p_{ij,k} \quad , \tag{2}$$

where $G_{ij,k}(z)$, $1 \leq i,j \leq L_k$, is the probability generating function describing the number of cells generated during a slot by a source from class k, given that the source is in state $S_{j,k}$ during the tagged slot and was in state $S_{i,k}$ during the preceding slot. For the Markov Modulated Bernoulli Process (MMBP), the number of cell arrivals generated by a source during any slot is either zero or equal to one, which is reflected by the property that each of the probability generating functions $G_{ij,k}(z)$ is a linear function of z, meaning that they can be written as

$$G_{ij,k}(z) = 1 - g_{ij,k} + z g_{ij,k} \quad , \tag{3}$$

for some parameters $g_{ij,k}$ satisfying $0 \leq g_{ij,k} \leq 1$. Although attention is focused on this specific arrival model, the theory developed here is far more general, and can also be applied when the $G_{ij,k}(z)$'s have a more complex form than given by (3).

The aggregate cell arrival process is fully determined, once the probability generating matrices $\mathbf{Q}_k(z)$, $1 \leq k \leq K$, have been specified for each individual traffic class. Let us define $e_k(n)$ as the total number of cell arrivals generated by the N_k sources of class k during slot n, and $a_{i,k}(n)$, $1 \leq i \leq L_k$, as the total number of sources of class k that are in state $S_{i,k}$ during slot n. Note that the latter random variables satisfy

$$\sum_{i=1}^{L_k} a_{i,k}(n) = N_k \quad , \tag{4}$$

for any value of n. We will denote by $\boldsymbol{x}_k$ the L_kx1 vector with elements $x_{i,k}$, $1 \leq i \leq L_k$. Let us also define the L_kx1 vector $\boldsymbol{B}_k(\boldsymbol{x}_k,z)$ with elements $B_{i,k}(\boldsymbol{x}_k,z)$, as the matrix product $\mathbf{Q}_k(z)\boldsymbol{x}_k$. Then, with the previous definitions, it is not difficult to show that the joint generating function of the random variables $e_k(n+1)$ and $a_{i,k}(n+1)$, $1 \leq i \leq L_k$, can be written in terms of the joint generating function of the random variables $a_{i,k}(n)$:

$$\mathbf{E}\left[z^{e_k(n+1)} \prod_{i=1}^{L_k} x_{i,k}^{\,a_{i,k}(n+1)} \right] = \mathbf{E}\left[\prod_{i=1}^{L_k} B_{i,k}(\boldsymbol{x}_k,z)^{a_{i,k}(n)} \right] \quad , \tag{5}$$

(where $\mathbf{E}[.]$ denotes the expected value of the argument) an important relation that describes the number of cells generated during consecutive slots by the N_k sources of class k.

We define $A_k(\boldsymbol{x}_k)$ as the joint probability generating function of the number of sources of class k in state $S_{i,k}$, $1 \leq i \leq L_k$, during an arbitrary slot in the steady state, i.e.,

$$A_k(\boldsymbol{x}_k) \triangleq \lim_{n \to \infty} \mathbf{E}\left[\prod_{i=1}^{L_k} x_{i,k}^{\,a_{i,k}(n)} \right] = A_k(\mathbf{Q}_k(1)\boldsymbol{x}_k) \quad , \tag{6}$$

assuming that the cell arrival process indeed reaches a stochastic equilibrium, and the latter limit exists. It readily follows from (4,5) that $A_k(\boldsymbol{x}_k)$ indeed should satisfy the above property. Furthermore, if we define $\sigma_{i,k}$ as the steady–state probability that a source of class k is in state $S_{i,k}$ during a slot, and $\boldsymbol{\sigma}_k$ as the L_kx1 column vector with $\sigma_{i,k}$ as its i–th element, which is the solution of the matrix equations

$$\boldsymbol{\sigma}_k^T = \boldsymbol{\sigma}_k^T \mathbf{Q}_k(1) \quad , \quad \boldsymbol{\sigma}_k^T \boldsymbol{1}_k = 1 \quad , \tag{7}$$

($\boldsymbol{1}_k$ is the L_kx1 column vector with all elements equal to 1, and $(.)^T$ represents the matrix transposition operation), then $A_k(\boldsymbol{x}_k)$ equals

$$A_k(\boldsymbol{x}_k) = (\boldsymbol{\sigma}_k^T \boldsymbol{x}_k)^{N_k} \quad , \tag{8}$$

and with (7), it is easily verified that this generating function indeed satisfies (6).

3 THE BUFFER OCCUPANCY : A FUNCTIONAL EQUATION

Due to the extremely low cell–loss ratios that will occur in B–ISDN communication networks, the multiplexer buffer could be considered having infinite storage–capacity, meaning that all arriving cells are accepted and temporarily stored to await their transmission. The multiplexer has c output lines via which cells are transmitted, thus allowing up to a maximum of c cells to leave the multiplexer buffer during each slot. Let us observe the system at the end of a slot (i.e., just after new arrivals, but before departures, if any), say slot n, and denote by the random variable v_n the buffer contents at that time instant; this is the number of cells in the multiplexer buffer, not including the cells that have been transmitted during slot n. From the previous, it is then clear that this quantity evolves according to the system equation

$$v_{n+1} = (v_n - c)^+ + \sum_{k=1}^{K} e_k(n+1) \quad , \tag{9}$$

where $(.)^+ \triangleq \max\{.,0\}$. Since we consider an infinite storage–capacity buffer, the system reaches a stochastic equilibrium only if the equilibrium condition, requiring that the mean number of cell arrivals per slot must be less than c, is satisfied. If we denote by p the mean number of cells carried by each output link per time slot, then, in view of the cell arrival model described in the previous section, it follows that this quantity equals

$$p = \frac{1}{c} \sum_{k=1}^{K} N_k \sum_{i=1}^{L_k} \sigma_{i,k} \sum_{j=1}^{L_k} G'_{ij,k}(1) p_{ij,k} \quad , \tag{10}$$

(where primes denote derivatives with respect to the argument) and $p<1$ is the necessary requirement for reaching a steady state.

The evolution of the (L+1)–th dimensional Markov chain $\{\boldsymbol{a}_k(n) \mid 1 \le k \le K\} \cup \{v_n\}$ (where $\boldsymbol{a}_k(n)$, $1 \le k \le K$, represents the set of random variables $\{a_{i,k}(n) \mid 1 \le i \le L_k\}$, and L is the sum of all L_k's) throughout consecutive slots completely determines the buffer behavior of the discrete–time queueing system previously described. Let us therefore define their joint generating function as

$$P_n(\boldsymbol{x}_1, \boldsymbol{x}_2, \ldots, \boldsymbol{x}_K, z) \triangleq \mathrm{E}\left[z^{v_n} \prod_{k=1}^{K} \prod_{i=1}^{L_k} x_{i,k}^{a_{i,k}(n)} \right] .$$

Combining this definition with system equation (9), together with (5), it follows that

$$P_{n+1}(\boldsymbol{x}_1, \boldsymbol{x}_2, \ldots, \boldsymbol{x}_K, z) = \mathrm{E}\left[z^{(v_n - c)^+} \prod_{k=1}^{K} \left[\prod_{i=1}^{L_k} B_{i,k}(\boldsymbol{x}_k, z)^{a_{i,k}(n)} \right] \right] .$$

Again, we assume that the system reaches a steady–state after a sufficiently long period of time (implying that the equilibrium condition $p<1$ must be satisfied), and that $P_n(.)$ has a steady–state limit, which will be denoted by P(.). Then, with the definition of $\boldsymbol{B}_k(\boldsymbol{x}_k, z)$, and using some standard probabilistic techniques, we find that this generating function must satisfy

$$P(\boldsymbol{x}_1, \boldsymbol{x}_2, \ldots, \boldsymbol{x}_K, z) = z^{-c}\{P(\boldsymbol{Q}_1(z)\boldsymbol{x}_1, \boldsymbol{Q}_2(z)\boldsymbol{x}_2, \ldots, \boldsymbol{Q}_K(z)\boldsymbol{x}_K, z) + R(\boldsymbol{Q}_1(z)\boldsymbol{x}_1, \boldsymbol{Q}_2(z)\boldsymbol{x}_2, \ldots, \boldsymbol{Q}_K(z)\boldsymbol{x}_K, z)\} \quad , \tag{11.a}$$

where $R(\boldsymbol{x}_1, \boldsymbol{x}_2, \ldots, \boldsymbol{x}_K, z)$ is given by

$$R(\boldsymbol{x}_1, \boldsymbol{x}_2, \ldots, \boldsymbol{x}_K, z) \triangleq \sum_{j=0}^{c-1} (z^c - z^j) \mathrm{E}\left[\prod_{k=1}^{K} \left[\prod_{i=1}^{L_k} x_{i,k}^{a_{i,k}} \right] \middle| v=j \right] \mathrm{Prob}[v=j] \quad . \tag{11.b}$$

In the right–hand side of this equation, the random variable v denotes the buffer contents at the end of an arbitrary slot, while $a_{i,k}$, $1 \le k \le K$, $1 \le i \le L_k$, is the number of sources of traffic class k that are in state $S_{i,k}$ during this slot. Equations (11.a,b) define a functional equation for P(.), (the joint probability generating function of these random variables), which contains all information concerning the buffer behavior of the queueing model with heterogeneous traffic under study. In the next section, we will describe a technique for solving this functional equation. Also note that the function R(.) contains a number of unknown probabilities, still to be determined. Throughout the following sections, it will also become clear how these unknowns can be computed.

4 SOLVING THE FUNCTIONAL EQUATION

4.1 Homogeneous Traffic

Let us first focus attention on the case where the cell arrival process is homogeneous, i.e., the N traffic sources generating cell arrivals all have the same traffic characteristics, and can be modeled as an L–state MMBP. In the following, the subscript k, $1 \le k \le K$, that reflected the traffic class type in the previous sections, will be omitted. The functional equation (11.a,b) then becomes

$$P(\boldsymbol{x},z) = z^{-c}\{P(\mathbf{Q}(z)\boldsymbol{x},z) + R(\mathbf{Q}(z)\boldsymbol{x},z)\} \quad , \tag{12.a}$$

where, $\mathbf{Q}(z)$ is the LxL probability generating matrix describing the arrival process per traffic source, given by (1), and $R(\boldsymbol{x},z)$ now becomes

$$R(\boldsymbol{x},z) \triangleq \sum_{j=0}^{c-1}(z^c-z^j)E\left[\prod_{i=1}^{L} x_i^{a_i} \,\middle|\, v=j\right]\mathrm{Prob}[v=j] = \sum_{j=0}^{c-1}(z^c-z^j)\sum_{\boldsymbol{\ell}}\left[\prod_{i=1}^{L} x_i^{\ell_i}\right]p(\boldsymbol{\ell},j) \quad , \tag{12.b}$$

where $\boldsymbol{a} \triangleq \{a_i \mid 1 \le i \le L\}$, is the set of ramdom variables representing the number of input sources in state S_i during an arbitrary slot, and where

$$p(\boldsymbol{\ell},j) \triangleq \mathrm{Prob}[\boldsymbol{a}=\boldsymbol{\ell},v=j] = \mathrm{Prob}[a_1=\ell_1, \ldots ,a_L=\ell_L,v=j] \quad . \tag{12.c}$$

In (12.b,c), $\boldsymbol{\ell}$ is the set of positive integers $\{\ell_i \mid 1 \le i \le L \text{ and } \ell_i \ge 0\}$ that satisfies

$$\sum_{i=1}^{L} \ell_i = N \quad , \tag{12.d}$$

and the sum for $\boldsymbol{\ell}$ includes all possible sets $\{\ell_i \mid 1 \le i \le L \text{ and } \ell_i \ge 0\}$.

The matrix $\mathbf{Q}(z)$ will be diagonizable under quite general circumstances, and let $\lambda_i(z)$ be the i–th eigenvalue of $\mathbf{Q}(z)$, and $\boldsymbol{w}_i(z)$ ($\boldsymbol{u}_i(z)$) the left row (right column) eigenvector of $\mathbf{Q}(z)$ with respect to $\lambda_i(z)$. Define the diagonal eigenvalue matrix

$$\Lambda(z) \triangleq \mathrm{diag}[\lambda_1(z), \lambda_2(z), \ldots ,\lambda_L(z)]$$

and eigenvector matrices

$$\mathbf{W}(z) \triangleq (\boldsymbol{w}_1(z)\ \boldsymbol{w}_2(z)\ \ldots\ \boldsymbol{w}_L(z))^T \quad , \quad \mathbf{U}(z) \triangleq (\boldsymbol{u}_1(z)\ \boldsymbol{u}_2(z)\ \ldots\ \boldsymbol{u}_L(z)) \quad . \tag{13.a}$$

From this definition, we have

$$\Lambda(z)\,\mathbf{W}(z) = \mathbf{W}(z)\,\mathbf{Q}(z) \quad , \quad \mathbf{U}(z)\,\Lambda(z) = \mathbf{Q}(z)\,\mathbf{U}(z) \quad . \tag{13.b}$$

For each value of i, $1 \le i \le L$, equation (13.b) determine the left row eigenvector and the right column eigenvector of $\mathbf{Q}(z)$ corresponding to $\lambda_i(z)$ upon some constant factor, which is uniquely determined when requiring that

$$U(z)\,\mathbf{1} = \mathbf{1} \quad \text{and} \quad \mathbf{W}(z)\,\mathbf{U}(z) = \mathbf{I} \Rightarrow \mathbf{W}(z)\,\mathbf{1} = \mathbf{1} \quad , \tag{13.c}$$

where **I** is the LxL identity matrix, and as defined before, $\mathbf{1}$ is the Lx1 column vector with all elements equal to 1. Equation (13.c) implies that $\mathbf{Q}(z)$ can be written as

$$\mathbf{Q}(z) = \mathbf{U}(z)\,\mathbf{\Lambda}(z)\,\mathbf{W}(z) = \sum_{i=1}^{L} \lambda_i(z)\,\boldsymbol{u}_i(z)\,\boldsymbol{w}_i(z) \quad . \tag{13.d}$$

Let us now go back to equation (12.a), from which we can derive that

$$P(\boldsymbol{x},z) = z^{-Hc}P(\mathbf{Q}(z)^H\boldsymbol{x},z) + \sum_{h=1}^{H} z^{-hc}R(\mathbf{Q}(z)^h\boldsymbol{x},z) \quad , \tag{14}$$

From (13.c,d), it is clear that $\mathbf{Q}(z)^h\boldsymbol{x} = \mathbf{U}(z)\mathbf{\Lambda}(z)^h\mathbf{W}(z)\boldsymbol{x}$. Then, letting H approach infinity, in a similar way as was explained in Steyaert (1992), the right hand side of the above equation can be further worked out, leading to an expression for $P(\boldsymbol{x},z)$, the joint generating function of the buffer occupancy at the end of an arbitrary slot, and $\boldsymbol{a}$, the set of random variables describing the L–state MMBP arrival process

$$P(\boldsymbol{x},z) = \sum_{\boldsymbol{m}} \frac{\prod_{i=1}^{L}\left\{\lambda_i(z)\,\boldsymbol{w}_i(z)\,\boldsymbol{x}\right\}^{m_i}}{z^c - \prod_{i=1}^{L}\lambda_i(z)^{m_i}} \sum_{\boldsymbol{\ell}} F_{\boldsymbol{\ell m}}(z) \sum_{j=0}^{c-1}(z^c - z^j)\,p(\boldsymbol{\ell},j) \quad . \tag{15}$$

The functions $F_{\boldsymbol{\ell m}}(z)$ are defined by the relation

$$\prod_{i=1}^{L}\left[\sum_{j=1}^{L} u_{ij}(z)\,x_j\right]^{\ell_i} \triangleq \sum_{\boldsymbol{m}} F_{\boldsymbol{\ell m}}(z)\left[\prod_{i=1}^{L} x_i^{m_i}\right] \quad , \tag{16}$$

where, similarly to $\boldsymbol{\ell}$, $\boldsymbol{m}$ represents a set of positive integers $\{m_i \mid 1 \le i \le L \text{ and } m_i \ge 0\}$ that satisfy (12.d)), and where $u_{ij}(z)$ is the i–th element of the column vector $\boldsymbol{u}_j(z)$. These functions can be calculated in terms of the $u_{ij}(z)$'s by identifying the appropriate coefficients in both hand sides of the above equation. It is worth noting that, in Section 5.2, we propose an approximation for the boundary probabilities, which in the mean time avoids the calculation of these functions.As in most applications, we are mainly interested in the distribution of the buffer occupancy, or, equivalently, the probability generating function V(z) of the buffer occupancy. Since V(z) equals $P(\mathbf{1},z)$, we find

$$V(z) = \sum_{\boldsymbol{m}} \frac{\prod_{i=1}^{L}\lambda_i(z)^{m_i}}{z^c - \prod_{i=1}^{L}\lambda_i(z)^{m_i}} \sum_{\boldsymbol{\ell}} F_{\boldsymbol{\ell m}}(z) \sum_{j=0}^{c-1}(z^c - z^j)\,p(\boldsymbol{\ell},j) \quad , \tag{17}$$

where we have used the property $\boldsymbol{w}_i(z)\mathbf{1} = 1$, which follows from (13.c). This expression for the probability generating function of the buffer occupancy at the end of an arbitrary slot still contains the unknown probabilities $p(\boldsymbol{\ell},j)$. These can be calculated by exploiting the property that V(z) is analytic inside the complex unit disk, which implies that the zeros inside the unit disk of the denominators in the right–hand side of (17) must also be zeros of the numerators. It can be shown that each denominator in (17) has c zeros inside the unit disk, and we thus find a total of $J = c.(N+L-1)i!/(N!(L-1)!)$ zeros inside the unit disk (including z=1, which leads to no additional equation for the unknowns).

Together with the normalization condition $V(1)=1$, in general, this is the number of linear equations we obtain for the same number of unknown probabilities, and this set of linear equations has a unique solution.

Once these unknowns have been calculated, all major characteristics concerning the buffer occupancy, such as mean value, variance, and tail distribution, can be calculated from (17). In this paper, we concentrate our efforts mainly on the tail distribution, which plays an important role in buffer dimensioning. However, as one observes from the value of J, the number of unknown probabilities can become quite large, as N and L increase, thus requiring solving a large set of linear equations. In order to avoid this, in Section 5, we discuss some techniques for approximating these unknown probabilities, that lead to accurate estimates of the tail probabilities, as will be shown by various numerical examples.

4.2 Heterogeneous Traffic

The derivation of the probability generating function in the case of heterogeneous traffic evolves along similar lines as in the case of homogeneous traffic, and adds no particularly new insights to the analysis. The final result for $V(z)$, the probability generating function of the buffer occupancy at the end of an arbitrary slot, can be written as

$$V(z)=\sum_{\boldsymbol{m}_1}\cdots\sum_{\boldsymbol{m}_K}\left[\frac{\prod_{k=1}^{K}\prod_{i=1}^{L_k}\lambda_{i,k}(z)^{m_{i,k}}}{z^c-\prod_{k=1}^{K}\prod_{i=1}^{L_k}\lambda_{i,k}(z)^{m_{i,k}}}\right]\sum_{\boldsymbol{\ell}_1}\cdots\sum_{\boldsymbol{\ell}_K}\left\{\prod_{k=1}^{K}F_{\boldsymbol{\ell}_k\boldsymbol{m}_k}(z)\right\}\sum_{j=0}^{c-1}(z^c-z^j)p(\boldsymbol{\ell}_1\cdots\boldsymbol{\ell}_K,j) \ . \tag{18}$$

As before, $\boldsymbol{\ell}_k$ ($\boldsymbol{m}_k$), $1 \le k \le K$, represents the set of positive integers $\{\ell_{i,k} | 1 \le i \le L_k$ and $\ell_{i,k} \ge 0\}$ ($\{m_{i,k} | 1 \le i \le L_k$ and $m_{i,k} \ge 0\}$) and the sums in (18) for $\boldsymbol{\ell}_k$, ($\boldsymbol{m}_k$) include all such sets that, as a consequence of (4), satisfy

$$\sum_{i=1}^{L_k}\ell_{i,k} = N_k \ , \quad \sum_{i=1}^{L_k} m_{i,k} = N_k \ . \tag{19.a}$$

Similarly to the homogeneous cell–arrivals case, $\{\lambda_{i,k}(z) \mid 1 \le i \le L_k\}$ is the set of eigenvalues of $\mathbf{Q}_k(z)$ (defined in (1)), and $\boldsymbol{\Lambda}_k(z)$ is the $L_k \times L_k$ diagonal matrix with $\lambda_{i,k}(z)$ on the intersection of the i–th row and the i–th column. In addition, $\boldsymbol{u}_{i,k}(z)$, $1 \le i \le L_k$, represent the right column eigenvectors corresponding to $\lambda_{i,k}(z)$, which is the i–th column of $\mathbf{U}_k(z)$, the $L_k \times L_k$ matrix that can be calculated from

$$\mathbf{U}_k(z)\,\boldsymbol{\Lambda}_k(z) = \mathbf{Q}_k(z)\,\mathbf{U}_k(z) \quad \text{and} \quad \mathbf{U}_k(z)\,\boldsymbol{1}_k = \boldsymbol{1}_k \ . \tag{19.b}$$

Finally, extending (16), the $F_{\boldsymbol{\ell}_k\boldsymbol{m}_k}(z)$'s that occur in (18) are implicitly defined by

$$\prod_{i=1}^{L_k}\left[\prod_{j=1}^{L_k} u_{ij,k}\,x_j\right]^{\ell_{i,k}} \triangleq \sum_{\boldsymbol{m}_k} F_{\boldsymbol{\ell}_k\boldsymbol{m}_k}(z)\left[\prod_{i=1}^{L_k} x_i^{m_{i,k}}\right] \ , \tag{19.c}$$

(with $u_{ij,k}(z)$ the j–th element of $\boldsymbol{u}_{ij,k}(z)$) and can be obtained in terms of the $u_{ij,k}(z)$'s by identifying the appropriate coefficients in both hand sides of this expression. The unknown probabilities

$$p(\boldsymbol{\ell}_1\cdots\boldsymbol{\ell}_K,j) \triangleq \text{Prob}[\boldsymbol{a}_1=\boldsymbol{\ell}_1,\cdots,\boldsymbol{a}_K=\boldsymbol{\ell}_K,v=j] \ , \tag{19.d}$$

(where $\boldsymbol{a}_k$ represents the set of random variables $\{a_{i,k} \mid 1 \le i \le L_k\}$, $a_{i,k}$ being the number of traffic sources of class k in state $S_{i,k}$ during an arbitrary slot) that occur in the right–hand side of (18), can be calculated by expressing that the zeros inside the unit disk of the denominators must also be zeros of the numerators. In general, this will involve solving a set of

$$J \triangleq c \prod_{k=1}^{K} \begin{bmatrix} N_k+L_k-1 \\ L_k-1 \end{bmatrix} , \tag{20}$$

linear equations for the same number of unknowns.

5. TAIL DISTRIBUTION OF THE BUFFER OCCUPANCY

In this section, we consider the tail distribution of the buffer occupancy, a performance measure of considerable interest for dimensioning purposes. We try to establish an approximation for the tail distribution of the buffer occupancy, that is both accurate, and easy to calculate, from a computational point-of-view.

5.1 The Multiple Poles Approximation

It has been observed in many cases that approximating the tail distribution of the buffer contents by a geometric form is quite accurate, if the poles of V(z) have a different modulus and multiplicity equal to one, which is, in general, the case. As in Steyaert (1992), we improve this kind of approach by considering a mixture of geometric terms in the approximation for the tail distribution of the buffer contents. In particular, in order to obtain accurate results, we claim that, in a first approximation, it is sufficient to merely consider multiple real and positive poles of V(z) in the series expansion of this function. Approximating the distribution of the buffer contents by a mixture of geometric terms (say M) corresponds to approximating V(z) by

$$V(z) \cong \sum_{m=1}^{M} \frac{\theta_m}{z - z_{o,m}} = - \sum_{m=1}^{M} z_{o,m}^{-1} \theta_m \sum_{s=0}^{\infty} \left[\frac{z}{z_{o,m}}\right]^s , \tag{21.a}$$

where we are particularly interested in sufficiently large values of s. In all cases considered further on, $z_{o,m}$, $1 \leq m \leq M$, are the M real and positive poles of V(z) with smallest modulus (which, of course, lay outside the unit disk).

The poles of V(z) correspond to the zeros outside the unit disk of the denominators in the right-hand side of expression (18) for V(z). Depending on the arrival model, the exact number of zeros outside the unit disk of each of the denominators varies, and calculating all the zeros can become a complicated numerical task. Nevertheless, for a wide variety of arrival models (among which those considered in Section 5.3), in all cases it has been observed that, the denominators which occur in the right-hand side of (18), in general, have a real and positive zero outside the unit disk, which, of course is a pole of V(z). The above expression for V(z) leads to the following approximation for the tail distribution of the system contents :

$$\text{Prob}[v>s] \cong - \sum_{m=1}^{M} \frac{\theta_m z_{o,m}^{-s-1}}{z_{o,m} - 1} , \quad s \geq 0 , \tag{21.b}$$

and this approximation improves for increasing values of s and M. Furthermore, using the residue theorem, the quantity $\theta_{\{m\}}$ in (21.a,b) can be shown to be equal to

$$\theta_m = \lim_{z \to z_{o,m}} (z - z_{o,m}) V(z) . \tag{21.c}$$

From expression (18) for V(z) and using de l'Hôpitals rule, these quantities can be easily calculated. The accuracy of the approximation for the buffer contents distribution presented here will be confirmed in Section 5.3 through comparison with the exact distribution.

5.2 Boundary Probabilities Approximation

As became clear in Section 4, a drawback of the technique presented here is the potentially huge number of boundary probabilities that must be calculated. Therefore, it is essential to find good approximations for these quantities. One possible approach for this problem in the single–server case is presented in this section. Denote by e the random variable describing the number of cell arrivals during a slot, whereas v, as before, indicates the buffer contents at the end of this slot. Using similar notations as in (19.d), let us also define the joint probability

$$q(\ell_1 \cdot\cdot \ell_K, 0) \triangleq \mathrm{Prob}[\boldsymbol{a}_1=\ell_1, \cdot\cdot , \boldsymbol{a}_K=\ell_K , e=0] \ .$$

Obviously, v=0 implies that there have been no cell arrivals during the tagged slot, i.e., $v=0 \Rightarrow e=0$. Consequently, it is clear that the following inequality between the latter quantities and the unknown probabilities $p(\boldsymbol{\ell},0)$ holds :

$$q(\ell_1 \cdot\cdot \ell_K, 0) > p(\ell_1 \cdot\cdot \ell_K, 0) \ .$$

In the next section, we will show through some numerical examples that approximating the conditional unknown probabilities by

$$p(\ell_1 \cdot\cdot \ell_K, 0)/(1-p) \cong q(\ell_1 \cdot\cdot \ell_K, 0)/\mathrm{Prob}[e=0] \ , \tag{22}$$

when calculating the tail probabilities of the buffer contents, yields an excellent upper bound for the latter quantities.

The values of the q(.,0)'s could be calculated from the traffic parameters. Indeed, combining the steady–state limit of (5) with (8), and using the statistical independence of different sources, we obtain that

$$\sum_{\ell_1} \cdot\cdot \sum_{\ell_K} \left\{ \prod_{k=1}^{K} \prod_{i=1}^{L_k} x_{i,k}^{\ell_{i,k}} \right\} q(\ell_1 \cdot\cdot \ell_K, 0) = \prod_{k=1}^{K} (\sigma_k^T \, Q_k(0) \, \boldsymbol{x}_k)^{N_k} \ , \tag{23.a}$$

and where Prob[e=0] which occurs in (22) obviously satisfies

$$\mathrm{Prob}[e=0] = \prod_{k=1}^{K} (\sigma_k^T \, Q_k(0) \, \boldsymbol{1}_k)^{N_k} \ . \tag{23.b}$$

An additional advantage of the approximation proposed in this section, is that it avoids the explicit calculation of the $q(.\boldsymbol{\ell})$'s, as well as the calculation of the functions $F_{\ell_k \boldsymbol{m}_k}(z)$ from (16) that occur in expression (18) for V(z) (which will be reflected in the calculation of the constants θ_m in (21.c)). From definition (19.c), it is not difficult to show, with the q(.,0)'s satisfying (23.a), that

$$\sum_{\ell_1} \cdot\cdot \sum_{\ell_K} \left\{ \prod_{k=1}^{K} F_{\ell_k \boldsymbol{m}_k}(z) \right\} q(\ell_1 \cdot\cdot \ell_K, 0) = \prod_{k=1}^{K} C_{\boldsymbol{m}_k}^{N_k} \prod_{i=1}^{L_k} (\sigma_k^T Q_k(0) \boldsymbol{u}_{i,k}(z))^{m_{i,k}} \ , \tag{24.a}$$

where $\boldsymbol{u}_{i,k}(z)$, as already mentioned, is the right column eigenvector with respect to $\lambda_{i,k}(z)$, which is obtained from solving (19.b), and where

$$C_{\boldsymbol{m}_k}^{N_k} \triangleq \frac{N_k}{m_{1,k}! \ \cdots \ m_{L,k}!} \ . \tag{24.b}$$

It is clear that (24.a) considerably reduces the numerical calculations, when using approximation (22) for the boundary probabilities in the evaluation of (21.c).

5.3 Numerical Examples

In this subsection, we compare the tail approximations derived above with the exact buffer contents distribution. The numerical examples to be shown below are based on the MMBP traffic model, especially the well–studied 2–state MMBP. The exact buffer contents distribution is obtained by just using the simple repeated substitution algorithm.

The homogeneous MMBP traffic model with L states and one–step transition probabilities p_{ij}, $1 \leq i,j \leq L$, has been described in Section 2. When a source is in state S_j, it will generate either one or no cell, with probabilities g_{ij} and $1-g_{ij}$ respectively. It is thus clear that the sojourn time (in slots) of a source in state S_{ij} is geometrically distributed, with mean value $T_i = 1/(1-p_i)$, $1 \leq i \leq L$.

For simplicity, we consider the following special case in our numerical examples : (1) upon leaving state S_i, the source will transit to the other states with equal probability, i.e., $p_{ij} = (1-p_{ii})/(L-1)$, if $j \neq i$; (2) the number of cells sent by the source during a slot only depends on the source state in this slot and is independent of the source state in the previous slot, i.e., $g_{ij} \equiv g_j$. It is further assumed without loss of generality that $g_j \geq g_i$ if $j \geq i$, for all $1 \leq i,j \leq L$.

So the MMBP traffic model we are going to use can be completely described by the mean sojourn time T_i and the average cell arrival rate g_i in state S_i ($1 \leq i \leq L$). In this case the steady–state probability of a traffic source being in state S_i is equal to

$$\sigma_i = T_i / \left[\sum_{i=1}^{L} T_i \right] .$$

From (10), the average traffic load on each outgoing link can be written as

$$p = \frac{N}{c} \left[\sum_{i=1}^{L} \sigma_i g_i \right] .$$

Note that in the following examples, we always take $g_1=0$, which implies that no cells are sent during the state 1 period. For the 2–state MMBP with $g_1=0$, g_2 is usually called the "mean peak rate" and T_2 is the "average burst length".

Now consider a queueing system fed by N identical MMBP traffic sources as described above. It is clear that when the number of states of each source L=2, the queueing performance is determined by the parameter set (N, p, c, g_2, T_2), as $g_1=0$. Let us first concentrate on the single–server case (c=1) and look at the impact of different parameters on the tail approximations. Fig. 1 compares the exact buffer contents distribution with its tail approximations for the traffic load p=0.4 and 0.8. The tail approximations are calculated using (21.b,c), where the boundary probabilities can be derived by solving the set of linear equations, obtained when expressing that the zeros of the denominators in the right hand side of (17) inside the unit disk are also zeros of the numerators. For this arrival model, we found that there are in total N–1 positive poles of V(z), the probability generating function of the buffer contents. One can observe from Fig. 1 that for high traffic load, the tail distribution can be well approximated by the geometric term of the smallest pole of V(z) (i.e., M=1), which will be referred to as the asymptotic queueing behavior. However, for low traffic loads, it is necessary to add more geometric terms corresponding to larger poles of V(z) in order to approximate the tail distribution more accurately. Fig. 1 illustrates that M=5 geometric terms are sufficient in approximating the tail distribution in the region of low probabilities (e.g., <10E–6) of interest. Of course, increasing the number of geometric terms M will eventually lead to better approximation in the high probability region (see M=10). In general, we found that the whole distribution of the buffer contents, except for small buffer contents (e.g., < 10 cells), can be accurately approximated by taking into account all the positive poles of V(z) (in this case M=19).

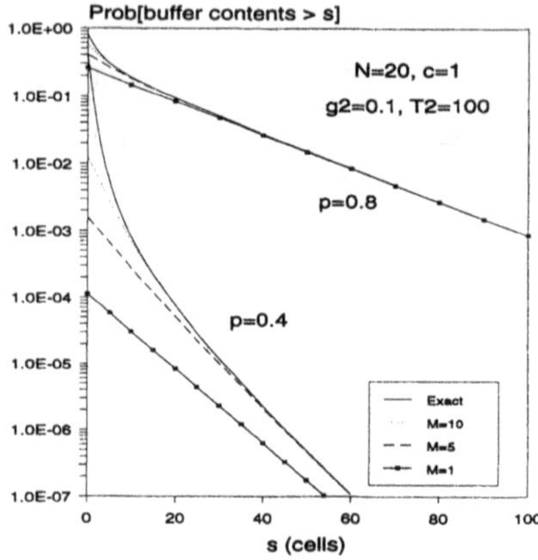

Fig. 1 : Buffer–contents distr. and approx.; traffic load p=0.4 and 0.8.

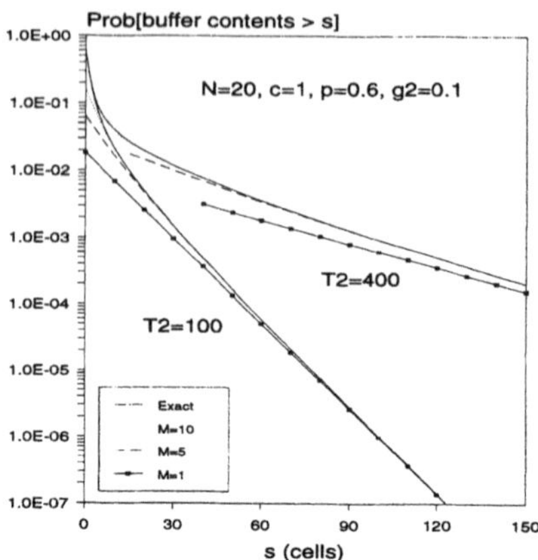

Fig. 2 : Buffer–contents distr. and approx.; mean burst length T_2=100 and 400.

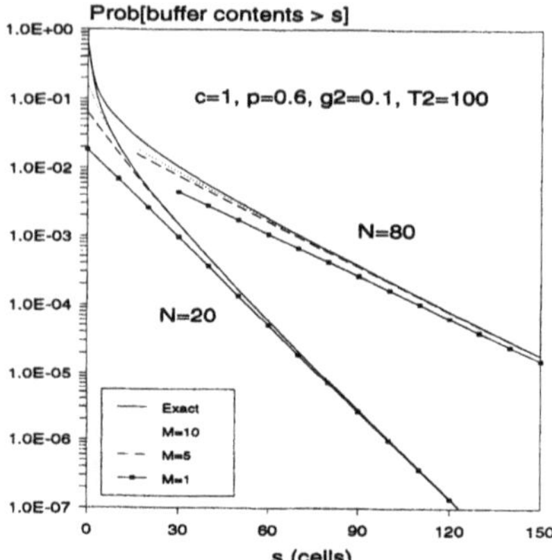

Fig. 3 : Impact of the number of traffic sources on the tail approximations.

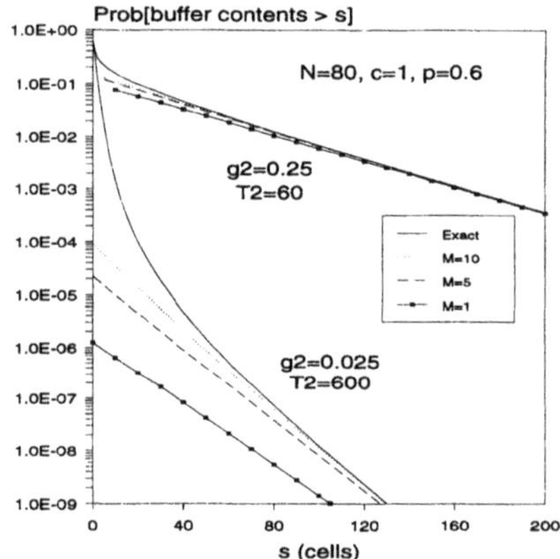

Fig. 4 : Impact of the mean peak rate on the tail approximations.

Similar results can also be observed when changing the values of the parameters. Fig. 2 shows an example where the average burst length T_2=100 and 400 slots, respectively. Fig. 3 gives another example in which the number of traffic sources N=20 and 80. It is interesting to see from the latter figure that although the total number of positive poles of V(z) (= N–1) increases linearly with N, the number of sources, the tail distribution of the buffer contents seems be dominated by a few geometric terms related to the smallest positive poles of V(z). The impact of the mean peak rates g_2 on the tail approximations is illustrated in Fig. 4. This figure reveals that for rather low source peak rate, more geometric terms might be required to get accurate approximations for the tail distribution. From Figs. 1–4, we also see that the tail distribution of the buffer contents cannot always be well approximated by only taking into account its asymptotic behavior.

The above results based on the 2–state MMBP's and single server case (c=1) also hold for an L–state MMBP's (L>2) and the multiple servers case (c>1). Fig. 5 shows an example when multiplexing of N=10 identical 3–state MMBP traffic sources. The buffer–contents distribution of a multiserver (c=4) queueing system with 2–state MMBP

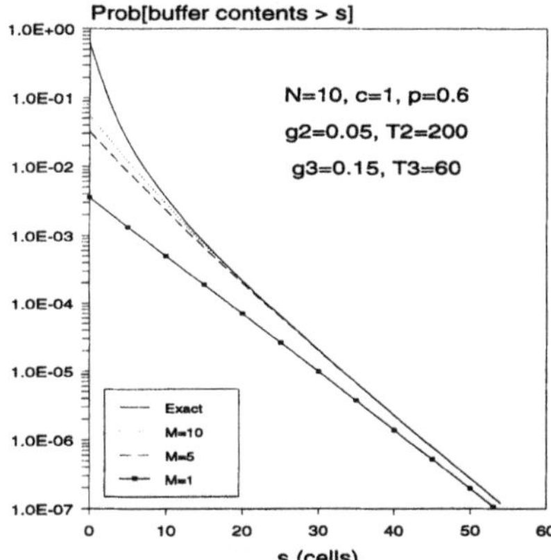

Fig. 5 : Tail approx. for a multiplex of 3–state MMBP traffic sources.

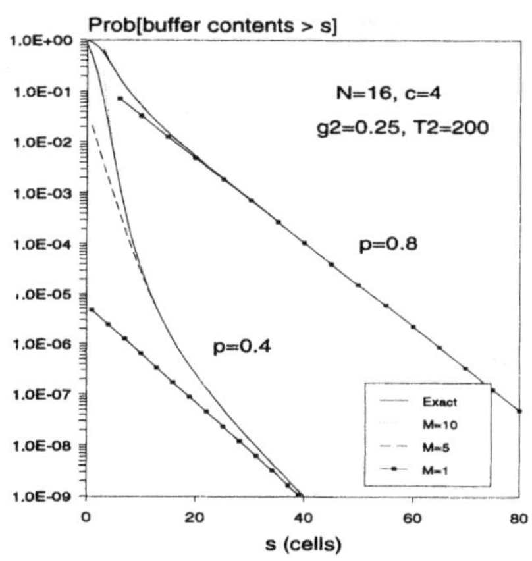

Fig. 6 : Tail approximations in the multiserver case.

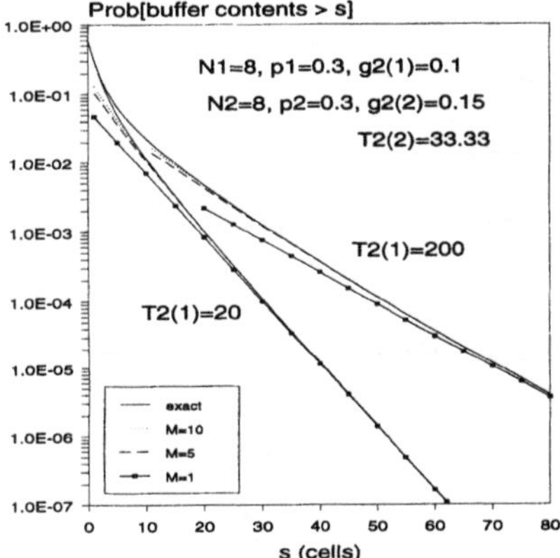

Fig. 7 : Heterogeneous traffic : Buffer–contents distribution and its approximations for $T_2(1) = 20$ and 200.

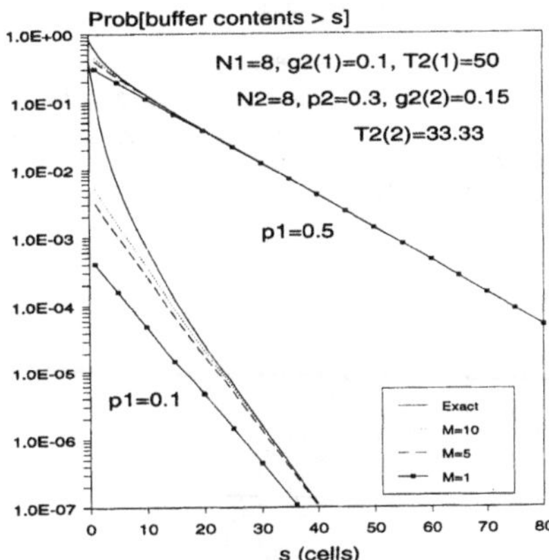

Fig. 8 : Heterogeneous traffic : Buffer–contents distribution and its approximations for $p_1 = 0.1$ and 0.5.

traffic sources is compared with its tail approximations in Fig. 6. Note that this multiserver queueing system can be used to model an output port of an ATM switching element, which has been used as a building block to construct a large ATM switching network (Henrion (1990, 1993)).

The above traffic descriptors for the homogeneous traffic case, are also well suited for describing heterogeneous traffic. Focusing attention on the case of a single–server queue fed by 2–state MMBP heterogeneous arrivals, the parameter set (N_k, p_k, $g_2(k)$, $T_2(k)$), $1 \leq k \leq K$, can then be used for characterizing each of the K traffic classes. Setting K=2, we have plotted some results in Figs. 7–8 for changing values of the average burst length (Fig. 7) and the offered load (Fig. 8) of the first traffic class, while keeping the traffic parameters of the other class constant. The conclusions that can be drawn here are basically identical to the case of homogeneous traffic : (1) considering even a relatively small number of terms (for instance M=5 or 10) in the geometric–tail approximation already leads to very accurate results as far as the tail behavior of the buffer contents

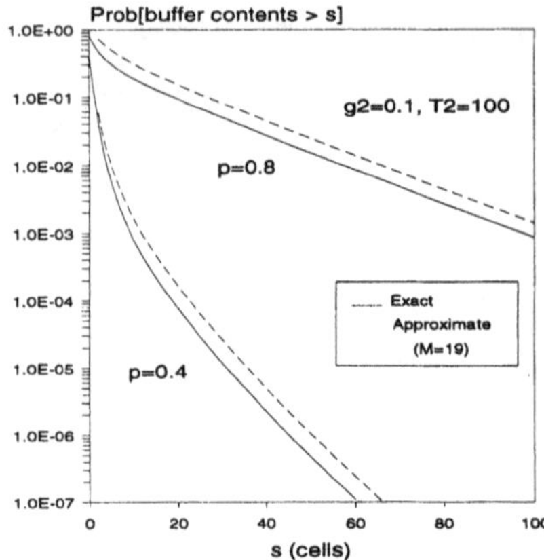

Fig. 9 : Upper—bound tail approximations for different traffic loads.

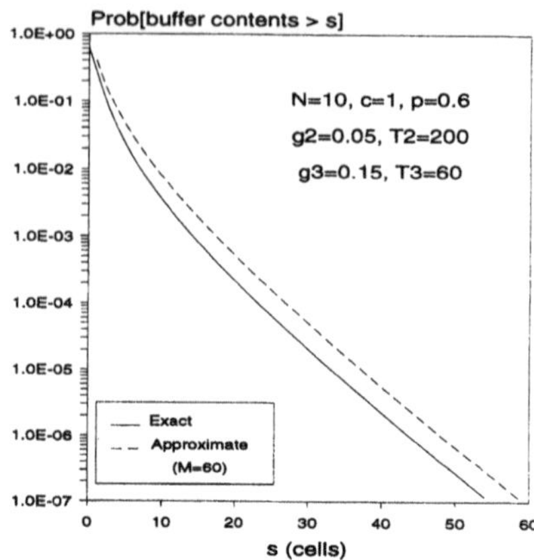

Fig. 10 : Upper—bound tail approximation for 3—state MMBP traffic sources.

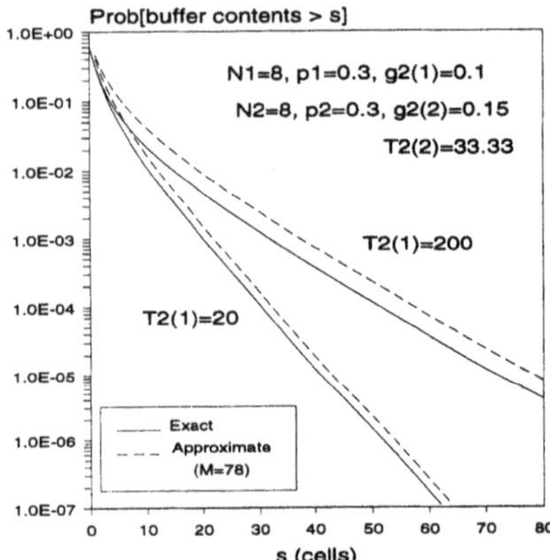

Fig. 11 : Heterogeneous traffic : upper—bound tail approximation.

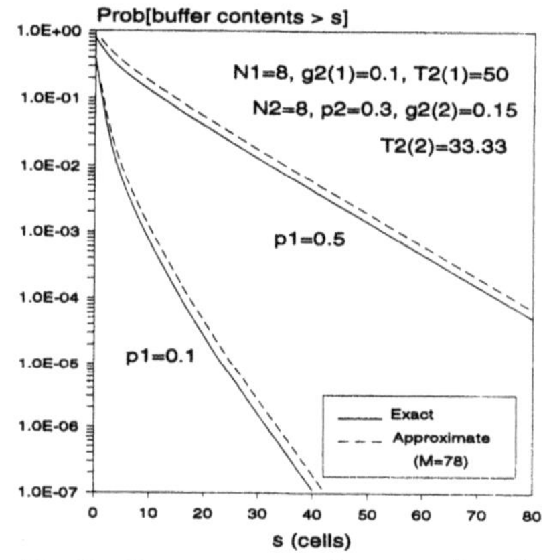

Fig. 12 : Heterogeneous traffic : upper—bound tail approximation.

distribution is concerned; (2) it is advisable to consider multiple terms in the above mentioned tail approximation, especially when the offered load is low.

As we discussed before, for large real systems with heterogeneous traffic, it is infeasible to obtain the unknown boundary probabilities due to the huge memory space requirement. Finding good approximations for the boundary probabilities is thus very important in ATM queueing analysis. In Section 5.2, we proposed a simple approximation for the boundary probabilities, from which all the geometric terms can be easily calculated. We found via numerous numerical results that this approximation leads to a good upper bound for the tail distribution of the buffer contents. Examples for homogeneous 2–state and 3–state MMBP's are shown in Figs 9 and 10, which use the same parameters as in Figs. 1 and 5 respectively; the heterogeneous arrivals case is illustrated in Figs. 11 and 12, with parameter sets that are identical as in Figs 7 and 8. In these figures, all the geometric terms related to the positive poles of V(z) are taken into account. An important property one observes from these curves is that the slopes of the tail distribution and its upper–bound approximation are identical. This is because they both contain the same positive poles of V(z), and merely differ in the value of the θ_m's (see (21.a–c)). Furthermore, the observed differences between exact and approximate results are small, thus leading to the conclusion that the approximation method proposed in Section 5.2 yields sufficiently accurate results.

6 CONCLUDING REMARKS

In this paper, we have presented an alternative solution technique for analyzing discrete–time queueing systems with general heterogeneous Markov–modulated arrival processes, which is relatively simple and easy to use compared to the matrix spectral decomposition method. We found via numerous numerical results that the tail distribution of the buffer contents can be well approximated by using only a few geometric terms related to the smallest positive poles of V(z), the probability generating function of the buffer contents. Moreover, an approximation for the boundary probabilities is given in the single server case, from which a good upper bound for the tail distribution is obtained, which is one of the main contributions of the paper. This upper bound is certainly quite useful in practical engineering (e.g., buffer dimensioning), and the calculation of this result is not limited by the system size and/or the number of traffic types.

Regarding the tail approximations, the main difficulty in this solution technique (as well as in the other methods) is the calculation of the poles of V(z) when the number of states L of each multiplexed source gets large (e.g., $L>3$). Finding an efficient way to calculate the poles of V(z) is one of the issues currently under study. As an initial result, we found a simple algorithm to calculate the smallest pole of V(z) for large value of L (Xiong (1994)). This smallest pole determines the asymptotic behavior of the tail distribution. Another issue that needs further investigation is finding efficient approximations for the boundary probabilities, in particular in the multiple servers case.

REFERENCES.

Anick, D.; Mitra, D. and Sondhi, M.M (1982) "Stochastic theory of a data– handling system with multiple sources", *Bell Syst. Tech. J.*, vol. 61, no. 8, pp. 1871–1894.

Blondia, C. and Casals, O., (1992) "Statistical multiplexing of VBR sources : A matrix–analytic approach", *Performance Evaluation*, vol. 16, pp. 5–20.

De Prycker, M. (1991) *Asynchronous Transfer Mode – Solution for Broadband ISDN*, Ellis Horwood Ltd. (England).

Henrion, M.; Schrodi, K.; Boettle, D.; De Somer, M. and Dieudonné, M. (1990) "Switching network architecture for ATM based broadband communications", Proc. of *ISS'90* (Stockholm), vol. V, pp. 1–8.

Henrion, M.; Eilenberger, G.; Petit, G. and Parmentier, P. (1993) "A multipath self–routing switch", *IEEE Communications Magazine*, pp. 46–52.

Hirano, M. and Watanabe, N. (1990) "Traffic characterization and a congestion control scheme for an ATM network", *Int. J. Digit. and Analog Commun. Syst.*, vol. 3, pp. 211–217.

Le Boudec, J.–Y. (1991) "An efficient solution method for Markov models of ATM links with loss priorities", *IEEE J. Select. Areas Commun.*, vol. 9, no. 3, pp. 408–417.

Li, S. and Sheng, H. (1991a) "Discrete queueing analysis of multi–media traffic with diversity of correlation and burstiness properties", Proc. of *IEEE INFOCOM'91*, pp. 368–381.

Li, S. (1991b) "A general solution technique for discrete queueing analysis of multimedia traffic on ATM", *IEEE Trans. Commun.*, vol. 39, no. 7, July 1991, pp. 1115– 1132.

Neuts, M.F. (1989) *Structured Stochastic Matrices of M/G/1 Type and Their Applications*, Marcel Dekker Inc. (New York).

Sohraby, K. (1992) "Heavy traffic multiplexing behavior of highly–bursty heterogeneous sources and their admission control in high–speed networks", Proc. of *IEEE GLOBECOM'92* (Orlando), pp. 1518–1523.

Stavrakakis, I., (1991) "Efficient modeling of merging and splitting processes in large networking structures", *IEEE J. Select. Areas Commun.*, vol. 9, no. 8, pp. 1336–1347.

Stern, and Elwalid, A.I. (1991) "Analysis of separable Markov–modulated rate models for information–handling systems", *Adv. Appl. Prob.*, vol. 23, pp. 105–139.

Steyaert, B. and Bruneel, H. (1992) "Analysis of the buffer behavior of an ATM switch with cell arrivals generated by a general correlated process : a novel approach", Proc. of *Digital Communication Networks Management Seminar* (St. Petersburg), pp. 199–217.

Xiong, Y. and Bruneel, H. (1993) "A tight upper bound for the tail distribution of the buffer contents in statistical multiplexers with heterogeneous MMBP traffic sources", Proc. of *IEEE GLOBECOM'93* (Houston), pp. 767–771.

Xiong, Y. and Bruneel, H. (1994) "On the asymptotic behavior of discrete–time single–server queueing systems with general Markov–modulated arrival processes", Proc. of *ITC'14* (Antibes Juan–les–Pins), pp. 179–189.

Zhang, Z. (1991) "Analysis of a discrete–time queue with integrated bursty inputs in ATM networks", *Int. J. Digit. and Analog Commun. Syst.*, vol. 4, pp. 191–203.

BIOGRAPHY

Bart Steyaert was born in Roeselare, Belgium, in 1964. He received the degrees of Licentiate in Physics and Licentiate in Computer Science in 1987 and 1989, respectively, all from the University of Ghent, Gent, Belgium. Since 1990, he has been working as a PhD student at the Laboratory for Communications Engineering as a member of the SMACS Research Group. His main research interests include performance evaluation of discrete–time queueing models, applicable in B–ISDN networks.

Yijun Xiong received the B.S. and M.S. degrees from Shangai Jiao–Tong University, China, in 1984 and 1987 respectively, and the Ph.D. degree from the University of Ghent, Belgium, in 1994, all in electrical engineering.

He is now a postdoctoral fellow at INRS–Telecommunications, Montreal, Canada. Before this, he worked at the Laboratory for Communications Engineering, University of Ghent, from Sept. 1992 to June 1994. From Aug 1989 to Aug 1992, he was with the Research Center of Alcatel Bell Telephone, Antwerp, Belgium. His current research interests include multicast ATM switches, traffic control, and recourse management of high speed networks.

12

Discrete-time analysis of a finite capacity queue with an 'all or nothing policy' to reduce burst loss

Johan M. Karlsson
Department of Communication Systems, Lund Institute of Technology
P.O.Box 118, S-221 00 LUND, Sweden, johan@tts.lth.se
Hideaki Yamashita
Faculty of Business Administration, Komazawa University
1-23-1 Komazawa, Setagaya-ku, Tokyo 154, Japan, i38666@m-unix.cc.u-tokyo.ac.jp

abstract

Many new and old applications have to split the information into smaller units while transmitted through a network. If not the whole packet is able to get through to the destination the fraction transmitted is of no value. Several areas within the tele- and data communication field where this applies are pointed out. Further a discrete time model with bursty arrivals is introduced and analyzed. The result shows the advantage of the 'All or Nothing Policy' for the burst loss probability and the waiting time.

keywords

Discrete-time, burst arrival, multi-server, finite capacity queue, all or nothing policy

1 INTRODUCTION

The needs for tele- and data communications have evolved almost from the first day of computing. Communications on which all forms of distributed systems are built, are concerned with the different techniques that are utilized to achieve the reliable transfer of information between two distant devices. The length of the physical separation may vary, but the issue is however the same, to exchange information in the most efficient way using existing equipment in the networks. For business, governments, universities and other organizations these information exchanges have become indispensable. The importance of efficiently utilize the network becomes essential, since we deal with limited resources. One important factor is the size of the buffers within the network nodes or in other connected equipment, since these are a relative expensive part. The evolution of telecommunications is towards a multi-service network fulfilling all user needs for voice, data and video communications in an integrated way. These services are also many times real time applications, which means that there are no time for retransmissions of lost information. Several papers show a substantial delay using different ARQ-schemes, which confirm our opinion (Anagnostou, Sykas and Protonotarios, 1984). Retransmissions are mostly due to lost information in full buffers somewhere along the route from sender to destination.

2 MODEL DESCRIPTION

The model under consideration could be applied on several areas in the communication field. A few of those are outlined below. The general problem solved by this model is to utilize the buffer as efficient as possible. The traffic arriving to this system emanate from several sources, which all generate bursts with a randomly distributed length. The interarrival time for bursts from one source are also randomly distributed. A burst is a unity, which means that the individual packets constituting the burst are of no use single-handed. If we in advance could discard packets, belonging to a burst, that are not able to enter the common queue due to space limitations, we have gained a lot. The first packet of a burst carries a length indicator, which displays the total length of the burst. If not all packets, arriving in succession, of a burst have opportunity to enter the queue or the server(s) all of them are lost. Since it is of no use to waste queueing space and/or processing time on packets that are of no value for the receiver our proposed policy tries to minimize the burst loss probability. This *'All or Nothing Policy'* will be explored on below.

3 APPLICATIONS

The mentioned policy could be applied to several areas within the tele- and data communication field. A few of those are briefly discussed below, however there are many more which we hope that the reader will discover and be able to use this general model for performance and dimensioning studies on.

3.1 Video Coding

Many of the new services that are going to evolve and that already exists are using images and/or high quality sound, which demand high bandwidth. To reduce the amount of information that has to be transferred, different coding schemes are evaluated to obtain efficient techniques and algorithms. If we try to focus on some of these services, they would correspond to services like HDTV (50-100 Mbps), picture telephony (64-128 kbps (CCITT H.261)) , Hi-Fi sound and group 4 telefax (64 kbps) and some other services related to office based communication devices. Coding techniques are going to be essential for all graphic, image and video information services. Some already standardized by JPEG (Joint Photographic Experts Group) and MPEG (Motion Picture Experts Group), sponsored by ISO and CCITT, provide for compression ratios up to 1:200. For further information about line transmission of non-telephone signals, the reader is referred to (CCITT, 1988). In the case of MPEG (see International Organization for Standardization, Joint Technical Committee 1, Subcommittee 29, WG11), a new more efficient version called MPEG-2 is now used. In this technique the code is dependent not only on the current image, but also on the previous as well as the succeeding images, as shown in Figure 1. This means that if we have a buffer there is no use capturing part of a packet,

Figure 1 The principle of the MPEG-2 coding scheme.

i.e. we could as well discard the entire packet making space for future needs.

3.2 Intermediate Systems

An Intermediate System (IS) is a device to interconnect two or more systems. Depending on the services the IS has to perform they are divided into three categories;

- **Repeater** connects two identical networks, it just regenerates the signal. This means that even collisions or disturbed signals would be regenerated. A repeater operates on OSI layer 1.
- **Bridge** connects two homogeneous networks, i.e. two LANs. The bridge acts like as an address filter, picking up packets from one LAN that are intended for a destination on another LAN. The bridge operates at layer 2 of the OSI model.
- **Router** connects several networks that may or may not be similar. It has the capability of connecting more than two networks, which means that it has to have some sort of routing algorithm implemented to decide to which output port the packet should be directed. The router operates on OSI layer 3.

In such a device, many different traffic streams are merged together and share the same buffer. Different connections use the same intermediate system, for example several connectionless services could be routed through a Token Ring (IEEE 802.4) and a CSMA/CD (IEEE 802.3) using a interconnecting bridge. The packets have to be queued in the bridge waiting for access to the CSMA/CD network, in which no access is granted within a certain amount of time. In this cases using real time data, there is no time for the receiver to resequent or demand retransmissions of parts of a packet, i.e. if we loose part of a packet we could as well discard the whole packet in the IS.

3.3 IP Traffic

The IP (Internet Protocol) is intended for communication through several networks. It provides a connectionless delivery system for hosts connected to networks with the protocol implemented. The IP has to be implemented on the hosts constituting the OD pair as well. The connection is unreliable and on a best effort basis. The interface to other layers, shown in Figure 2 are to transport and network layers. The transport layer is

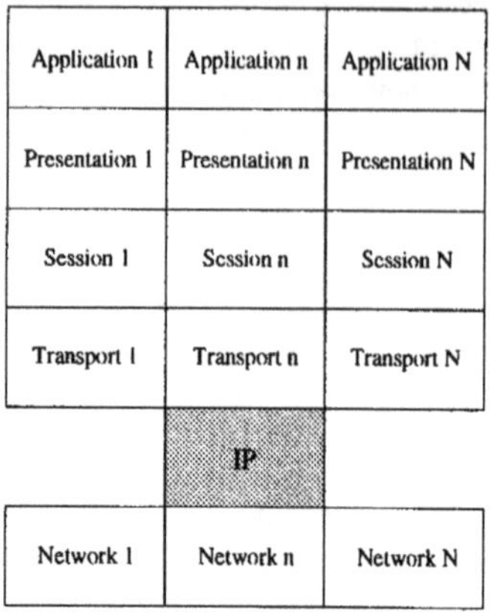

Application 1	Application n	Application N
Presentation 1	Presentation n	Presentation N
Session 1	Session n	Session N
Transport 1	Transport n	Transport N
	IP	
Network 1	Network n	Network N

Figure 2 The seven layer protocol stack, showing the position of IP.

usually implemented as the Transmission Control Protocol (TCP), but other protocols could be used on this level. On the link layer we could have interfaces towards different LAN access protocols or others like SMDS. IP is also to be used over ATM networks and on ATM Local Area Networks (Chao, Ghosal, Saha and Tripathi, 1994). The basic unit for transfer is specified as a datagram. The datagram consists of a data field and a header, the different fields are shown in Figure 3 and a more specific explanation of each field could be found in (Comer, 1991). Parts of certain interest to our studies are the "Total Length" and "Fragment Offset" fields, described below:

- **Total Length** To identify the number of octets in the entire datagram. (Usually less than 1500 octets, which is the maximum packet length of Ethernet.) IP specifications sets a minimum size of 576 octets that must be handled by routers without fragmentation.
- **Fragment Offset** The field represents the displacement (in octets) of this segment from the beginning of the entire datagram. Since the datagrams may arrive out of sequence, this field is used to assemble the collection of fragments into the original datagram.

The datagram could during transmission be duplicated, lost, delayed or out of order. This is to permit nodes with limited buffer space to handle the IP datagram. In some error situations datagrams are discarded, while in other situations error messages are sent. If we have to discard information it is more efficient to discard information belonging to the same IP datagram since fractions are of no use.

4 ANALYTICAL MODEL

The queueing model under consideration is a discrete-time, multi-server, finite capacity queue with burst arrivals, shown in Figure 4. Once the first packet of a burst arrives at

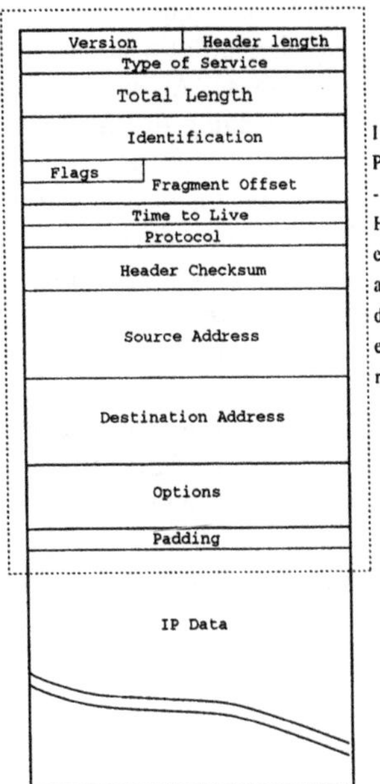

Figure 3 The IP datagram.

the queue, the successive packets will arrive on every time slot until the last packet of the burst arrives. The number of packets of the nth burst is denoted by S^n, which is assumed to be independent and identically distributed (*i.i.d.*) with a general distribution. We assume that there exists a positive number S_{max} such that $Pr[S^n > S_{max}] = 0$ and that we can know the value of S^n when the first packet of the nth burst arrives. The interarrival time between the nth and $(n+1)$st burst is denoted by T^{n+1}, which is assumed to be *i.i.d.* with a general distribution. We allow that T^n may take the value 0, i.e. the first packet of more than one burst may arrive on the same slot. There are m servers

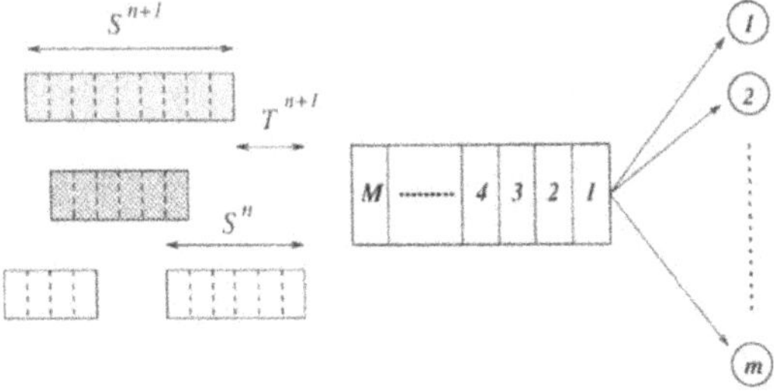

Figure 4 Queueing model of the analyzed system.

which are synchronized so that they start and end service at the same time. The service time of a packet is assumed to be equal to one slot. The packets arrive at the queue at the beginning of a slot and leave the queue at the end of a slot. The capacity of the shared buffer is finite, say M, excluding the service space. Arriving packets are allowed to enter the queue only when none of the packets belonging to the same burst are lost. At the arrival instant of the first packet of the nth burst, the system tries to reserve buffer space for all packets which belong to the nth burst, if it finds all servers busy, so that all packets of the nth burst can enter the queue. If this is not possible, all packets of the nth burst are lost. This *'All or Nothing Policy'* minimizes burst loss probabilities. The packets in the queue are served in a FIFO discipline on a burst basis. That is, the packets of the nth burst have higher priority than any packets of the (n+1)st, or higher numbered, burst whenever they arrive.

Indeed, if $T^{n+1} + t < S^{n+1}$, the tth packet of the $(n+1)$st burst has already arrived at the arrival instant of the $(T^{n+1} + t + 1)$st packet of the nth burst. In this case, the tth packet of the $(n+1)$st burst is served earlier than the $(T^{n+1} + t + 1)$st packet of the nth burst only if any server is idle.

In the following two sections, we propose an efficient numerical method to analyze the queueing model described above.

5 EMBEDDED MARKOV CHAIN

In this section, we construct a finite state embedded Markov chain, which will be useful for obtaining some stationary performance measures of the queue described in the previous section. First of all, let us consider an embedded Markov chain by giving attention to all active bursts, i.e., accepted bursts with remaining packets (which have not yet arrived). If we keep track of the number of remaining packets of each active burst and the number of packets in the buffer (the queue length) at the arrival instant of bursts, the process has a Markov property. It might be possible to obtain some stationary performance measures, e.g., the burst loss probability, the queue length distribution and the waiting time distribution from the steady state probability distribution of this process. However, the process becomes intractable as the number of active bursts increases. Therefore, it is important to construct an embedded Markov chain in order to efficiently obtain some performance measures such as a packet loss probability.

A similar model was analyzed and an effective embedded Markov chain was proposed by (Yamashita, 1994). We extend this methodology for the system with the *'All or Nothing Policy'*. The basic idea of the method is as follows: Let us consider the embedded point of the nth burst arrival instant. When the number of active bursts is equal to or less than m, we keep track of the number of remaining packets of every burst. However, when the number of active bursts is greater than m, we choose m bursts in decreasing order of the number of remaining packets and keep track of the numbers of remaining packets of these m bursts. The number of packets in the buffer never decreases while at least m bursts are active whenever the $(n+1)$st burst arrives. Therefore, if we also keep track of the number of packets in the buffer on the last slot when at least m bursts are active, we can deside whether the $(n+1)$st burst should be accepted or not.

Let v_i^n denote the ith largest number of remaining packets among active bursts at the arrival instant of the nth burst. In other words, v_i^n means the number of time slots with i arriving packets counting from the arrival instant of the nth burst , excluding the $(n+1)$st burst and all the bursts after $(n+1)$st. Note that $v_1^n \geq v_2^n \geq \cdots \geq v_m^n$. $v_{i-1}^n > v_i^n = 0$ means that only $(i-1)$ bursts are active at the arrival instant of the nth burst. $\boldsymbol{v}^n$ denotes the vector $(v_1^n, v_2^n, \cdots, v_m^n)$. Further, let w^n be the number of packets in the buffer on the v_m^nth slot counting from the arrival instant of the nth burst, excluding the $(n+1)$st burst and all the bursts after $(n+1)$st even if they have arrived already. w^n takes into account the arrival packets due to the other active bursts not included in the vecor $\boldsymbol{v}^n$.

Now, let us obtain the relationship between w^n and w^{n+1} given $\boldsymbol{v}^n$, T^{n+1}, and S^{n+1}. If $v_m^n \leq T^{n+1}$, then w^{n+1} represents the number of packets in the buffer on the T^{n+1}th slot counting from the arrival instant of the nth burst and is less than w^n since the packets in the buffer, if any, will be served after the v_m^nth slot. The ith server is capable of serving $(T^{n+1} - v_i^n)^+$ packets during the T^{n+1} slots, where $(N)^+ = \max(0, N)$. If $v_m^n > T^{n+1}$, on the other hand, w^{n+1} represents the number of packets in the buffer on the $\max(v_m^n, T^{n+1} + S^{n+1})$th slot counting from the arrival instant of the $(n+1)$st burst, and is greater than w^n since the number of packets increases on every slot by one from the T^{n+1}th to $\min(v_m^n, T^{n+1} + S^{n+1})$th slots counting from the arrival instant of the nth burst, as long as there is enough space in the buffer. If any arriving packets is not able to enter the buffer due to lack of space, all packets of the nth burst are rejected to enter the buffer. Then, the ith server will serve the packets in the buffer on every slot by one from $(v_i^n + 1)$st to T^{n+1}th slots (if $v_i^n < T^{n+1}$) counting from the arrival instant of the nth burst. From the above two discusions, we have

$$w^{n+1} = \begin{cases} [w^n - \sum_{i=1}^{m}(T^{n+1} - v_i^n)^+]^+, & \text{if } v_m^n - T^{n+1} \leq 0, \\ w^n + \min(v_m^n - T^{n+1}, S^{n+1}), & \text{if } 0 < \min(v_m^n - T^{n+1}, S^{n+1}) \\ & \quad \leq M - w^n, \\ w^n, & \text{if } \min(v_m^n - T^{n+1}, S^{n+1}) \\ & \quad > M - w^n. \end{cases} \tag{1}$$

Similarly, we can obtain the relationship between $\boldsymbol{v}_i{}^n$ and $\boldsymbol{v}_i{}^{n+1}$ given w^n, T^{n+1}, and S^{n+1}. When the $(n+1)$st burst is accepted, the last packet of the $(n+1)$st burst arrives on the $(T^{n+1} + S^{n+1})$th slot counting from the arrival instant of the nth burst. If $v_i^n \leq T^{n+1} + S^{n+1} < v_{i-1}^n$, then the ith largest number of remaining packets among active bursts at the arrival instant of the $(n+1)$st burst, v_i^{n+1}, is S^{n+1}. At the same time, $v_{i-1}^{n+1} = v_{i-1}^n - T^{n+1}$ and $v_{i+1}^{n+1} = v_i^n - T^{n+1}$. On the other hand, when the $(n+1)$st burst is rejected because of the *'All or Nothing Policy'*, no new burst arrives during T^{n+1} slots. Accordingly, we have the following relations:

$$v_i^{n+1} = \begin{cases} (v_{i-1}^n - T^{n+1})^+, & \text{if } \min(v_m^n - T^{n+1}, S^{n+1}) \le M - w^n, \\ & \quad v_{i-1}^n \le T^{n+1} + S^{n+1}, \\ S^{n+1}, & \text{if } \min(v_m^n - T^{n+1}, S^{n+1}) \le M - w^n, \\ & \quad v_i^n \le T^{n+1} + S^{n+1} < v_{i-1}^n, \\ v_i^n - T^{n+1}, & \text{if } \min(v_m^n - T^{n+1}, S^{n+1}) \le M - w^n, \\ & \quad T^{n+1} + S^{n+1} < v_i^n, \\ & \text{or } \min(v_m^n - T^{n+1}, S^{n+1}) > M - w^n. \end{cases} \tag{2}$$

where $i = 1, 2, \cdots, m$ and we define $v_0^n = \infty$ for $n = 1, 2, \cdots$.

$(\boldsymbol{v}^n, w^n)$ has the Markov property, because $(\boldsymbol{v}^{n+1}, w^{n+1})$ depends only on $(\boldsymbol{v}^n, w^n)$ given T^{n+1} and S^{n+1}. Let us denote the relationship by:

$$(\boldsymbol{v}^{n+1}, w^{n+1}) = f(\boldsymbol{v}^n, w^n, S^{n+1}, T^{n+1}).$$

Since $\boldsymbol{v}^n$'s are bounded by S_{max}, $(\boldsymbol{v}^n, w_1^n)$ is a finite state embedded Markov chain at the arrival instant of bursts with less than $(S_{max}+1)^m(M+1)$ states, i.e., $O(S_{max}^m M)$. $T^{n+1} \ge S_{max} + M$ is a sufficient condition for $v_1^{n+1} = S^{n+1}$, $v_2^{n+1} = \cdots = v_m^{n+1} = 0$, and $w^{n+1} = 0$. Therefore, it is sufficient to consider the case $T^{n+1} = 0, 1, \cdots, S_{max} + M$, $S^{n+1} = 1, 2, \cdots, S_{max}$ for every state $(\boldsymbol{v}^n, w^n)$ when we calculate the coefficients of the equilibrium equations using (1) and (2). That is, it requires $O(S_{max}^{m+1} M(S_{max} + M))$ time units to calculate. Once we calculate the coefficients of the equilibrium equations, we can get the steady state probability distribution of $(\boldsymbol{v}^n, w^n)$, denoted by $P(\boldsymbol{v}, w)$, by solving the system of stationary equilibrium equations:

$$P(\boldsymbol{v}, w) = \sum_{S=1}^{S_{max}} \sum_{T=0}^{S_{max}+M} P(S)P(T) \sum_{(\boldsymbol{v}', w') \in \Lambda(\boldsymbol{v}, w, S, T)} P(\boldsymbol{v}', w')$$

for all possible states $(\boldsymbol{v}, w)$, where

$$\Lambda(\boldsymbol{v}^{n+1}, w^{n+1}, S^{n+1}, T^{n+1})$$

$$= \{(\boldsymbol{v}^n, w^n) \mid (\boldsymbol{v}^{n+1}, w^{n+1}) = f(\boldsymbol{v}^n, w^n, S^{n+1}, T^{n+1})\},$$

and $P(S)$ and $P(T)$ denote the probability that the number of packets of a burst is S and the probability that the interarrival time between bursts is T, respectively.

In particular, when $m = 1$, we get,

$$P(v, 0) = \sum_{v'=1}^{S_{max}} \sum_{w'=0}^{M} Pr[S = v] \sum_{k=v'+w'}^{\infty} Pr[T = k] P(v', w')$$

$$+\Psi[v \geq M+1] \sum_{v'=v}^{S_{max}} \sum_{j=M+1}^{S_{max}} Pr[S=j]Pr[T=v'-v]P(v',0).$$

$$P(v,w) = \sum_{v'=1}^{S_{max}} \sum_{w'=\max(0,w-v',w-v)}^{M} Pr[S=v]Pr[T=v'+w'-w]P(v',w')$$

$$+\sum_{v'=v}^{S_{max}} \sum_{w'=\max(0,w-v+1)}^{w-1} Pr[S=w-w']Pr[T=v'-v]P(v',w')$$

$$+\Psi[v \geq M-w+1] \sum_{v'=v}^{S_{max}} \sum_{j=M-w+1}^{S_{max}} Pr[S=j]Pr[T=v'-v]P(v',w),$$

$$w = 1, 2, \cdots, M.$$

where $\Psi[\cdot]$ is an indicator function which takes 1 or 0.

We note that this method is still much more efficient than the straightforward way mentioned at the beginning of this section, though the process $(\boldsymbol{v}^n, w^n)$ becomes intractable as the number of servers increases.

6 PERFORMANCE MEASURES

In this section, we get the performance measures using the steady state probability distribution $P(\boldsymbol{v}, w)$ obtained in the previous section. We first calculate the burst loss probability, i.e., that one or more packets in the burst is rejected. As discussed in the previous section, when $\min(v_m^n - T^{n+1}, S^{n+1}) > M - w^n$ the $(n+1)$st burst is lost because of the *'All or Nothing Policy'*. Then the burst loss probability denoted by P_{loss}^{burst} is represented by

$$P_{loss}^{burst} = \sum_S \sum_T P(S)P(T) \sum_{v_m} \sum_{w > M - \min(v_m - T.S)} P(v_m, w), \tag{3}$$

where $P(v_m, w)$ is the marginal probability of $P(\boldsymbol{v}, w)$. Let us define the packet loss probability by the ratio between the average number of packets that are lost and the average number of packets that arrive in a burst. Similar arguments give the expression for the packet loss probability denoted by P_{loss}^{packet} as follows:

$$P_{loss}^{packet} = \sum_S \sum_T SP(S)P(T) \sum_{v_m} \sum_{w > M - \min(v_m - T,S)} P(v_m, w) / \sum_S SP(S). \tag{4}$$

Now, we get the waiting time distribution W, assuming FIFO discipline on a burst basis, that is, the packets of the nth burst have the higher priority than any packets of the (n+1)st burst whenever they arrive. We define the waiting time distribution so that it satisfies the following equation:

$$\sum_{k=0}^{\infty} P[W=k] + P_{loss}^{packet} = 1.$$

Accordingly, we suppose the nth burst is not rejected. We first consider the waiting time of the jth packet of the $(n+1)$st burst, denoted by W_j^{n+1}, given $\boldsymbol{v}_n$, w_n, and T^{n+1}. Note that any packets of the (n+1)st burst can not be served at least until the v_m^nth slot counting from the arrival instant of the nth burst, if $T^{n+1} \le v_m^n$. The jth packet of the $(n+1)$st burst arrives at the queue on $(T^{n+1}+j)$th slot counting from the arrival instant of nth burst. $W_j^{n+1} = k$ means that the number of packets in the buffer which should be served before the jth packet of the $(n+1)$st burst first becomes 0 at the $(T^{n+1}+j+k)$th slot counting from the arrival instant of the nth burst. The number of packets in the buffer which should be served before the jth packet of the $(n+1)$st burst at the $(T^{n+1}+j+k)$th slot counting from the arrival instant of the nth burst is equal to :

[queue length on the v_m^nth slot excluding all the bursts after nth: w_n]

+ [# arrived packets of the $(n+1)st$ burst at the arrival instant of the jth packet: j]

+ [# arrived packets of all the bursts before $(n+1)$st from the (v_m^n+1)st slot: $\sum_{i=v_m^n+1}^{T^{n+1}+j+k} e_i^n$]

− [# served packets from the (v_m^n+1)st slot: $(T^{n+1}+j+k-v_m^n)m$],

where e_i^n is the largest number which satisfies $v_{e_i^n}^n \ge i$, $(e_i^n = 0, 1, \cdots, m)$. In other words, e_i^n is the number of active bursts on the ith slot counting from the arrival instant of the nth burst, excluding all the burst after nth. Using

$$(T^{n+1}+j+k-v_m^n)m - \sum_{i=v_m^n+1}^{T^{n+1}+j+k} e_i^n = \sum_{i=1}^{T^{n+1}+j+k} (m-e_i^n),$$

we have the following relations:

$$W_j^n = \begin{cases} 0, & \text{if } w^n + j \le \sum_{i=1}^{T^{n+1}+j}(m-e_i^n), \\ \\ k, & \text{if } \sum_{i=1}^{T^{n+1}+j+k-1}(m-e_i^n) < w^n + j \le \sum_{i=1}^{T^{n+1}+j+k}(m-e_i^n). \end{cases}$$

Therefore, taking into account the condition that the burst is accepted, the waiting time distribution can be obtained by

$$Pr[W=0] = \sum_S \sum_T P(S)P(T) \sum_v \sum_{w \le M-\min(v_m-T,S)} P(v_m, w)$$

$$\sum_{j=1}^{S} \Psi[w+j \le \sum_{i=1}^{T+j}(m-e_i)] / \sum_S SP(S),$$

$$Pr[W=k] = \sum_S \sum_T P(S)P(T) \sum_v \sum_{w \le M-\min(v_m-T,S)} P(v_m, w)$$

$$\sum_{j=1}^{S} \Psi[\sum_{i=1}^{T+j+k-1}(m-e_i) < w+j \le \sum_{i=1}^{T+j+k}(m-e_i)] / \sum_S SP(S),$$

$$k = 1, \cdots, S_{max} + \lfloor M/m \rfloor,$$

where the summation of $\boldsymbol{v}$ extends over all possible states of $(v_1, v_2, \cdots, v_M)$ and e_i means e_i^n given $\boldsymbol{v}$.

We can now calculate the mean waiting time using the waiting time distribution. Then, we get the mean queue length $\bar{L}$ using Little's law [?], i.e.

$$\bar{L} = \bar{W}\bar{S}/\bar{T}, \tag{5}$$

where $\bar{S}$, $\bar{T}$, and $\bar{W}$ denote the first moments of S, T, and W, respectively. Though we assumed FIFO discipline on a burst basis when we derived the waiting time distribution, (3) $\sim$ (5) hold true for other service disciplines, e.g., FIFO on a packet basis.

7 NUMERICAL EXAMPLES

In this section, we present some numerical examples and demonstrate the advantage of the *'All or Nothing Policy'* for the burst loss probability and the waiting time. We consider two examples, single and double server queues. For both examples, we assume that the interarrival time between bursts is uniformly or binomially distributed from 1 to 15, with mean 8.0 and squared coefficient of variation 0.2917 or 0.06944. We also assume that the number of packets in a burst is uniformly or binomially distributed from 1 to 11, with mean 6.0 and squared coefficient of variation 0.2778 or 0.05469. The number of states of the embedded Markov chain $(\boldsymbol{v}^n, w^n)$ for each example are shown in Table 1.

We compare two systems, with and without the *'All or Nothing Policy'*. The packet loss probability, the burst loss probability, and the mean waiting time of packets are illustrated as a function of the buffer capacity in Figures 5 $\sim$ 10. In each example, we can find that the system with the *'All or Nothing Policy'* is superior to the one without it for the burst loss probability and the mean waiting time of packets. On the other hand, using the *'All or Nothing Policy'*, the packet loss probability increases, however the packet loss probability is not important in applications discussed in section 3.

Table 1 The number of states $(\boldsymbol{v}^n, w^n)$

Capacity of the buffer	1	3	5	7	9
$m = 1$	22	44	66	88	110
$m = 2$	152	304	456	608	760

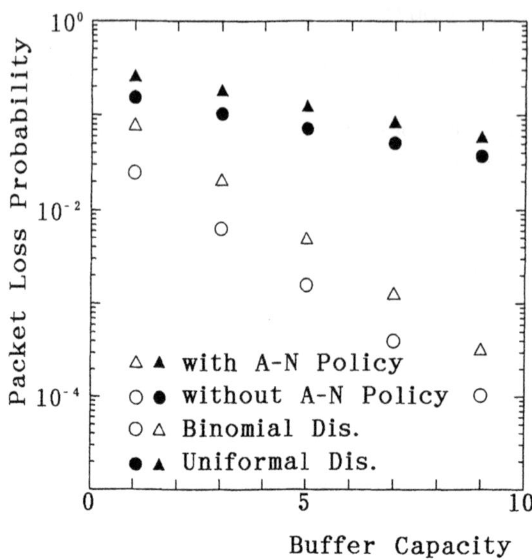

Figure 5 Packet Loss Probability ($m = 1$).

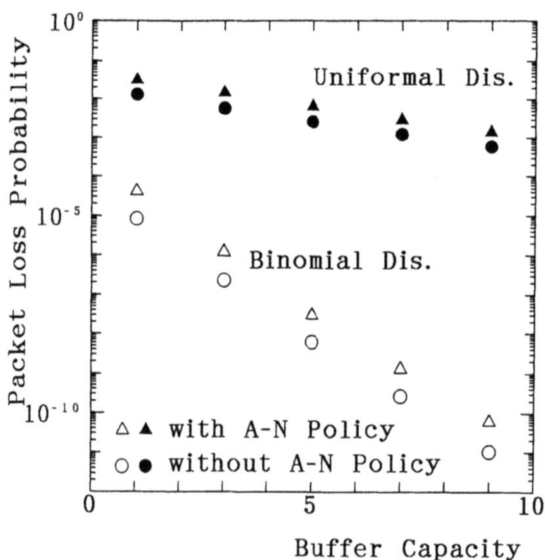

Figure 6 Packet Loss Probability ($m = 2$).

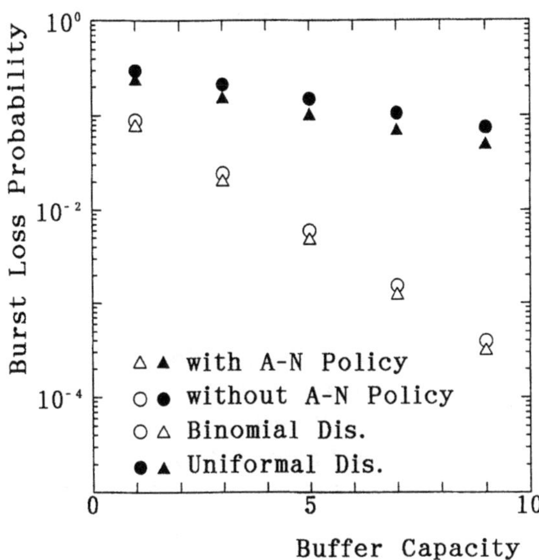

Figure 7 Burst Loss Probability ($m = 1$).

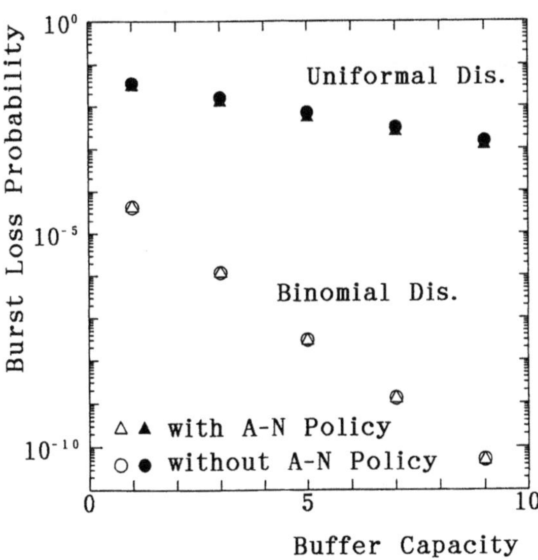

Figure 8 Burst Loss Probability ($m = 2$).

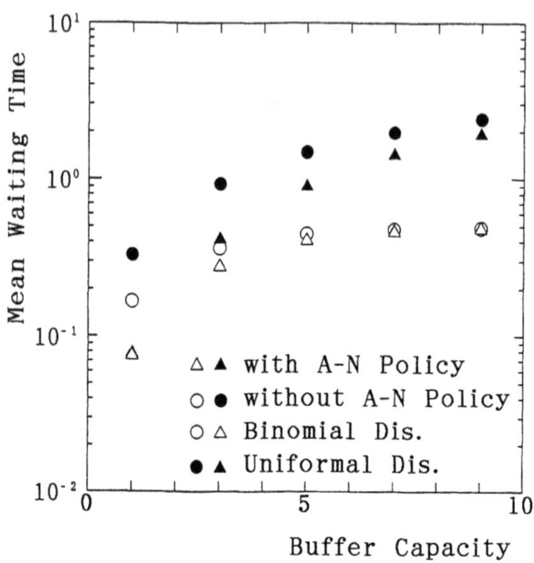

Figure 9 Mean Waiting Time ($m = 1$).

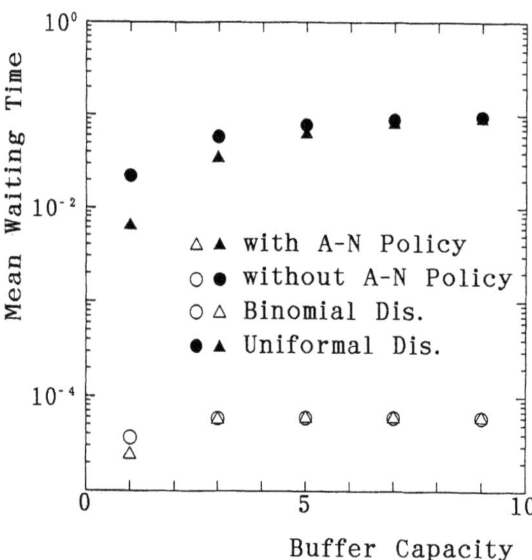

Figure 10 Mean Waiting Time ($m = 2$).

8 REFERENCES

Anagnostou, M. E., Sykas, E. D. and Protonotarios, E. D. (1984), Steady-state and transient delay analysis of ARQ protocols, *Computer Communications,* **7**, No.1.

CCITT (1988), *CCITT Recommendation,* H and J series, November.

Chao, H. J., Ghosal, D., Saha D. and Tripathi, S. K. (1994), IP on ATM Local Area Networks, *IEEE Communications Magazine,* August.

Comer, D.E. (1991), *Internetworking with TCP/IP, vol.I,* Prentice-Hall.

Little, J. D. C. (1961), A Proof of the Queueing Formula $L = \lambda W$, *Operations Research* **9**.

Yamashita, H. (1994), Numerical analysis of a discrete-time finite capacity queue with a burst arrival, *Annals of Operations Research,* **49**.

9 BIOGRAPHY

Johan M. Karlsson has been Associate Professor of the Department of Communication System, Lunt Institute of Technology, Sweden, since 1993.

Hideaki Yamashita received the B.S., M.S., and Ph.D. degrees in mechanical engineering from Sophia University in 1982, 1984, and 1987, respectively. He is currently Lecturer (faculty member) of the Faculty of Business Administration, Komazawa University, Tokyo, since 1995. As a Visiting Faculty in Computer Science, he stayed at the North Carolina State University, Raleigh, during 1989 - 1991. His research interests include the modeling and analysis of queueing networks and their applications to production systems and high speed networks. He is a member of the Operations Research Society of Japan, the Institute of Electronics, Information and Communication Engineers, and the Japan Society of Mechanical Engineers.

13

Study of the Impact of Temporal / Spatial Correlations on a Tagged Traffic Stream[1]

Marco Conti and Enrico Gregori
CNR - Istituto CNUCE
Via Santa Maria 36, 56126 Pisa, Italy
Email: {man@vm. and e.gregori@}cnuce.cnr.it , Fax: +39 50 904052

Ioannis Stavrakakis
Northeastern University
Electrical & Computer Engineering, Boston, MA 02115, USA.
Email: ioannis@cdsp.neu.edu , Fax: (617) 373-8970

Abstract

The problem of evaluating the end-to-end (multi-node) performance of a tagged traffic stream in an ATM environment is known to be difficult and largely open. To simplify the analysis, past studies have assumed that the co-existing (background) traffic is uncorrelated and is diverted after a single multiplexing stage; for this reason, temporal and spatial correlations were assumed to be insignificant and were not investigated. The objective in this work is to consider such correlations and evaluate their impact on the end-to-end performance of a tagged traffic stream. Such correlations can be significant due to temporal correlation in the background traffic or partial commonality in the routing path (background traffic is not necessarily diverted). A binary Queueing Activity Indicator (QAI) is proposed in this work to provide for a simple mechanism to capture these correlations. Results derived for the delay distribution of a tagged traffic stream traversing two consecutive nodes show the substantial impact of spatial/temporal correlations and indicate that these correlations are effectively captured by the proposed QAI. In addition, it is shown that increasing the temporal/spatial correlation results in increasing inaccuracy of the results obtained by ignoring them.

Keywords

Network traffic, jitter, correlations, end-to-end performance.

[1]M. Conti's and E. Gregori's work was supported by CNR in the framework of a bilateral project. I. Stavrakakis' research was performed in part while visiting CNR Istituto CNUCE and has been supported in part by the Advanced Research Project Agency (ARPA) under Grant F49620-93-1-0564 monitored by the Air Force Office of Scientific Research (AFOSR); I. Stavrakakis is on leave from the University of Vermont.

1 INTRODUCTION

Asynchronous Transfer Mode (ATM) is considered to be the prevailing switching and multiplexing technique for the implementation of high-speed Broadband - Integrated Services Digital Networks (B-ISDN). Increased network utilization in the presence of diversified applications is expected to be achieved through the adopted statistical multiplexing and the on-demand bandwidth allocation (Onvural, 1993).

Although the maximum network utilization can be achieved by allowing for uncontrolled access to the network of all users, it is clear that in this case congestion will occur and the Quality of Service (QoS) delivered to the applications will suffer significantly. As a result, an optimal compromise between network utilization and delivered QoS is desirable. The Call Admission Control (CAC) and Traffic Regulation (TR) functions are being developed in an effort to achieve this optimal balance (Onvural, 1993).

Unless a very conservative approach is followed – limiting severely the network utilization – statistical multiplexing will surely result in modification of the source traffic profile, which may be severe at periods of network overload. This modification to the source traffic profile (distortion) represents a measure of the reduction of the QoS that the application will experience. When more than one multiplexing stages are along the path from source to destination, the distortion can be significantly increased.

The queueing problems induced by the statistical multiplexing process are well known and studied. Most of the past studies have focused on the determination of the impact of the multiplexing process on a *random* cell (information unit). That is, the underlying assumption has been that all multiplexed applications have the same QoS requirements. A measure of the traffic distortion due to the multiplexing process – the correlation in the *random* cell departure process – has been considered in (Stavrakakis, 1991a), (Stavrakakis, 1991b), (Lau, 1993), where an approximate description of the resulting random cell profile is presented and used for an approximate end-to-end study.

In an ATM environment in which applications may have quite diverse QoS requirements, studies like the above can be of limited usefulness, (Kurose, 1993), (Nagarajan, 1992). It is important that the magnitude of the distortion to a specific traffic profile – in the presence of other multiplexed applications – be determined. Thus, an application needs to be tagged and observed at the output of the multiplexer to evaluate the distortion to its traffic profile.

A number of recent works have focused on the evaluation of the distortion of a tagged traffic profile due to the statistical multiplexing process (Guillemin, 1992), (Bisdikian 1993), (Matragi, 1994a), (Matragi, 1994b), (Roberts, 1992), (Landry, 1994). The delay jitter has been adopted as a measure of the distortion and has been determined by developing analytic (Bisdikian 1993), (Matragi, 1994a), (Matragi, 1994b) and numerical (Landry, 1994) approaches. These works evaluate the magnitude of the distortion to a tagged traffic profile and point to the fact that this distortion may be unacceptable for certain applications. In (Landry, 1994) it is attempted to restore in part the original traffic profile by proposing a modification to the FIFO (First-In First-Out) multiplexing discipline and evaluating the reduced distortion to the tagged traffic profile. A typical approach followed for the study of a tagged traffic stream considers all non-tagged traffic to form a single traffic stream called the background traffic.

The above mentioned past work is based on a number of assumptions. First, the background traffic at a single node is assumed to be a (time) uncorrelated process. The second major assumption is that of the almost complete nodal decomposition in the approximate end-to-end performance study. Although the tagged interarrival process to node $n+1$ is described in terms of the tagged interdeparture process from node n, consecutive interdepartures are identically distributed and independent. In addition, the background traffic in node $n-1$ is assumed to follow a different path from that of the tagged traffic and, thus, it is not forwarded to node $n+1$. As a result, the background traffic in node n is fresh and independent from any other process. In this discussion, $n-1$, n and $n+1$ denote three consecutive nodes along the path from the source to the destination of the tagged traffic.

In this work, the major assumptions outlined above will be relaxed to a certain degree. At first, the background traffic will be assumed to be a correlated process, which may be better reflecting a realistic environment. One of the objectives will be to determine the impact of correlations in the background process on the distortion to the tagged traffic stream (temporal correlation). A second departure from the past work is regarding the background traffic of node $n-1$. By definition, this is the non-tagged traffic which is present at node $n-1$ and is forwarded to node n together with the tagged traffic. A portion of this traffic will be assumed to be forwarded to node $n+1$ along with the tagged traffic. This carried-on traffic – which will be allowed to be correlated – will be added to the fresh background traffic at node n. The consideration of the carried-on traffic contributes to a coupling between the queueing processes in nodes $n-1$ and n (spatial correlation). Again, the existence of carried-on traffic may be better capturing a realistic networking environment.

The objective in this paper is to assess the significance of the temporal/spatial correlation discussed above and evaluate the resulting accuracy in calculating measures of the distortion of the tagged traffic profile. For this reason, results for a system of two consecutive nodes will be derived in the present study. The end-to-end performance when more than two nodes are present can be derived by following the approach proposed in this paper. The present study will help identify the processes which impact significantly on the measures of interest of the distortion of the tagged traffic profile and provide guidance for the larger scale end-to-end study in the presence of arbitrarily many nodes.

In the next section the system to be studied is described; a model is developed and the quantities of interest are defined. An outline of the queueing analysis is presented in section 3, followed by numerical results and discussion in section 4.

2 SYSTEM DESCRIPTION AND MODELING

Consider two consecutive nodes (nodes $n-1$ and n) along the path from the source to the destination of the tagged traffic. Suppose that: (a) the background traffic is a (time) uncorrelated process and (b) the background traffic at node $n-1$ is diverted after node n, that is, it is not forwarded to node $n+1$ along with the tagged traffic.

Let $\{G_k^n\}_k$ denote a process describing some traffic descriptor (such as delay or interdeparture) associated with tagged cell k and node n. Due to the memory present in the queueing process, $\{G_k^n\}_k$ will be a (time) correlated process. Since the queueing activity at node $n-1$ modulates $\{G_k^{n-1}\}_k$ which shapes the tagged cell arrival process to node n, $\{G_k^{n-1}\}_k$ and $\{G_k^n\}_k$ will be (space) correlated processes. All the past work has ignored such temporal

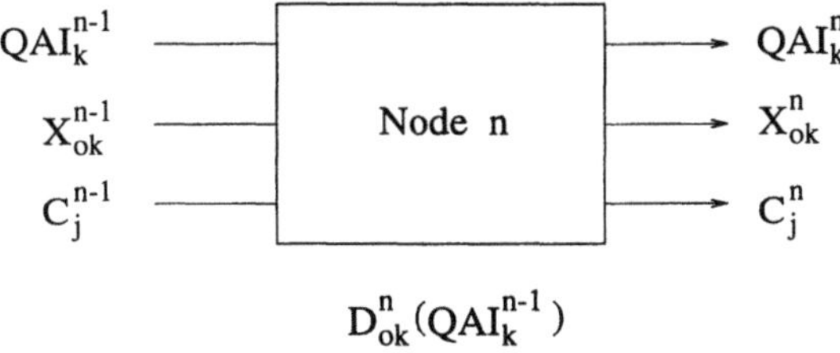

Figure 1: Input - output quantities for a generic node.

and spatial correlations. Some results suggest that - at least under certain conditions - these correlations may not be significant, (Landry, 1994).

In this work, the above assumptions (a) and (b) are being relaxed. In addition to the memory in the queueing process, (temporal) correlation in the background traffic will be present, potentially increasing substantially the temporal correlation in the process $\{G_k^n\}_k$. For instance, if a 2-state background arrival process remains in one of two substantially different (in terms of cell arrivals) states, a bimodal queueing behavior may be induced. Ignoring such correlations may result in very inaccurate performance calculations. By relaxing (b), the spatial correlation will be increased through the contribution of the carried-on traffic which is shaped by the queueing activity in the previous node.

The objective in this work is to study the impact of spatial and temporal correlations (coupling) on the characteristics of a tagged traffic stream. The term *spatial correlation* is adopted in this work to refer to the dependence of the queueing activity at some node n from the queueing activity at node $n-1$. The term *temporal correlation* refers to the (temporal) correlation in a traffic descriptor associated with two consecutive tagged cells. A simple, binary Queueing Activity Indicator (QAI) is adopted in this work, to provide for a simple mechanism to (approximately) capture these spatial and temporal correlations. The QAI will modulate the carried-on traffic as well as the tagged traffic stream, and provide for a limited coupling of queueing processes associated with consecutive cells and consecutive nodes.

The queueing study of a generic node n will be based on the consideration of the input triplet $\{QAI_k^{n-1}, X_{ok}^{n-1}, C_j^{n-1}\}$ - see Figure 1 - where the involved quantities denote the QAI associated with node $n-1$ and tagged cell k, the tagged cell interdeparture distribution associated with node $n-1$ and tagged cell k, and the background traffic at the output of node $n-1$, respectively.

Process $\{QAI_k^{n-1}\}_k$ will be approximated by a 2-state, first-order Markov process with parameters matched to those of the exact process. The state of this process - which will be part of the state description of node n - will modulate X_{ok}^{n-1} as well as C_j^{n-1}; that is, the input traffic to node n will be shaped by the queueing activity at node $n-1$, through the consideration of QAI_k^{n-1}. Furthermore, measures of the queueing behavior at node n - such as the delay D - will be shaped by the QAI associated with node $n-1$. Given the current state of QAI_k^{n-1}, the conditional probabilities X_{ok}^{n-1} given QAI_k^{n-1}, C_j^{n-1} given QAI_k^{n-1} and D_{ok}^n given QAI_k^{n-1} will be derived. Finally, the queueing analysis of node n will determine the output triplet $\{QAI_k^n, X_{ok}^n, C_j^n\}$ which will be considered in the study of node $n+1$.

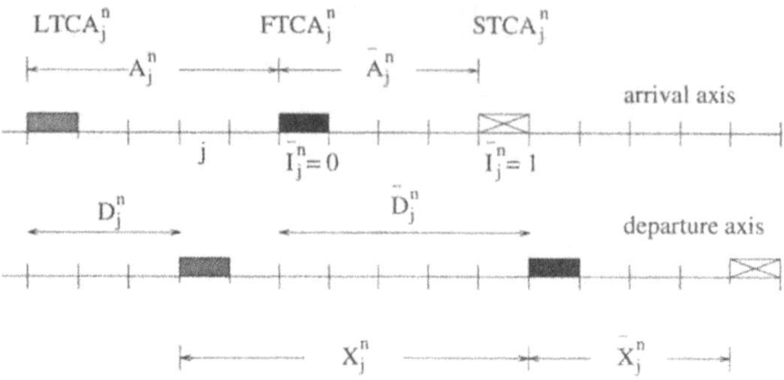

Figure 2: Arrival and departure time axes.

The following definitions will be used in the description and analysis of the system to be studied in this work. Although results for two nodes will be derived in the present work, the general notation n will be used to denote the node under study. A subscript j denotes a quantity associated with time slot j. A subscript ok denotes a quantity associated with the kth tagged cell arriving in the associated node; for simplicity, subscript k will be used instead of ok if the associated quantity is defined only with respect to a tagged cell arrival and not at time slot j. A superscript n denotes a quantity associated with node n. For simplicity, let $LTCA_j^n$ denote the Latest Tagged Cell Arrival to node n which occurred before or at time (slot) j. Similarly, let $FTCA_j^n$ ($STCA_j^n$) denote the First (Second) Tagged Cell Arrival following time (slot) j. Some of the quantities introduced below are depicted in Figure 2.

$A_j^n(\bar{A}_j^n)$: Interarrival time between $LTCA_j^n$ and $FTCA_j^n$ ($FTCA_j^n$ and $STCA_j^n$); $1 \leq A_j^n, \bar{A}_j^n \leq A_{max}^n$; let $f_a^n(k) = Pr\{A_j^n = k/A_j^n > k-1\}$, $1 \leq k \leq A_{max}^n$; let $\bar{f}_a^n(k) = Pr\{A_j^n > k/A_j^n > k-1\} = 1 - f_a^n(k)$; let $E\{A_j^n\} = \frac{1}{\lambda^n}$, where $E\{\cdot\}$ denotes the expectation operator.

$X_j^n(\bar{X}_j^n)$: Interdeparture time between $LTCA_j^n$ and $FTCA_j^n$ ($FTCA_j^n$ and $STCA_j^n$).

$D_j^n(\bar{D}_j^n, \tilde{D}_j^n)$: Delay of $LTCA_j^n$ ($FTCA_j^n$, $STCA_j^n$).

$I_j^n(\bar{I}_j^n, \tilde{I}_j^n)$: Indicator function assuming the value 1 if $LTCA_j^n$ ($FTCA_j^n$, $STCA_j^n$) finds the queue non-empty of *tagged* cells upon arrival to node n; for instance, $\bar{I}_j^n = 1_{\{A_j^n \leq D_j^n\}}$. Process $\{I_j^n\}_j$ will be modeled in terms of a 2-state Markov process embedded at times of tagged cell arrivals. The parameters of this approximate process are derived by matching its transition probabilities to those of the true process; the latter are determined by considering the evolution of process $\{W_j^n\}_j$ (described in section 3). $I_j^n(\bar{I}_j^n, \tilde{I}_j^n)$ is the QAI associated with node n modulating the output process of node n and employed in the study of node $n+1$.

J_j^{n-1} : Indicator function assuming the value 1 if $LTCA_j^n$ finds the queue of node $n-1$ non-empty of tagged cells upon arrival to node $n-1$. Notice that J_j^{n-1} is not

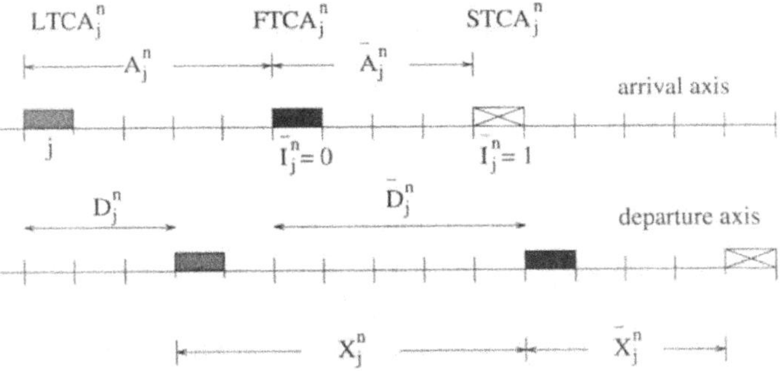

Figure 3: Arrival and departure time axes.

identical to I_j^{n-1}, since the former refers to $LTCA_j^n$ and the latter to $LTCA_j^{n-1}$, which are different cells. J_j^{n-1} is the QAI associated with node $n-1$, modulating the output process of node $n-1$ and employed in the study of node n.

Q_j^n : Queue occupancy.

B_j^n : State of the Markov background traffic; let $S_b = \{0,1\}$, $\pi_b^n(i)$ and $p_b^n(i,j)$, $i,j \in S_b^n$ denote the state space and the stationary and transition probabilities, respectively.

$\hat{B}_j^n(B_j^n)$: Number of background cells generated when the state of the background process is (B_j^n); let $f_b^n(k,l) = Pr\{\hat{B}_j^n(l) = k\}$, $0 \le k \le M_b$, $l = 0,1$.

T_j^n : Time (in slots) that has elapsed since the arrival time of $LTCA_j^n$; $T_j^n = 1$ if a tagged cell arrives at slot j, as it is the case in Figure 3; $1 \le T_j^n \le A_{max}^n$.

$$T_{j+1}^n = \begin{cases} T_j^n + 1 & \text{if } \{A_j^n > T_j^n\} \\ 1, & \text{if } \{A_j^n = T_j^n\} \end{cases} \tag{1}$$

It is easy to establish that the conditional probability distribution for T_j^n – denoted by $p_t^n(l,k) = Pr\{T_j^n = k / T_{j-1}^n = l\}$ – is given in terms of that of A_j^n as shown below.

$$p_t^n(k-1,k) = Pr\{A_j^n > k-1 \ / \ A_j^n > k-2\} = \bar{f}_a^n(k-1)\ , \quad 2 \le k \le A_{max}^n \tag{2}$$
$$p_t^n(k,1) = Pr\{A_j^n = k \ / \ A_j^n > k-1\} = f_a^n(k)\ , \quad 1 \le k \le A_{max}^n \tag{3}$$
$$p_t^n(k,l) = 0 \text{ elsewhere} \tag{4}$$

Determination of the evolution of this process requires knowledge of the conditional interarrival time A_j^n given that $A_j^n > T_j^n - 1$ or, equivalently, knowledge of X_j^{n-1}, which depends on J_j^{n-1} (the QAI). The conditional probability distribution of X_j^n given J_j^{n-1} is derived from the joint probability $\{X_j^n, J_j^{n-1}\}$ which is derived by considering the evolution of process $\{W_j^n\}_j$ (section 3).

$C_j^n(J_j^{n-1}, T_j^n)$: Indicator function assuming the value 1 if a background cell of node $n-1$ arrives at node n at time j. Notice that

$$C_j^n(1, T_j^n) = 1_{\{T_j^n \neq 1\}} \tag{5}$$

$$C_j^n(0, 1) = 0 \tag{6}$$

$$C_j^n(0, T_j^n) = l, \quad l = 0, 1, \quad T_j^n \neq 1 \quad \text{(to be evaluated)} \tag{7}$$

Notice that $J_j^{n-1} = 1$ implies that the arrival of a tagged cell found the previous tagged cell in the queue and, thus, the interval between their transmission instants will be filled up with background traffic. The probabilistic description of $C_j^n(0, T_j^n)$ is derived assuming that J_j^{n-1} and T_j^n determine completely the (probabilistic) behavior of $C_j^n(J_j^{n-1}, T_j^n)$ and considering the evolution of the process $\{W_j^n\}_j$ (section 3).

F_j^n : State of the Markov splitting process associated with the carried-on traffic; let $S_f^n = \{0, 1\}$, $\pi_f^n(i)$ and $p_f^n(i, j)$, $i, j \in S_f^n$, denote its state space, and its stationary and transition probabilities, respectively. If a background cell from node $n-1$ arrives at node n at slot j ($C_j^n(J_j^{n-1}, T_j^n) = 1$), then this cell is forwarded to node $n+1$ (along the path of the tagged stream) if $F_j^n = 1$ – and this is called *carried-on* cell – and it is diverted otherwise. Notice that transitions of this Markov chain are assumed to occur at slots containing background traffic ($C_j^n(J_j^{n-1}, T_j^n) = 1$) from node $n-1$.

It should be noted that when the QAI J_j^{n-1} (or, I_j^n, $\bar{I}_j^n$, $\tilde{I}_j^n$) is equal to zero, light to moderate queueing activity may be assumed; when it is equal to one, moderate to serious queueing activity may be assumed. It is expected that this QAI biases the delay and inter-departure distributions. For this reason, families of these distributions will be obtained by considering the different values of the QAI, as indicated earlier in the presentation of Figure 1. By considering the correlation in the QAI process and capturing the dependence from the QAI associated with node $n-1$ of the input processes and the induced tagged cell delay at node n, an end-to-end performance based on limited nodal-coupling can be obtained. In addition to the previous, the particular selection of the QAI facilitates the description of the background traffic at the output of the corresponding node. When $I_j^n = 1$, every slot in the output link over the interdeparture interval X_j^n must contain a background cell from node n, as indicated in (5). When $I_j^n = 0$, the previous is not necessarily true, (7).

3 OUTLINE OF THE QUEUEING ANALYSIS

In this section, the queueing measures of interest are derived for node n and its output process is characterized. Details regarding the derivations are omitted due to space limitations.

Assuming that the departures occur before arrivals over the same slot, the evolution of the queue occupancy process is given by

$$Q_j^n = [Q_{j-1}^n - 1]^+ + \hat{B}_j^n(B_j^n) + 1_{\{C_j^n(J_j^{n-1}, T_j^n)=1, F_j^n=1\}} + 1_{\{T_j^n=1\}} \tag{8}$$

where the first term describes departures from the queue provided that it is non-empty and the second term describes fresh background arrivals. The third term describes background traffic from node $n-1$ to node n which is forwarded to node $n+1$ (carried-on traffic). The last term describes a tagged cell arrival. In view of equation (8) and the definitions of $\{B_j^n\}_j$ and $\{F_j^n\}_j$, the following multi-dimensional process:

$$\{W_j^n\}_j \equiv \{T_j^n, J_j^{n-1}, F_j^n, B_j^n, Q_j^n\}_j, \tag{9}$$

becomes a Markov process under the approximations associated with processes $\{J_j^{n-1}\}_j$, $\{T_j^n\}_j$ and $\{C_j^n(J_j^{n-1}, T_j^n)\}_j$, as outlined above. Let

$$\Phi_j^n \equiv \{T_j^n, J_j^{n-1}, F_j^n, B_j^n\}. \tag{10}$$

Then $\{W_j^n\}_j$ can be written as

$$W_j^n \equiv \{\Phi_j^n, Q_j^n\} \tag{11}$$

The transition probability matrix of (11) has the M/G/1 structure with Φ_j^n and Q_j^n corresponding to the phase and level processes, respectively. The numerical complexity in deriving the stationary probabilities $\pi_w(\cdot,\cdot,\cdot,\cdot,\cdot)$ of (11) is determined by the dimensionality of the space of the phase process (Neuts, 1989) which is equal to $2^3 A_{max}^n \times 2^3 A_{max}^n$. Due to the particular evolution of process T_j^n (it increases by 1 or returns to 1), the number of possible states of the phase process is significantly smaller than $2^3 A_{max}^n \times 2^3 A_{max}^n$.

Let j be the time of arrival of the kth tagged cell to node n (Figure 3). Then

$$\{T_j^n, J_j^{n-1}, F_j^n, B_j^n, Q_j^n\} \equiv \{1, J_{ok}^{n-1}, F_{ok}^n, B_{ok}^n, D_{ok}^n\} \tag{12}$$

Notice that the queue occupancy upon the kth tagged cell arrival is equal to the delay of this cell and that $T_j^n = 1$ when a tagged cell arrival occurs.

Let

$$\{\Psi_k^n\}_k \equiv \{J_{ok}^{n-1}, F_{ok}^n, B_{ok}^n, D_{ok}^n\}_k \tag{13}$$

with stationary probabilities given by $\pi_\psi(i_1, i_2, i_3, i_4) = \frac{1}{\lambda^n}\pi_w(1, i_1, i_2, i_3, i_4)$. Clearly, the joint probability of $\{J_{ok}^{n-1}, D_{ok}^n\}$ can be derived from the stationary probabilities of $\{\Psi_k^n\}_k$ and finally

$$P\{D_{ok}^n / J_{ok}^{n-1} = i\} = \frac{P\{J_{ok}^{n-1} = i, D_{ok}^n\}}{P\{J_{ok}^{n-1} = i\}} \quad , \quad i = 0, 1. \tag{14}$$

4 NUMERICAL RESULTS AND DISCUSSION

In this work results have been derived for the system of two consecutive nodes depicted in Figure 4. The tagged source S generates a traffic stream which is multiplexed with the background traffic B_1 and B_2 – at nodes 1 and 2, respectively – before it reaches the destination D. The FIFO service discipline and infinite queue capacity are assumed at nodes

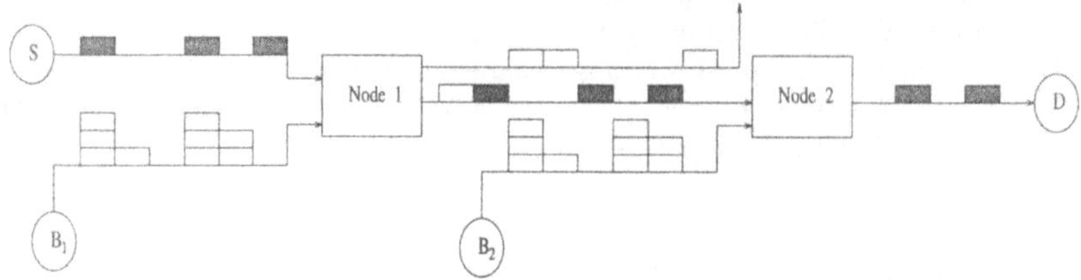

Figure 4: A system of two consecutive nodes.

1 and 2. A portion of the background traffic at node 1 is forwarded to node 2 while the remaining is diverted. The cumulative arrival process to node 2 contains tagged cells, fresh background cells from B_2 and carried-on traffic coming from B_1.

By definition, carried-on traffic is the background traffic that is coming from the previous node and is forwarded to the next node, together with the tagged traffic. As a consequence, carried-on traffic is not present in the first node and the phase and level processes for $\{W_j^1\}_j$ will be described as follows.

$$\{\Phi_j^1\}_j \equiv \{T_j^1, B_j^1\}_j \tag{15}$$

$$Q_j^1 = [Q_{j-1}^1 - 1]^+ + \hat{B}_j^1(B_j^1) + 1_{\{T_j^1=1\}} \tag{16}$$

Under the assumption that $\{A_j^1\}_j$ is an independent process and $\{B_j^1\}_j$ is a Markov process, it is easy to establish that $\{W_j^1\}_j$ is a Markov process. The block matrices in the transition matrix of $\{W_j^1\}_j$ (M/G/1 structure) are easily determined.

The phase process associated with node 2 is given by (10) and the evolution of the level process by (8). $\{B_j^2\}_j$ and $\{F_j^2\}_j$ are assumed to be Markov processes with given parameters (section 2). $\{J_j^1\}_j$ is approximated by a 2-state first-order Markov chain by utilizing the solution of $\{W_j^1\}_j$. Process $\{X_j^1(J_j^1)\}_j$ as well as the background traffic at the output of node 1 are calculated as indicated section 2. The results presented below have been derived for the system of 2 nodes under the following traffic parameters.

The tagged cell interarrival time at node 1 is constant and equal to 8; that is, $A_j^1 = 8$. The background traffic at node 1 has parameters $p_b^1(0,0) = p_b^1(1,1) = .99$, (and thus, $\pi_b^1(0) = \pi_b^1(1) = .5$); $f_b^1(k,l) = b(k,6;p_l)$, where $b(k,6;p_l)$ is the binomial probability with parameters 6 and p_l, $l = 0,1$; $\lambda_b^1 = (\pi_b^1(0)p_0 + \pi_b^1(1)p_1)6 = .7$; $\frac{p_1}{p_0} = 4$ (burstiness measure). Notice that the background traffic is very bursty. On the average, it stays in state 1 for about 100 slots delivering background cells at a rate greater than 1. The objective in selecting such burstiness of the background traffic at node 1 is to investigate how congestion in node 1 affects the performance figures at node 2, and determine the effectiveness of the QAI in characterizing this environment.

The traffic at node 2 consists of the tagged traffic ($\lambda^1 = \lambda^2 = .125$), the fresh background traffic and the carried-on traffic from node 1. The model considered for the background traffic at node 2 is identical to that at node 1 with the following differences in the parameters: $\frac{p_1}{p_0} = 1$ and $\lambda_b^2 = .1$. That is, this process is uncorrelated. The parameters of the splitting process

$\{F_j^2\}_j$ – determining the carried-on traffic – are: $\pi_f^2(0) = \pi_f^2(1) = .5$, $p_f^2(0,0) = p_f^2(1,1) = p_f$, $p_f \in \{.50, .90, .95, .99\}$. That is, half of the background traffic from node 1 becomes carried-on traffic and competes with the tagged traffic and the background traffic of node 2 for the same resources; p_f determines the burstiness of the carried-on traffic.

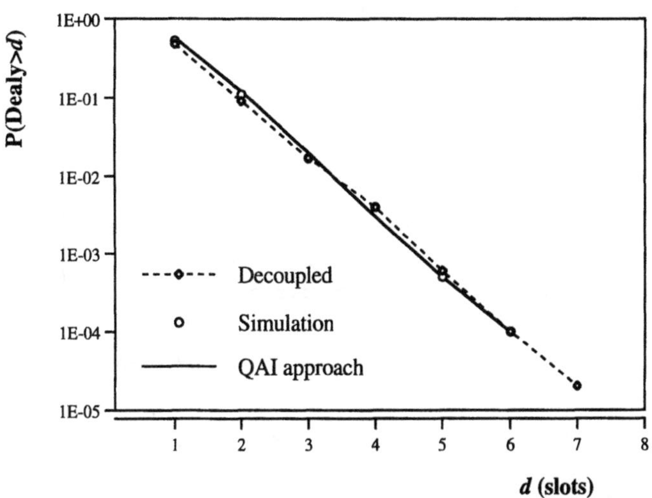

Figure 5: Tail of node 2 tagged cell delay distribution for $p_f = .50$.

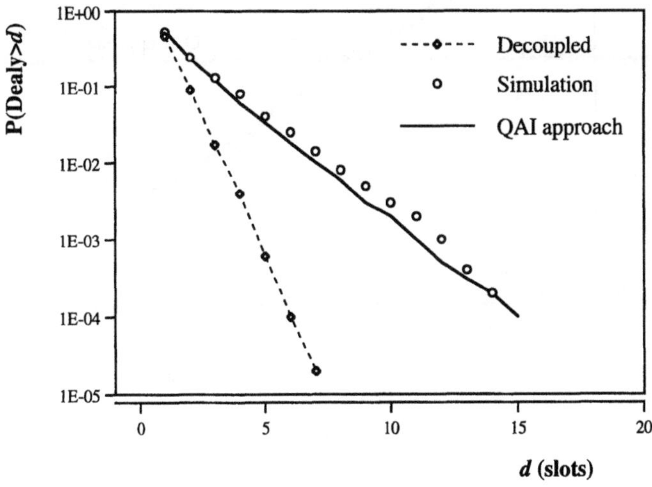

Figure 6: Tail of node 2 tagged cell delay distribution for $p_f = .90$.

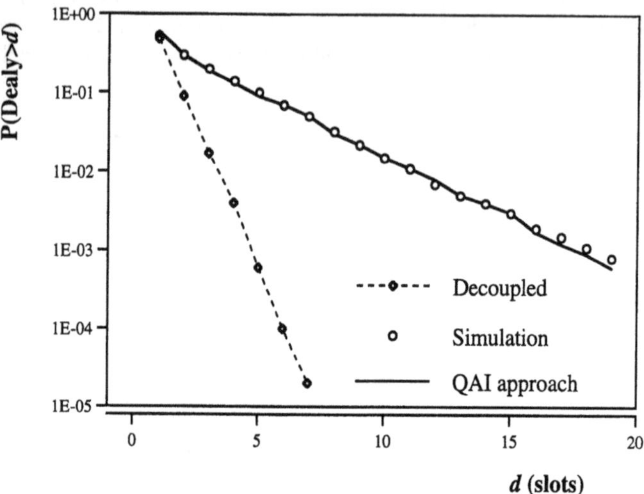

Figure 7: Tail of node 2 tagged cell delay distribution for $p_f = .95$.

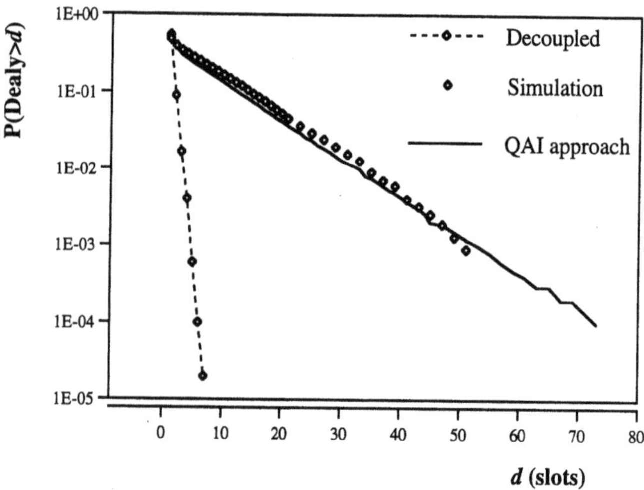

Figure 8: Tail of node 2 tagged cell delay distribution for $p_f = .99$.

Figures 5-8 present the tail distribution of the tagged cell delay at node 2 for different values of the burstiness p_f. The results under "QAI approach" are derived by applying the approach developed in this work. The results under "Decoupled" refer to simulation results of the system where the carried-on traffic has been replaced by a Bernoulli fresh background traffic at node 2 of the same rate (.35). The results under "Simulation" refer to the simulation of the real system with the carried-on traffic present. Notice that the simulative results have been obtained with a 90% confidence level; the width of the confidence interval is always lower than 2% and for this reason it never appears in the figures.

When $p_f = .50$, the carried-on traffic is generated through an uncorrelated splitting process of the background traffic coming from node 1. This traffic should be very similar to the additional fresh background traffic considered in the "Decoupled" case and, thus, the "Simulation" and "Decoupled" results should be very close. This can be observed in Figure 5 where $p_f = .50$. In addition, the "QAI approach" results are very close to the other ones, indicating that the approximations involved in this approach do not compromise its accuracy.

As expected, the queueing activity at node 2 increases as p_f increases. This is observed in Figures 6-8 for $p_f = .90$, .95, .99, respectively. Notice the increasing inaccuracy (as p_f increases) of the "Decoupled" results and the consistent accuracy of the "QAI approach" results. These results suggest that :

(a) The modulation of the output processes (tagged and carried-on traffic) by the QAI results in an accurate evaluation of the queueing behavior in the next node, as determined by the tagged cell delay tail distribution.

(b) Destination correlation – indicated here by a large value of p_f – can have a significant impact on the queueing behavior. Ignoring such correlation may result in very inaccurate performance evaluation.

(c) The QAI approach presented here seems to be capable of capturing the correlation among the queueing processes associated with consecutive nodes (spatial correlation). When node 1 is temporarily overloaded by what will become highly correlated carried on traffic – when strong spatial correlation is present due to a high vaue of p_f –, the increased queueing activity at node 1 induces increased queueing activity at node 2, as the results in Figures 5-8 indicate.

Since a (time) correlated background traffic is expected to induce increased queueing activity at node 1 leading to a sustainable value of the QAI equal to 1, it is expected that the Markov approximation to the QAI process will also exhibit similar level of correlation. Indeed, it was found that $P\{QAI_k^1 = 1/QAI_{k-1}^1 = 1\} = .96$ and $P\{QAI_k^1 = 0/QAI_{k-1}^1 = 0\} = .94$ under correlated background traffic at node 1 and $p_f = .99$ (Figure 8). That is, a temporal correlation in the input process to node 1 seems to be well captured by the temporal correlation of the QAI process. Thus, the resulting increased temporal correlation in the queueing process due to the temporal correlation in the arrival process seems to be well captured by the temporal correlation in the QAI process. This may be important in accurately evaluating the temporal correlation in the end-to-end tagged cell delay process which may be useful in identifying potential starvation problems when a large number of consecutive cells are delayed excessively.

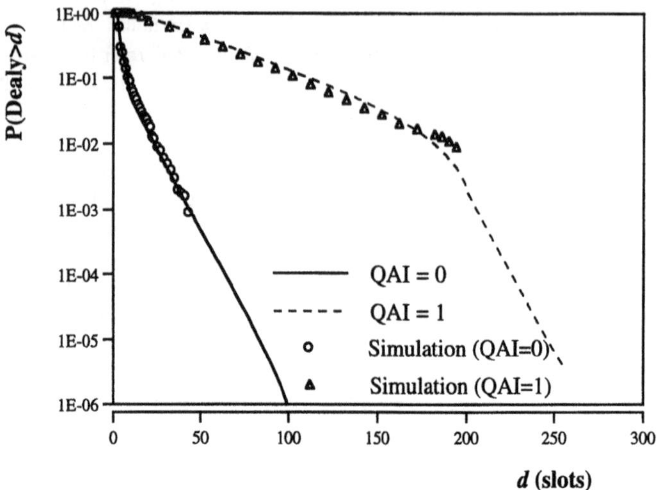

Figure 9: Tail of the end-to-end tagged cell delay distribution given the value of QAI in the first node.

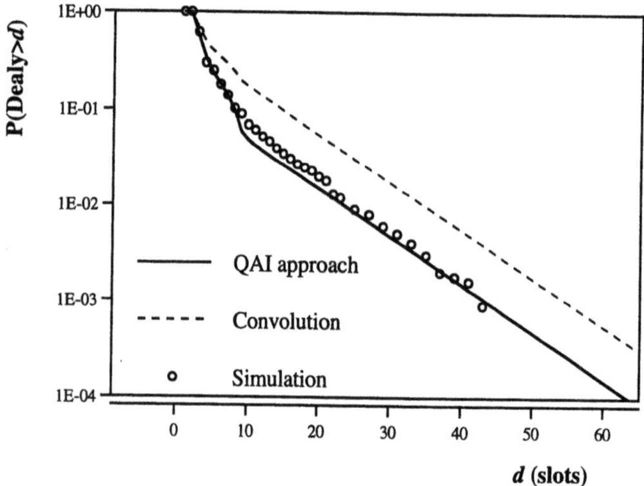

Figure 10: Tail of the end-to-end tagged cell delay distribution under independent delays in the nodes ("convolution") and considering the delay in node 2 as shaped by the QAI in node 1 ("QAI approach").

Figure 9 shows the end-to-end tagged cell delay tail distribution given that QAI (associated with node 1) is equal to 0 or 1. It can be observed that the QAI affects significantly the performance measure. Given that process QAI remains unchanged for a long time (associated change probabilities are .06 or .04), it may be concluded that a large number of consecutive tagged cells may all experience either small or large delays (temporal correlation).

Finally, Fig. 10 presents the end-to-end tagged cell delay tail distribution by assuming independent delays at nodes 1 and 2 under "Convolution" and considering the delay at node 2 given the QAI at node 1 under the "QAI approach". The "Convolution" involves the distributions of $\{D^1_{ok}/QAI^1_k = 0\}$ (derived from the Markov chain $\{W^1_j\}_j$) and $\{D^2_{ok}\}$ (derived from the Markov chain $\{W^2_j\}_j$). The "QAI approach" involves the conditional distributions associated with $\{D^1_{ok}/QAI^1_k = 0\}$ and $\{D^2_{ok}/QAI^1_k = 0\}$. Notice that the results under the "QAI approach" represent lighter queueing conditions at node 2, suggesting that the QAI process seems to effectively modulate the output processes from node 1 and induce lower queueing activity at node 2 when QAI=0 at node 1, as expected. Again, given that process QAI remains unchanged for a long time (associated change probabilities are .06 or .04), it may be concluded that a large number of consecutive tagged cells may all experience either small or large delays (temporal correlation).

Additional results to further substantiate the points made above are under derivation.

5 REFERENCES

Onvural, R. (1993) *Asynchronous Transfer Mode: Performance Issues.* Artec House, Boston.

Stavrakakis, I. (1991a) Queueing Behavior of Two Interconnected Buffers of a Packet Network with Application to the Evaluation of Packet Routeing Policies. *International Journal of Digital and Analog Communication Systems*, John Wiley and Sons, Vol. 4.

Stavrakakis, I. (1991b) Efficient Modeling of Merging and Splitting Processes in Large Networking Structures. *IEEE Journal on Selected Areas in Communications*, Vol.9 No. 4.

Lau, W.-C. and Li, S.-Q. (1993) Traffic Analysis in Large-Scale High Speed Integrated Networks: Validation of Nodal Decomposition Approach. *in IEEE Infocom'93*, San Francisco, CA.

Kurose, J. (1993) Open Issues and Challenges in Providing Quality of Service Guarantees in High Speed Networks. *Computer Communications Review*, Vol. 23, No. 1.

Nagarajan, R. and Kurose, J. (1992) On Defining, Computing and Guaranteeing Quality-of-Service in High-Speed Networks. *in IEEE Infocom'92*, Florence, Italy.

Guillemin, F. and Monin, W. (1992) Management of Cell Delay Variation in ATM Networks. *in IEEE Globecom'92*, Orlando, FL.

Bisdikian, C. and Matragi, W. and Sohraby, K. (1993) A Study of the Jitter in ATM Multiplexers. *in Fifth International Conference on Data Communication Systems and their Performance*, Raleigh, NC.

Matragi, w. and Bisdikian, C. and Sohraby, K. (1994a) Jitter Calculus in ATM Networks: Single Node Case. ", *in IEEE Infocom'94*, Toronto, Canada.

Matragi, W. and Bisdikian, C. and Sohraby, K. (1994b) Jitter Calculus in ATM Networks: Multiple Node Case. *in IEEE Infocom'94*, Toronto, Canada.

Roberts, J. and Guillemin, F. (1992) Jitter in ATM Networks and its Impact on Peak Rate Enforcement. *Performance Evaluation*, Vol. 16, No. 1-3.

Landry, R. and Stavrakakis, I. (1994) A Queueing Study of Peak-Rate Enforcement for Jitter Reduction in ATM Networks", *IEEE GLOBECOM'94*, San Francisco.

Neuts, M.f. (1989) *Structured Stochastic Matrices of the M/G/1 Type and their Applications*, Marcel Dekker, New York.

6 BIOGRAPHY

Marco Conti received the Laurea degree in Computer Science from the University of Pisa, Pisa, Italy, in 1987. In 1987 he joined the Networks and Distributed Systems department of CNUCE, an institute of CNR (the Italian National Research Council). He has worked on modeling and performance evaluation of Metropolitan Area Network MAC protocols. His current research interest include ATM, Wireless Networks, design, modeling and performance evaluation of computer communication systems.

Enrico Gregori graduated in Electronic Engineering from the University of Pisa in 1980. He joined CNUCE, an institute of the Italian National Research Council (CNR), in 1981. In 1986 he spent one year as a visiting at the IBM Research Laboratory in Zurich. He has worked in several projects on network architectures and protocols. His current research interests include high-speed network design and performance evaluation. He is author or co-author of numerous publications in this field. From 1989 to 1991, he gave a course on Metropolitan Area Networks and their performance evaluation in the Computer Science department of the University of Pisa. He is now lecturing on computer network architectures, protocols and their performance evaluation in the Faculty of Engineering of the University of Siena, Italy.

Ioannis Stavrakakis received the Diploma in Electrical Engineering from the Aristotelian University of Thessaloniki, Greece, 1983, and the Ph.D. degree in Electrical Engineering from the University of Virginia, 1988. In 1998, he joined the faculty of Computer Science and Electrical Engineering of the University of Vermont as an assistant and then associate professor. Since 1994, he has been an associate professor of Electrical and Computer Engineering of the Northeastern University, Boston. His research has been focused on the design and performance evaluation of communication networks. He is an active member of the IEEE Communications Society (Computer Communications) and has been involved in the technical program of conferences sponsored by the IEEE, ACM and IFIP societies.

PART FIVE

Source Modelling

14

On the prediction of the stochastic behavior of time series by use of Neural Networks - performance analysis and results

M. D. Eberspaecher
University of Stuttgart, Institute of Communications Switching and Data Technics
Seidenstrasse 36, 70174 Stuttgart, Germany, Phone: +49 711 1212482,
Fax: +49 711 1212477, E-mail: eberspaecher@ind.uni-stuttgart.de

Abstract
In time series theory, the prediction of future values is a widely discussed subject. There are manyfold methods to derive models from data. One of the main objectives is to obtain the model parameters. Some proposals use self adapting techniques like Neural Networks to estimate the model parameters. Most of these approaches predict one future value of a time series. Some simulation tasks require models for traffic sources that are closely related to time series prediction though there exist different requirements. One of them is that a simulated traffic source should show the same stochastic behavior as a reference source. In this paper a procedure is presented that automatically adapts to a given reference source in the sense described above.

Keywords
Time series, prediction, Neural Networks, source modelling

1 INTRODUCTION

General

The analysis of time series is an extensively developed area of mathematics. There are many approaches to model dynamic systems. They can be classified as follows: linear, linear stochastic, nonlinear and nonlinear stochastic dynamic systems. Nonlinear systems may show chaotic behavior, depending on system parameters. There is a sliding transition from systems with a random disturbance to systems with probabilistic transitions between states. In addition seasonal effects and trends may be observed.

Depending on the underlying system an appropriate model has to be found. ARMA and ARIMA models (Auto-Regressive-Moving-Average and Auto-Regressive-Integrated-Moving-Average, respectively) represent simple methods to model linear and linear stochastic systems in a suitable way. ARIMA models often are used when trends have to be considered. For other sequences, more sophisticated methods are required (Janacek, 1993, Harvey, 1993, Hamilton, 1994).

Nonlinear models are necessary to adequately model nonlinear and nonlinear stochastic systems. Here chaotic behavior of time series might occur and has to be identified since chaotic series must be treated differently than stochastic time series, see Scargle (1992).

However, many problems need their own specific solutions (Brillinger, 1992, Weigend, 1993). Most of these approaches share the following characteristics: firstly, as much information as possible about the characteristics of the underlying data is collected and, secondly, a model that covers the essential features is deduced.

Most procedures that deal with forecasting future values predict one (the next) value based on a set of N past values. The selection of N is not trivial since it determines the prediction quality and depends on the observed dynamic system. Some remarks how to determine useful values of N can be found in Scargle (1992).

All methods mentioned above share one disadvantage: they are inflexible in terms of changing stochastic behavior of the underlying data. These changes have to be taken into account by the model and increase its complexity very much.

Time series models that are able to deal with changing parameters should be based on an architecture that is inherently able to automatically adapt to these changes. This architecture could be based on a Neural Network. Until today there are not many approaches that use a Neural Network (NN) architecture. Their advantage primarily consists in their ability to learn a given behavior without the exact knowledge of the underlying system and without difficult analysis needed for modelling. One disadvantage is that no detailed and understandable model of the underlying system is built.

These models are most often used for prediction: Chakraborty (1992) presents a neural network to multivariate time series analysis, Deppisch (1994), Lowe (1994) and Hudson (1994) use neural network algorithms to predict chaotic time series. Mozer (1993) presents a general taxonomy of neural net architectures for processing time-varying patterns. In Tang (1994) a neural net approach is compared to the Box-Jenkins methodology. Wan (1993) presents a somewhat different method that uses a neural net with internal delay lines, i. e. a neural net with inherent memory.

Source modelling is an area where the generation of future values is frequently used.

Source modelling

In source modelling the generation of deterministic new values from given data such as forecasting exchange rates is often of no particular interest. In contrast to that, a random data sequence is generated by a stochastic model. Every new value is randomly chosen from a given distribution depending on the state of the model. State changes are most often defined by a state transition probability function that may be nonlinear and depend on past states. Source modelling is frequently used for traffic generation in the simulation of communication networks, in the simulation of manufacturing plants or in measurement technology. Since this paper concentrates on communications all the examples will relate to this area.

Multiple traffic sources built from the same model must be statistically independent from each other. A simple reproduction of measured data from a file (play back) is not sufficient. A shifted play back from a file where one traffic sequence starts at one point and another traffic sequence starts at another point of the file is not sufficient, too, because of the strong correlations (especially when the file is short or when many sources are needed). Even in the case of very long files problems might occur in large simulations. This leads to the conclusion that reproduction from files is inflexible.

Like in conventional time series analysis ARMA models can be used for source modelling but they don't fit very well because of nonlinearities in almost all systems. There are some approaches employing Neural Networks that avoid this problem but most of them don't fulfill all requirements for source modelling.

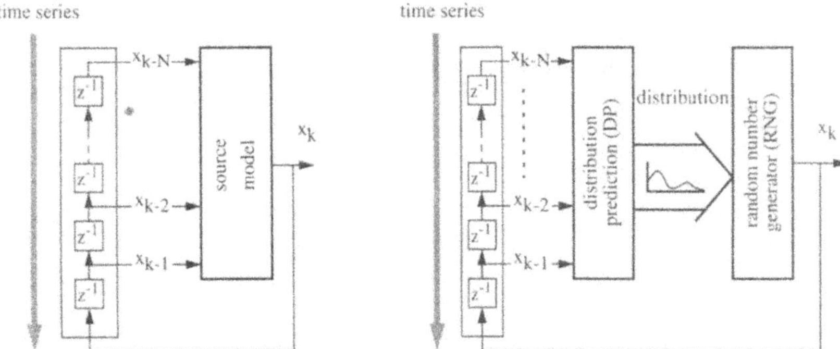

Figure 1 Principle of source model. **Figure 2** Extended principle of source model.

In Tarraf (1993, 1994) NNs are used for modelling an ATM cell stream (Asynchronous Transfer Mode) without adding a random component. This kind of modelling is sufficiently exact, but not very useful for simulation.

How can traffic models be obtained? This is a simple task when only distributions are of interest and no correlations. Distributions might stem from measurement. The task can become very difficult when correlations must be taken into account. In this case an ARMA process might be profitably used, but how to obtain the model parameters? Another problem is that ARMA models are normally driven by white noise with the consequence that negative values are possible in any case, even if they are not allowed. Gruenenfelder (1991) shows an example for source modelling using an ARMA model. Here the parameters are partly estimated from the measured data and partly calculated via the frequency domain.

Most of the cited approaches (Gruenenfelder (1991) is an exception) lead to a deterministic behavior of the model in the way that they calculate a new value based on some observations, without adding any noise. Neural Network based models that adapt to a given time series during a learning phase therefore do not learn the stochastic behavior but the conditional expectation of future values.

In the following parts of this paper a new method is presented that adapts a source model to many different random processes or time series. It uses Neural Networks to automatically learn the stochastic behavior of the underlying data and therefore avoids some of the problems of other models. The adaptation process is fully automated and only a few topological model parameters have to be estimated from the data.

Automatic source identification

The advantage of automatic source identification is a gain in productivity and saving of money since computational power is much cheaper than man power today. The disadvantage is the reduced possibility of interpretation of the generated source model. Only very few topological parameters have to be estimated from the data. All random processes that are weakly stationary and that have some seasonal effects can be modelled.

The objective of the identification procedure is to model the distribution and autocorrelation of a time series as good as possible. Some simple tools for evaluation are presented later in this paper, see section 3.

Figure 1 shows one principle of this approach. Suppose that the content of the box named "source model" is already adapted to the data. The scalar value x_k is the output value of the modelled traffic source. N time delayed output values form the input vector $\boldsymbol{i}_k$

$$\boldsymbol{i}_k = (x_{k-1}, x_{k-2}, \ldots, x_{k-N}) . \qquad (1)$$

The elements of $\boldsymbol{i}_k$ form the embedding space coordinates with dimension N. To determine a suitable value of N a simple approach is used. The sample autocorrelation function of the observed time series is calculated and evaluated. In the case of periodic signals N is chosen to be greater than the period length. In the case of vanishing autocorrelation N is chosen to be equal the lag where the absolute value of the autocorrelation falls below $1.96/\sqrt{T}$, where T is the sample size (see Figure 8,a). It is assumed that autocorrelation values less than this value are based on white noise, see Harvey (1993).

The shift register forms a memory of the past. This is the only memory of the model. The vector $\boldsymbol{i}_k$ is fed into the "source model," the output of which is the newly generated value. This output is fed back to the shift register, delayed by one step. So the number generating loop is closed.

Figure 2 provides a more detailed view. The source model is now divided into two parts, the distribution prediction (DP) and the random number generation (RNG). The RNG simply draws a new number according to the distribution density at its input. The general inverse-transform method is used as RNG algorithm, see Law (1991). For every input vector the DP block computes (predicts) the distribution of the following output value.

This model can be described by some equations. The internal state of the DP is a function of the input vector $\boldsymbol{i}_k$:

$$S_k = h(\boldsymbol{i}_k) = h(x_{k-1}, x_{k-2}, \ldots, x_{k-N}) . \tag{2}$$

The density function at the output can be expressed as a function of the internal state and therefore depends on the input vector $\boldsymbol{i}_k$, too:

$$G_k = g(S_k) = g(h(x_{k-1}, x_{k-2}, \ldots, x_{k-N})) . \tag{3}$$

The output value of the model, x_k, is chosen according to the density G_k.

In Section 2 the components of the distribution prediction including the learning process are described. In section 3 some measures for performance evaluation are introduced and in section 4 some examples of time series that are learned and predicted by applying the new source model are presented.

2 DISTRIBUTION PREDICTION

At first the scenario shown in Figure 3 is examined. N values of a time series preceding the current value x_k are fed to a black box named "distribution prediction." Inside this box the prediction of the distribution of the actual value is computed. For each vector $\boldsymbol{i}_k$ at the input there is a distinct distribution at the output.

Before continuing the theoretical model description the correspondence between distribution and correlation has to be clarified. The occurrences of vectors $\boldsymbol{i}_k$ obey an N-dimensional distribution $\hat{g}$ in the embedding space $\Re^N$. From $\hat{g}$ the autocorrelation of the series x_k can be derived, see Papoulis (1984). Therefore it is sufficient for the model to learn a good approximation of $\hat{g}$ to model the autocorrelation of the underlying time series.

The actual value x_k and the input vector $\boldsymbol{i}_k$ are now treated as one (N+1)-dimensional vector $\boldsymbol{x}_k = (x_k, \boldsymbol{i}_k)$ with an (N+1)-dimensional distribution function

$$F(x_k, x_{k-1}, \ldots, x_{k-N}) = P\{X_k \le x_k, \ldots, X_{k-N} \le x_{k-N}\} \tag{4}$$

and density function

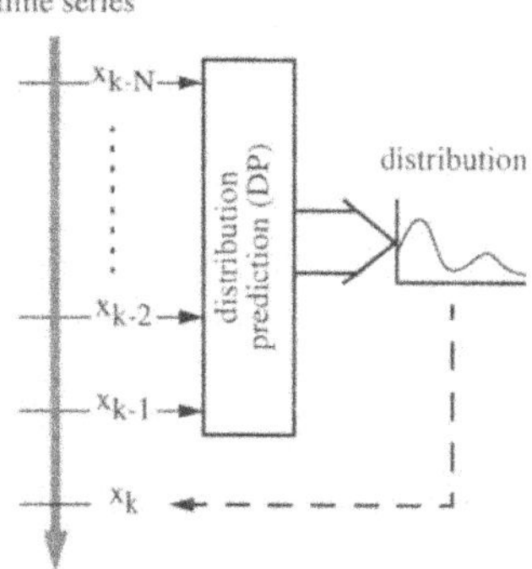

Figure 3 Principle of distribution prediction.

$$f(x_k, x_{k-1}, \ldots, x_{k-N}) = \frac{\partial^N}{\partial x_k \ldots \partial x_{k-N}} F(x_k, \ldots, x_{k-N}) \,. \tag{5}$$

Without loss of generality k is now set to 0 for simplicity. Then the density function becomes

$$f(x_0, x_{-1}, \ldots, x_{-N}) \,. \tag{6}$$

For distribution prediction the conditional distribution of x_0 for distinct values of the input vector $\boldsymbol{i}_k$ is needed. This is done by local approximations of parts of this distribution. For the local approximation the *N*-dimensional embedding space that belongs to the input vector is quantized into *M* discrete vectors

$$\hat{\boldsymbol{p}}_i = \left(\hat{x}^i_{-1}, \ldots, \hat{x}^i_{-N}\right), \qquad i = 1, \ldots, M. \tag{7}$$

This task is carried out by a vector quantizer (VQ).

Each vector $\hat{\boldsymbol{p}}_i$ points to the center of a region I_i of $\Re^N$. In these regions the density function of x_0, (6), is approximated by the function $\hat{f}_i(x_0)$. The union of all regions I_i forms the *N*-dimensional space $\Re^N$.

The approximation in region I_i is defined by the mean value of density (6) in this region:

$$\hat{f}_i(x_0) = \frac{1}{Vol(I_i)} \cdot \int_{I_i} \ldots \int f(x_0, t_1, \ldots, t_N)\, dt_1 \ldots dt_N \tag{8}$$

where

$$Vol(I_i) = \int_{I_i} \ldots \int dt_1 \ldots dt_N. \tag{9}$$

This leads to the following error inside region I_i (the error measure is the squared difference between $f(\ldots)$ and $\hat{f}_i(\ldots)$):

$$E_i = \int_{-\infty}^{\infty} \left[\int_{I_i} \ldots \int f^2 (x_0, t_1, \ldots, t_N)\, dt_1 \ldots dt_N \right] dx_0 - Vol(I_i) \cdot \int_{-\infty}^{\infty} \hat{f}_i^2 (x_0)\, dx_0 \tag{10}$$

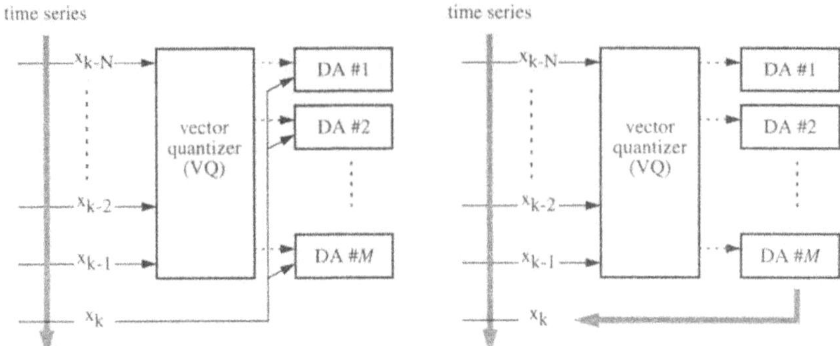

Figure 4 Learning of DP.

Figure 5 Prediction with VQ and DA's.

To compute the total error that results from quantization some features of the vector quantizer need to be known. This VQ adapts to the x_k in a way that all regions I_i are equally probable for the given time series. The algorithm used is partly taken from literature and is described in detail in Appendix A. The resulting total error is:

$$E_G = \sum_{i=1}^{M} E_i \tag{11}$$

E_G decreases for an increasing number M of regions.

The approximated density $\hat{f}_i(x_0)$ is obtained from the given time series, too. A new Neural Network algorithm was investigated that is able to form a distribution when a sequence of data is offered to it. This algorithm is presented in the next subsection.

Figures 4 and 5 show the relationship between vector quantizer and distribution approximation (DA). Whenever an input vector falls into a region of the VQ, the corresponding DA is chosen for learning or prediction. In other words, the VQ is responsible to detect all correlations, the DA's are responsible for representing distributions.

Figure 4 shows the learning case. The actual value x_k is needed here as input for the actual DA to adapt the distribution. In the case of prediction (Figure 5), when learning is completed, one DA is chosen by the VQ for prediction of x_k.

Distribution adaptation

Figure 6 shows how the distribution adaptation module works. A time series that obeys to a distinct distribution is given. Minimum and maximum values are not known a priori. The algorithm inside the black box shall form an approximation of the distribution of the given data. The values of the time series are fed to the input of the adaptation module. The density function at the output is represented by a piecewise constant function, see Figure 7.

Usually, a distribution is measured by dividing the whole interesting region into small regions of equal width. The local density inside these regions is calculated from the frequency. The approach presented here has two advantages compared to normal distribution measurement: firstly, there is no a priori knowledge needed concerning the minimum and maximum values of the time series, the algorithm adapts automatically to them. Secondly, the regions are not of equal width and are adapted in order to obtain an optimal split. To achieve a high approximation quality the density function is approximated finer where it is high and coarser where it is low. This is achieved by equal probable regions. The region probability is approxi-

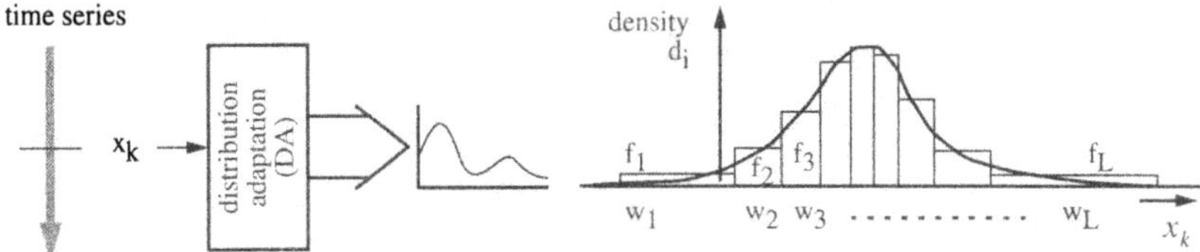

Figure 6 Principle of distribution adaptation. **Figure 7** Example of approximated distribution.

mated by the frequency f_i of each region. Figure 7 shows a sample approximation function with L regions. The values w_i denote the width of the regions and d_i the densities, respectively. The correspondence between width, density and region probability is $f_i = d_i \cdot w_i$ for i in $1..L$.

The algorithm is implemented as a non supervised Neural Network with L computing elements. There are no other inputs than the events from the time series for learning. The structure of the NN and the learning rule are described in more detail in Appendix B. In the recall phase, which is needed for distribution prediction, the output of the trained NN is used as the approximation of a density function.

3 PERFORMANCE ANALYSIS

To test the performance of derived models against the original data (i) the distribution and correlation diagrams can be qualitatively compared or (ii) some quantitative tests can be applied.

In this section some statistical performance measures are introduced that are used for quantitative tests.

3.1 Distribution test

To test the distribution the Kolmogorow-Smirnow test is used. This test compares the empirical distribution functions of the underlying time series and the time series generated by the source model. The hypothesis that the source model models the distribution according to a given significance level is accepted or rejected based on a measure that involves the maximal difference between the empirical distribution functions.

In all examples in this paper a significance level of 0.05 is used which leads to a critical value of $\sqrt{T} \cdot 1.923$ for acceptance of the hypothesis, where T is the sample size of both time series. For the used sample size of 10000 the critical value becomes 192.

3.2 Autocorrelation test

To test the autocorrelation a pragmatic approach is used: the mean square error (MSE) between empirical autocorrelation functions of the underlying time series and the time series generated by the source model. Here once again a distinction has to be made between periodic systems and pure stochastic systems since in case of periodic systems the autocorrelation function does not vanish for higher lags.

Vanishing autocorrelation

The maximal lag for calculating MSE is determined as the lag where the absolute value of the autocorrelation of the underlying time series falls below $1.96/\sqrt{T}$, where T is the sample size,

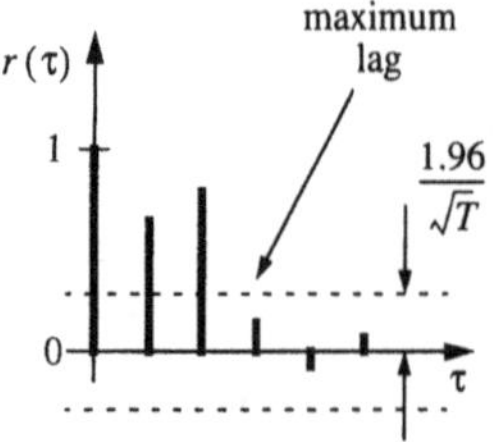

a.) Determination of maximum lag.

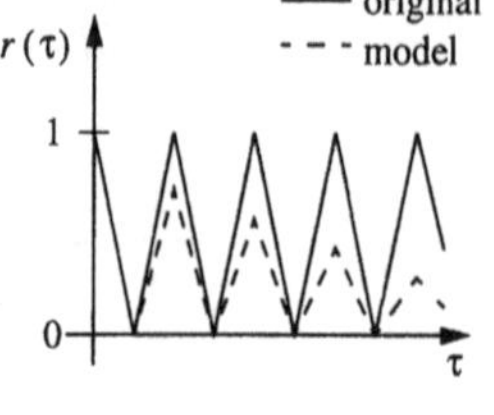

b.) Original and model autocorrelation.

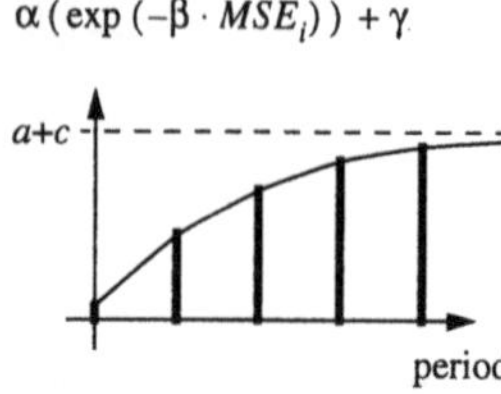

c.) Exponential approximation.

Figure 8 Tests for correlation.

see Figure 8,a. It is assumed that autocorrelation values less than this value are based on white noise, see Harvey (1993).

In the examples in this paper a threshold of 0.01 for MSE was used for acceptance of autocorrelation.

Periodic time series

In case of periodicity of the underlying time series the autocorrelation does not vanish. On the other hand the autocorrelation function of the generated series decreases approximately exponentially, see Figure 8,b.

That is a feature of the source model. To handle this the MSE is calculated for a number of periods and an exponential function fitted to the resulting error series according to the function

$$\alpha\left(\exp\left(-\beta \cdot MSE_i\right)\right) + \gamma, \tag{12}$$

see Figure 8,c, where MSE_i is the MSE of period i and α, β, γ are parameters to be fitted. The only interesting parameter is β which is used as criterion.

In the examples in this paper a threshold of 0.1 for β was used for acceptance of autocorrelation.

4 SOURCE MODELLING - EXAMPLES

In this section some examples of modelling traffic sources are presented. The comparison between the original time series (reference) and the modelled time series is done by calculating the correlogram for both of them as well as the statistical tests described above.

Markov modulated poisson process - MMPP

The first example is a MMPP process (Markov Modulated Poisson Process) with two states. The MMPP is a frequently used traffic model in telecommunications and represents a source with two activity states. Figure 9 shows a state-transition diagram. The process switches

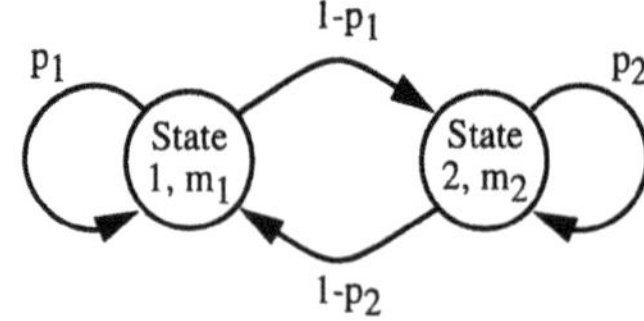

Figure 9 MMPP with 2 states.

between the two states with probabilities $1-p_1$ and $1-p_2$, respectively. The expectation of event values in the states are m_1 and m_2. The parameters for the example are $p_1 = p_2 = 0.8$, $m_1 = 0.1$ and $m_2 = 15$.

For distribution prediction N=5 VQ inputs and M=2 VQ units were used, thus having 2 distinct distributions approximated. L=100 segments were used for the approximation of each distribution. Figure 10 shows the resulting distribution functions of the approximation. Note that these distribution functions are not the same than those of the two states of the underlying system since they include the probabilities of state changes, too.

The correlograms in Figure 11 differ to a slight extent, because only two distribution approximation units have been used in this example. See Table 1 for results.

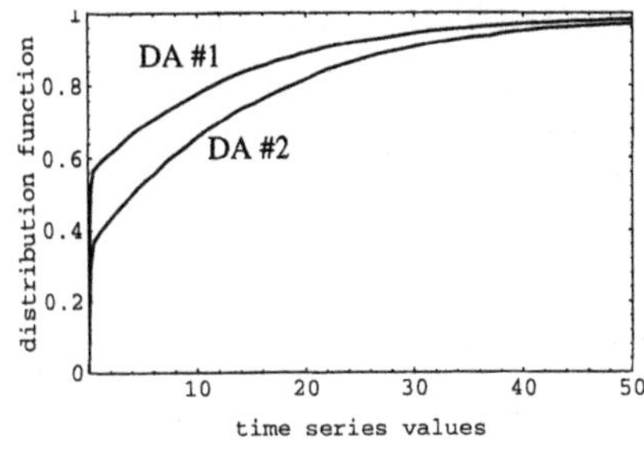

Figure 10 MMPP: Distribution functions.

Figure 11 MMPP: Correlogram.

Second order moving average process - MA(2)

The reference data for this test was produced from $y_k = \varepsilon_k + \Theta \cdot \varepsilon_{k-2}$ with $\Theta = 0.2$ and ε being white noise with mean 0 and variance 1.

See Table 1 for model parameter and results. In Figure 12 the good correspondence of the autocorrelations can be seen.

MPEG coded video frames

This example is a real world case. A MPEG coded video sequence from the movie "Star Wars" is used. The sequence consists of the amount of data per video frame after compression.

Figure 13 shows the sequence generated by the MPEG scheme in principle. The large frames are so-called I-frames and comprise a whole picture. The medium-sized frames are P-frames,

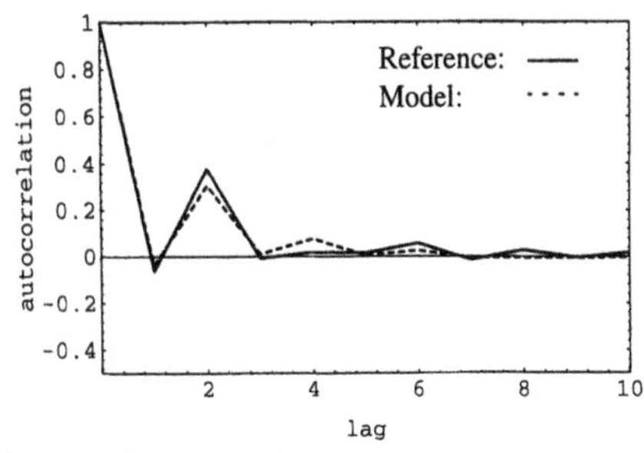

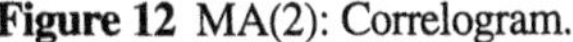

Figure 12 MA(2): Correlogram.

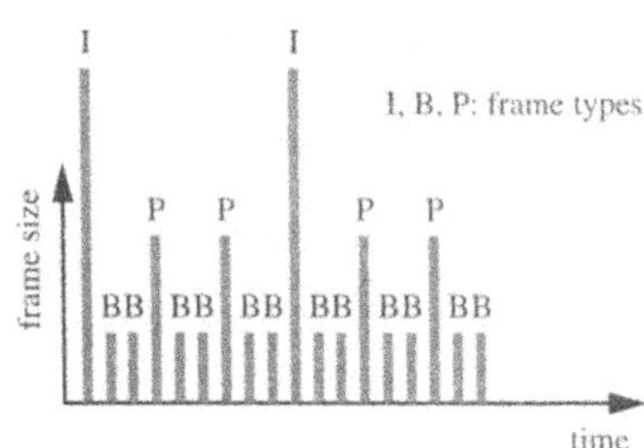

Figure 13 Sequence of MPEG frames.

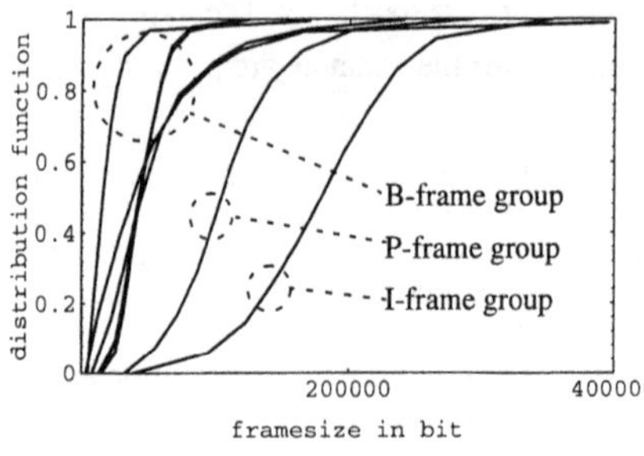

Figure 14 MPEG: Distribution functions.

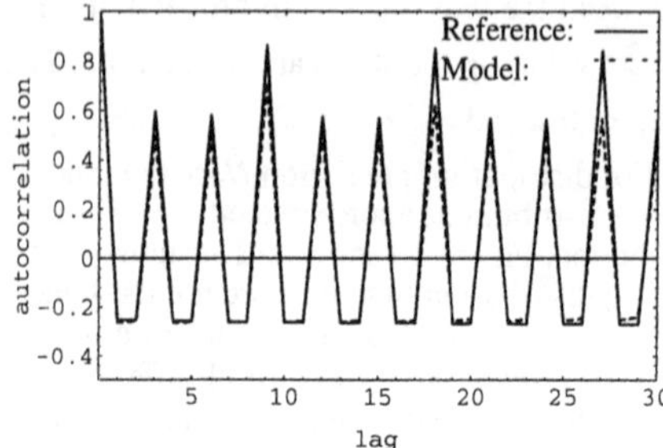

Figure 15 MPEG: Correlation between frames.

the small frames B-frames. The sequence I-B-B-P-B-B-P-B-B- is repeated cyclically and is defined by the MPEG parameters (more detailed information can be found in Le Gall (1991)). The size of the I-, B-, and P-frames is distributed according to the underlying video scenes. Due to this cyclical behavior the correlogram of MPEG-coded video data has a characteristic form as shown in Figure 15.

Figure 14 shows the resulting distribution approximations that are learned by the VQ and the DA's during the adaptation process. It can be seen that there are different groups of distributions. Each group represents a different frame type (I, B or P) and the occurrence of the different distribution types obeys the same rule than the I-B-P-sequence above.

In Figure 15 it can be seen that the model fits the reference quite well. Model parameters and further results are contained in Table 1.

The long term correlation is poor in this example but it could be further improved by increasing the number N of model inputs.

Table 1 Examples

	N	M	L	*distribution test (value should be < 192)*	*correlation test*
MMPP	5	2	100	185	0.002 (MSE)
MA(2)	3	25	10	189	0.0026 (MSE)
video frames	9	15	10	165	0.04 (periodic)

5 FURTHER DEVELOPMENT

This work is still in a preliminary state and extensions are under development. Future extensions will include:

- Control inputs for source models. This allows hierarchical models for different time scales. In case of video data a higher model may be responsible to model video scenes whereas a lower model is responsible to model the frame sizes.
- Multivariate time series.

This should be easily done by extending the input vector with control lines and values from other time series. During the adaptation procedure the distribution approximation can be adapted independently for every input stream.

6 SUMMARY

In this paper a new algorithm is presented that identifies arbitrary time series that may contain seasonalities. It consists of a vector quantizer that reduces the complexity of the input data and a special Neural Network type that is able to learn distributions.

The model parameters for both the vector quantizer and the Neural Network are automatically derived by the learning process if an adequate topology is chosen.

A performance analysis is presented and some examples demonstrate the usability of the method.

7 REFERENCES

Bagchi, A. (1993) *Optimal Control of Stochastic Systems*. Prentice Hall.

Brillinger, D. (1992) *New Directions in Time Series Analysis, Part I+II*. Springer.

Chakraborty, K., Mehrotra, K., Mohan, C. K. and Ranka, S. (1992) Forecasting the Behavior of Multivariate Time Series Using Neural Networks. *Neural Networks*, vol. 5, 961-70.

Deppisch, J., Bauer, H. U. and Geisel, T. (1994) Hierarchical training of neural networks and prediction of chaotic time series, in *Artificial Neural Networks: Forecasting Time Series* (ed. V. R. Vemuri, R. D. Rogers), IEEE Computer Society Press, 66-71.

Gruenenfelder, R., Cosmas J. P. and Odinma-Okafor, A. (1991) Characterization of Video Codecs as Autoregressive Moving Average Processes and Related Queueing System Performance. *IEEE Journal on Selected Areas in Communications*, vol. 9, no. 3, 284-93.

Hamilton, J. D. (1994) *Time Series Analysis*. Princeton University Press.

Harvey, A. C. (1993) *Time Series Models*. Harvester Wheatsheaf.

Hecht-Nielsen, R. (1989) *Neurocomputing*. Addison-Wesley.

Hudson, J. L., Kube, M., Adomaitis, R. A., Kevrekidis, I. G., Lapedes, A. S. and Farber, R. M. (1994) Nonlinear Signal Processing and System Identification: Applications to Time Series from Electrochemical Reactions, in *Artificial Neural Networks: Forecasting Time Series* (ed. V. R. Vemuri, R. D. Rogers), IEEE Computer Society Press, 36-42.

Janacek, G. (1993) *Time Series*. Ellis Horwood.

Law, A. M. and Kelton, W. D. (1991) *Simulation Modelling & Analysis*. McGraw-Hill.

Le Gall, D. (1991) MPEG: A Video Compression Standard for Multimedia Applications, *Communications of the ACM*, vol. 34, no. 4, 46-58.

Lowe, D. and Webb, A. R. (1994) Time series prediction by adaptive networks: a dynamical systems perspective, in *Artificial Neural Networks: Forecasting Time Series* (ed. V. R. Vemuri, R. D. Rogers), IEEE Computer Society Press, 12-9.

Mozer, M. C. (1993) Neural Net Architectures for Temporal Sequence Processing, in *Time Series Prediction: Forecasting the Future and Understanding the Past* (ed. A. S. Weigend), Addison-Wesley, 243-64.

Papoulis, A. (1984) *Probability, Random Variables, and Stochastic Processes*. McGraw-Hill.

Scargle, J. D. (1992) Predictive Deconvolution of Chaotic and Random Processes, in *New Directions in Time Series Analysis, Part I* (ed. D. Brillinger), Springer, 335-56.

Tang, Z., de Almeida, C. and Fishwick, P. A. (1994) Time series forecasting using neural networks vs. Box-Jenkins methodology, in *Artificial Neural Networks: Forecasting Time Series* (ed. V. R. Vemuri, R. D. Rogers), IEEE Computer Society Press, 20-7.

Tarraf, A. A., Habib, I. W. and Saadawi, T. N. (1993) Neural Networks for ATM Multimedia Traffic Prediction. *Proceedings of IWANNT '93*, 85-91.

Tarraf, A. A., Habib, I. W. and Saadawi, T. N. (1994) A Novel Neural Network Traffic Enforcement Mechanism for ATM Networks. *IEEE Journal on Selected Areas in Communications*, vol. 12, no. 6, 1088-96.

Wan, E. A. (1993) Time Series Prediction Using a Connectionist Network with internal Delay Lines, in: *Time Series Prediction: Forecasting the Future and Understanding the Past* (ed. A.

S. Weigend), Addison-Wesley, 195-217.
Weigend, A. S. (1993) *Time Series Prediction: Forecasting the Future and Understanding the Past.* Addison-Wesley.

APPENDIX A: VECTOR QUANTIZER

The VQ algorithm described here is closely related to the Kohonen learning rule as proposed in Hecht-Nielsen (1989).

The vector quantizer consists of M units that receive the following input (Euclidean distance between $\boldsymbol{i} = (i_1, i_2, \ldots, i_N)$ and internal weight vectors $\boldsymbol{w}_i = (w_{i1}, w_{i2}, \ldots, w_{iN})$, i=1..M) that belong to the units, see Figure 16:

$$d_i = D(\boldsymbol{w}_i, \boldsymbol{i}) = \sqrt{(w_{i1} - i_1)^2 + \ldots + (w_{iN} - i_N)^2} \tag{13}$$

A competition takes place between the units. The unit with the lowest value $d_i - b_i$ is the winner and its output z_i is set to 1. The outputs of all other units are set to 0. The value b_i is a bias term that ensures that the frequency of winning the competition becomes $1/M$ for all units after some learning.

The rule for calculating the bias is

$$b_i = \gamma \cdot \left(\frac{1}{M} - f_i\right), \quad i=1..M \ (\gamma \text{ typically } 10) \tag{14}$$

After the competition the weight vector of the winner unit and the f_i of all units are modified:

$$\boldsymbol{w}_i^{new} = \boldsymbol{w}_i^{old} + \alpha\left(\boldsymbol{i} - \boldsymbol{w}_i^{old}\right) \tag{15}$$

$$f_i^{new} = f_i^{old} + \beta\left(z_i - f_i^{old}\right), \quad i=1..M \ (\beta \text{ typically } 0.0001) \tag{16}$$

The learning rate α decreases exponentially during the learning process (typically from 0.05 to 0.001).

APPENDIX B: NEURAL NETWORK ALGORITHM

In this Appendix the learning rule of the new Neural Network algorithm is shortly described. For the meaning of some variables refer to Figure 7 in Section 2 and to Table 2.

The first step is to determine the segment i the event e_k falls into. Next segment frequency and width are adapted.

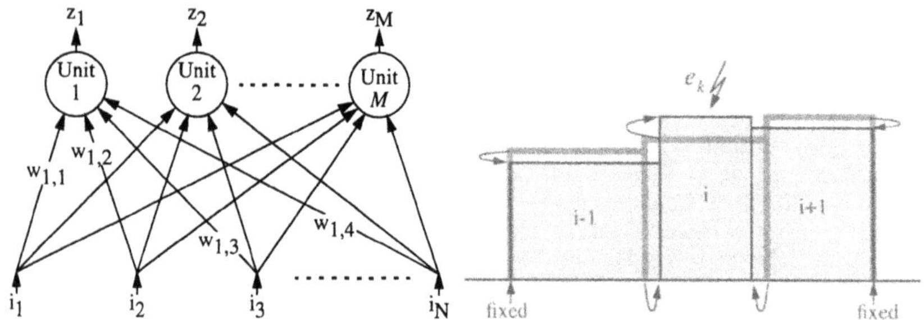

Figure 16 Vector quantizer. **Figure 17** Event in segment i.

Principle

Events have only an effect in one segment and its direct left and right neighbors (segments i, i-1, i+1). Therefore the algorithm has a high locality and can be easily implemented in parallel.

Table 2 Variables for Neural Network algorithm

variable	*description*
e_k	Event at time k
$w_{i,k}$:	Width of segment i
$d_{i,k}$:	Density of segment i
$f_{i,k}$:	Frequency of segment i
L	Number of segments for approximation

Adaptation of segment width and position

The adaptation of segment width consists of the distribution of an "amount of frequency" of segment i to its neighboring segments. The density of segment i is kept constant as well as the borders of the neighbors. Adaptation takes place by means of the change of the width of segments i, i-1, i+1 and the change of density of segments i-1 and i+1. The width of segment i is reduced to enhance the resolution in areas of higher event frequency. This procedure leads to nearly the same probability $1/L$ of each segment.

The marginal segments (segments 1 and L) have to be treated in a special way. Their outer borders have to adapt to the minimum and maximum values of the underlying distribution.

The "amount of frequency" mentioned above is governed by a learning parameter λ that decreases exponentially during learning (typically from 0.1 to 0.001).

Adaptation of segment frequency

The adaptation of segment frequency firstly corrects some errors possibly made in the adaptation of segment width and secondly provides a fine tuning of the density function. The adaptation of segment frequency is performed for more cycles than the adaptation of segment width. So even if the first step leads to non-optimal density function this is corrected in the second step.

The frequency of events in the distinct segments is calculated as exponentially weighted mean value of all events falling in one segment:

$$f_{i,k} = \alpha \cdot \sum_{j=0}^{k} (1-\alpha)^j \cdot t_{i,k-j} \tag{17}$$

with

$$t_{i,k} = \begin{cases} 1 & \text{if event } e_k \text{ in area } i \\ 0 & \text{if event } e_k \text{ not in area } i. \end{cases} \tag{18}$$

Since the $t_{i,k}$ are zero for k negative equation (17) can be rewritten as a recursion:

$$f_{i,k} = \alpha \cdot t_{i,k} + (1-\alpha) \cdot f_{i,k-1} \tag{19}$$

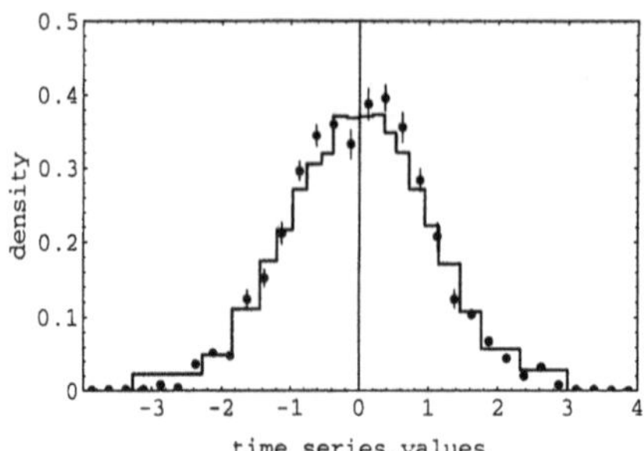

Figure 18 Density approximation of normal distributed time series.

The factor α determines the contribution of the actual event $t_{i,k}$ to the sum. Like for the adaptation of the segment width a slightly modified method is necessary to adapt the segment frequency of boundary segments. The value of α depends on the number of samples available for learning. The combined effects of both adaptations can be seen in Figure 17.

Example

An example for the adaptation capabilities of the Neural Network is shown in the sequel. A normally distributed time series with mean zero and variance 1 is used as training data. Figure 18 shows the original density function (dots) with 95% confidence intervals (irregularity in shape stems from non-ideal random number generator and not much adaptation cycles) and the approximation with L=20 segments (solid line). Note the decreasing width with increasing density.

Error approximation

The resulting distribution approximation has an approximation error. The error is computed as the integral over the squared difference of original distribution f and density approximation d_i. After some simplification and assuming that all regions have the same probability $1/L$ this leads to

$$E = \int_{-\infty}^{\infty} f^2(x)\,dx - \frac{1}{L} \cdot \sum_{i=1}^{L} d_i . \tag{20}$$

8 BIOGRAPHY

Markus D. Eberspächer was born in 1963 and studied Electrical Engineering at the University of Stuttgart where he received the Dipl.-Ing. degree in 1991. Since 1991, he is a member of the scientific staff at the Institute of Communications Switching and Data Technics, University of Stuttgart (Prof. Kühn). His interests include Neural Networks and Fuzzy Control in telecommunication networks.

15

The Entropy of Cell Streams as a Traffic Descriptor in ATM Networks

*N. T. Plotkin**
SRI International
333 Ravenswood Avenue
Menlo Park, CA 94025, USA
ninatp@erg.sri.com
and
C. Roche
Laboratoire MASI
Université Paris VI
75252 Paris Cedex 05, France
roche@masi.ibp.fr

Abstract

We examine the properties of a promising new traffic descriptor for ATM networks, namely the entropy of cell streams. The entropy is a measure of the disorganization among cells within a traffic stream; alternatively we can say that entropy captures the amount of randomness in cell scattering. We study the entropy of ON-OFF sources with respect to the typical queue parameters of interest: average queue size, queue variance and equivalent buffers. The equivalent buffer is defined as the minimum buffer size needed to achieve a specific loss probability. We demonstrate that the average queue size and the variance of queue size are monotonically decreasing with increasing entropy in streams with the same fixed load. We find that the inverse of the entropy is closely linear, to within a good approximation, to the equivalent buffer. This simple relation demonstrates the appeal of the entropy estimator. In addition to a measure of cell scattering, our results suggest another interpretation of entropy as a measure of smoothness. Traffic streams with higher entropy (i.e. smoother) have less buffering needs in terms of average, variance and equivalent buffers. Based on this observation we introduce a traffic shaping mechanism whose goal is to boost the entropy of a stream.

Keywords

Traffic Characterization, Entropy, ATM Networks

---

*This work was carried out while this author was a visiting researcher at the Laboratoire MASI at Université Paris VI.

1 INTRODUCTION

Control of congestion in ATM networks is expected to be carried out through the use of preventive mechanisms. The overall approach to network control is based on call admission control and real-time source policing. When the user submits a connection request to a network agent, the agent uses a call admission control mechanism to decide whether to accept or reject the call. This decision is made based on the expected consumption of network resources needed by the traffic stream, and the available resources in the network. When a call is accepted, a traffic contract is established between the user and the network. This traffic contract should specify the anticipated characteristics of the traffic flow. During the lifetime of the connection, policing mechanisms at the entry point of the network ensure that the actual characteristics of the traffic stream match those specified in the contract. The goal is to avoid the occurrence of a congested state inside of the network.

Obviously the proper operation of such network control depends heavily on a correct characterization of the traffic flow upon establishment of the traffic contract. One way to characterize a flow is through a well chosen set of traffic descriptors [CCITT, 1992]. A good traffic descriptor should measure important properties of the traffic stream which influence the performance of the network. Traffic descriptors which exhibit simple relations to network performance parameters are desirable. Moreover, for traffic descriptors to be useful for source policing, they must be measurable "on-the-fly", or in real-time [Eckbern, 1992].

Common traffic descriptors include the peak rate, mean rate, utilization factor, and autocorrelation factors. A variety of descriptors based on the concept of a *burst* have also been proposed. The study of superposed Ethernet traffic reported in [Leland, 1993] hints that, although some of these quantities may be useful to characterize simple traffic streams (e.g., single source traffic), they prove inadequate for more complex traffic streams that appear in actual networks, in particular superposed streams. For example, the notion of burst length is ill-defined for highly-superposed LAN traffic. This motivates us to pursue the study of novel traffic descriptors that do not suffer from such limitations.

In this paper we examine the properties of a promising traffic parameter known as the *entropy* of a traffic stream. The idea of using entropy as a traffic metric was first introduced in [Plotkin, 1994], where entropy is used as a tool for studying properties of departure processes in a tandem queueing system. Here we are interested in exploring the relation of entropy to standard ATM QoS parameters, in order to assess its applicability to these networks. We consider traffic streams of discrete-time slotted systems which carry fixed sized cells, such as in ATM networks. Each slot either carries a cell or is empty, therefore a particular sample stream can be viewed as a binary sequence. Intuitively entropy measures the degree of randomness with which cells are dispersed over slots, within a given stream. If there is little randomness in cell placement among slots, then the cells will exhibit repeating patterns. In our binary sequence representation this corresponds to shorter bit patterns that appear repeatedly within a longer binary sequence. If there are few detectable repeating patterns, then the cells are considered more disorganized. The entropy measure grows with increasing disorganization. Since a deterministic stream has an exact organization, its entropy is zero. The terms cell *dispersion*, *disorganization*, and *scattering* can be used interchangeably. Entropy can thus be used as a measure of cell scattering in ATM networks.

For a parameter to be useful as a traffic descriptor it must be both measurable in real-time and meaningful to the performance of the network. In ATM network complex traffic streams arise because as single streams traverse the network they are merged, superposed, interleaved, queued, split apart and so on. Some of these effects will cause cell scattering or dispersion. Others will cause clustering or clumping. Entropy may be useful to capture this naturally occurring phenomenon of cell scattering. The second interesting point is the possibility to estimate the entropy of actual traffic streams. In [Taft-Plotkin, 1994] they illustrated an efficient and implementable method for entropy measurement in real-time systems. The algorithm developed can be used inside a traffic monitor to estimate the entropy of traffic streams on the fly, i.e., while watching cells bypass the monitor. This algorithm uses a variation on Lempel-Ziv data compression techniques.

In this paper we explore the meaning of entropy to discrete-time ATM-network queues. We consider a basic system in which an ON-OFF traffic stream is fed into a queue with a single deterministic server. In particular, we study the relationship between the entropy and each of the three typical queue parameters of interest: average queue size, variance of queue size, and equivalent capacity. We show that both the average queue size and the variance of the queue size are monotonically decreasing with increasing entropy, for fixed loads. We then examine the entropy of the input in relation to the *equivalent buffer*, i.e. the minimum buffer size needed to achieve a specific loss probability. We find that the equivalent buffer is closely linear, to within a good approximation, to the inverse of the entropy.

These results inadvertently suggest that entropy can be used as a measure of traffic "smoothness". Traffic streams which - for a fixed load - have smaller equivalent buffers, find smaller average queue sizes and see smaller queue variance, can be considered *smoother* than those which have larger equivalent buffers, find larger average queue sizes and see larger queue variance. This interpretation is intuitive since traffic streams which require less buffers can be considered easier to handle, and hence smoother. This understanding of smoothness is based on the same concept of smoothness given in [Low, 1993]. With this interpretation, our results for ON-OFF streams imply that traffic streams with higher entropy are smoother than those with lower entropy. We therefore design a traffic smoother whose goal is to boost the entropy of a stream, and call this smoother an *entropy booster*. Simulation tests demonstrate that our entropy booster can be quite efficient in traffic smoothing and leads to performance improvement.

In section 2 we specify the class of input sources that we consider. Section 3 covers basic entropy definitions and presents the entropy of our sources in a useful form. Section 4 illustrates the relation between the first two moments of our queue system and the entropy of the source. Section 5 introduces our entropy booster and provides preliminary results from simulation testing of this smoothing mechanism. In section 6 we examine the influence of entropy on the equivalent buffer. Our conclusions and ideas for future research directions are discussed in section 7.

2 CHARACTERIZATION OF AN ON-OFF TRAFFIC PROCESS

An ON-OFF traffic process can be used to describe a single variable bit rate source with a bounded peak rate. It is given by a two-state discrete time Markov chain as depicted in Figure 1. In the OFF state the process generates 0 cells/slot, and in the ON state the process generates m cells/slot. The amount of time spent in the ON state, at each visit to the ON state, is given by a geometric random variable B with parameter α. Similarly the lengths of visits to the OFF state are given by a geometric random variable I with parameter β. The probability of transitioning to the OFF (ON) state given that we are in the ON (OFF) state is $1-\alpha$, ($1-\beta$, respectively). Therefore, the fraction of time the source is ON, also called the utilization factor, is given by

$$\sigma = \frac{E[B]}{E[B]+E[I]} = \frac{\frac{1}{1-\alpha}}{\frac{1}{1-\alpha}+\frac{1}{1-\beta}} = \frac{1-\beta}{2-\alpha-\beta} \tag{1}$$

In this model the peak rate is m and the average input rate is σm. The k-th order autocorrelation function for a single on-off input is (the derivation is provided in the Appendix)

$$\gamma(k) = (\alpha+\beta-1)^k \tag{2}$$

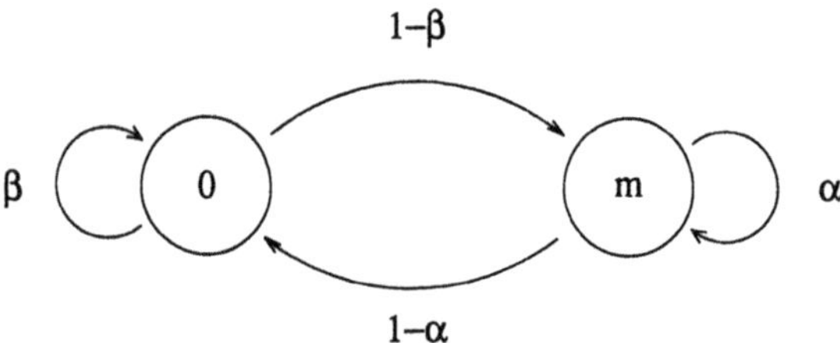

Figure 1 ON-OFF Source Model

A specific sample stream generated by this model can be represented not by a binary sequence of 0's and 1's, but rather by a binary sequence of 0's and m's. This on-off source is entirely described by the triple (α, β, m). This source is equivalently described by the triple (σ, γ, m), where $\gamma = \gamma(1)$, by using the transformation

$$1-\alpha = (1-\sigma)(1-\gamma) \tag{3}$$
$$1-\beta = \sigma(1-\gamma) \tag{4}$$

The advantages of this transformation are that the parameters σ and γ have a natural physical meaning to the traffic stream generated by the source, and that it leads to simpler forms for the equations we develop below.

3 ENTROPY

The entropy of a random variable X with a probability mass function $p(x)$ defined in an information theoretic sense is given by

$$\mathcal{H}(X) = -\sum p(x)\log_2 p(x) \tag{5}$$

We use logarithms to base 2, and thus measure the entropy in bits. The entropy is a measure of the uncertainty of a random variable. For a Bernoulli process, with parameter $0 \leq p \leq 1$, where $Pr[X_i = 1] = p$, the entropy is

$$\mathcal{H} = -p\log p - (1-p)\log(1-p) \stackrel{def}{=} H(p) \tag{6}$$

and is plotted in Figure 2. We see here that the entropy is symmetric with respect to $p = 1/2$, i.e. $H(p) = H(1-p)$. Consider a binary sequence generated by a Bernoulli process. The symmetry means that the entropy doesn't distinguish between a given sample path and the same path in which the 0's and 1's are interchanged. Entropy is only concerned with the organization of the sample path. This example is simply that of coin tossing and we see that the entropy achieves its maximum for the case of an unbiased coin, i.e. $p = 1/2$. This is the case for which the random variable X_i is the most uncertain. It is known that $0 \leq \mathcal{H} \leq \log|\aleph|$ where $|\aleph|$ denotes the number of elements in the range of X_i. Since in this example $X_i \in \{0,1\}$ we have $0 \leq \mathcal{H} \leq 1$. (For a complete coverage of entropy definitions and properties see [Cover, 1991].)

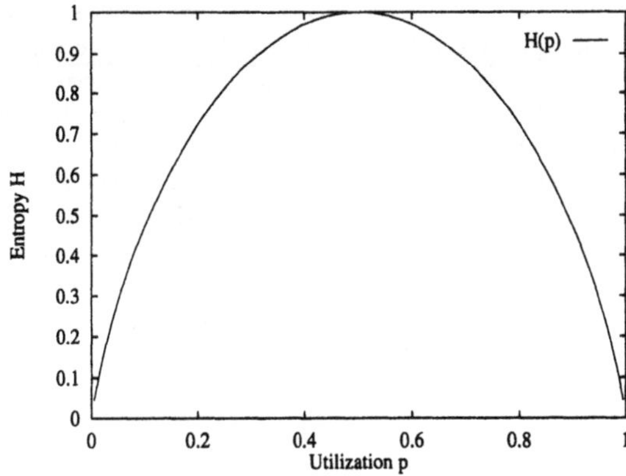

Figure 2 Entropy of a Bernoulli Process with parameter p.

The *entropy rate* (in bits per unit time) of a stochastic process $\{X_i\}$, $i = 0, 1, 2\ldots$ is defined by $\mathcal{H} = \lim \frac{1}{n}\mathcal{H}(X_1, X_2, \ldots, X_n)$, when the limit exists, where $\mathcal{H}(X_1, X_2, \ldots, X_n)$

is defined as in (5) with $p(x)$ replaced by the joint probability distribution $p(x_1, x_2, \ldots, x_n)$. For a stationary Markov chain $\{X_i\}$ with invariant distribution μ and transition matrix P, the entropy rate is

$$\mathcal{H} = \mathcal{H}(X_2/X_1) = -\sum_{ij} \mu_i P_{ij} \log P_{ij} \tag{7}$$

According to this last equation, the entropy of our bursty on-off traffic source in terms of α and β is given by

$$\mathcal{H} = \frac{1-\beta}{2-\alpha-\beta} H(\alpha) + \frac{1-\alpha}{2-\alpha-\beta} H(\beta) \tag{8}$$

Using the transformation in equations (3) and (4), and the fact that $H(p) = H(1-p)$, the entropy can be rewritten in a more useful form in terms of σ and γ as follows

$$\mathcal{H} = \sigma\, H\Big((1-\sigma)(1-\gamma)\Big) + (1-\sigma)\, H\Big(\sigma(1-\gamma)\Big) \tag{9}$$

In this form, the equation clearly reveals the symmetry of the entropy of an ON-OFF source in the utilization factor σ. This symmetry indicates that the entropy should not be used alone as a traffic descriptor, but rather that it should be coupled with the utilization. Essentially we are studying the traffic descriptor (σ, H) since including the utilization factor allows us to distinguish between two streams with the same entropy but different queueing behavior. In the case of $\gamma = 0$, the right-hand term of equation (9) collapses to $H(\sigma)$ as it should since in this case the ON-OFF process reduces to a simple Bernoulli process with parameter σ. For our sources $X_i \in \{0, m\}$ hence $|\aleph| = 2$ and the range of the entropy values is again given by $0 \leq \mathcal{H} \leq 1$. Note that the value m does not appear in the formula for the entropy, since the entropy is only concerned with the organization of unoccupied slots and occupied slots; (this is true as long as the occupied slots always contain a fixed number of cells).

The interpretation of these definitions in terms of cell dispersion is as follows. Let the Markov Chain $\{X_i\}$ $i = 0, 1, 2\ldots$ be given by an ON-OFF source. In our binary sequence representation an example of a sample path generated by the source might is given by $X_0 = 0, X_1 = m, X_2 = m, X_3 = 0, X_4 = 0, X_5 = 0, X_6 = m$ or simply $0mm000m$. The sample stream $0m00mmm0mm00$ is considered more disorganized than the stream $mm0mm0mm00m0$ since in the latter stream the pattern $mm0$ occurs repeatedly.

4 ON-OFF/D/1 QUEUE

We begin by examining the basic case of a infinite FIFO queue with a deterministic server and an ON-OFF input source. ON-OFF streams are suitable for this study since the ON-OFF traffic model is a common model for ATM networks and because it allows us to vary the amount of cell scattering by tuning its characteristic parameters. These processes also include Bernoulli processes which are a degenerate case of ON-OFF processes when $\alpha + \beta = 1$, or $\gamma = 0$. In [Bruneel, 1994] they determine the z-transform of the steady state buffer occupancy process for an ON-OFF/D/1 queue, whose input is described by the triple (α, β, m). This generating function is given by

$$U(z) = \sum_{k=0}^{\infty} p(k) z^k = \frac{(1-\lambda)(z-1)[\beta + (1-\alpha-\beta)z^{m-1} + (1-\beta)z^m]}{z - \beta - (1-\alpha-\beta)z^{m-1} - \alpha z^m} \quad (10)$$

where $p(k)$ denotes the probability that there are k cells in the queue and where the average input rate λ is given by

$$\lambda = \frac{1-\beta}{2-\alpha-\beta} m \quad (11)$$

We now use this function to determine the mean and variance of the corresponding queue occupancy process. The average number in the queue is obtained from

$$E(N) = \frac{dU(z)}{dz}|_{z=1} \quad (12)$$

After the transformation into the parameter set (σ, γ, m) we obtain

$$E(N) = \frac{m\sigma[2(m-1) - (1-\gamma)(2m\sigma + m - 3)]}{2(1-\gamma)(1-m\sigma)} \quad (13)$$

We now have, with equations (9) and (13), formulas for both the average queue size and the entropy of the input in terms of the same triple (σ, γ, m). By fixing the load (i.e. m and σ) and varying the autocorrelation factor γ we can trace a parametric plot of the average queue length in terms of the entropy. This plot is given in Figure 3. We see that the average queue length increases when the entropy decreases to zero, which corresponds to $\gamma \to 1$. The backtracking behavior that we observe at the bottom of the curves corresponds to $\gamma \to -1$ and the change of direction occurs at the point $\gamma = 0$. For bursty traffic models, the interesting values of γ are those that correspond to positive correlations, i.e. $\gamma > 0$. In this region, $E(N)$ is a monotonically decreasing function of H.

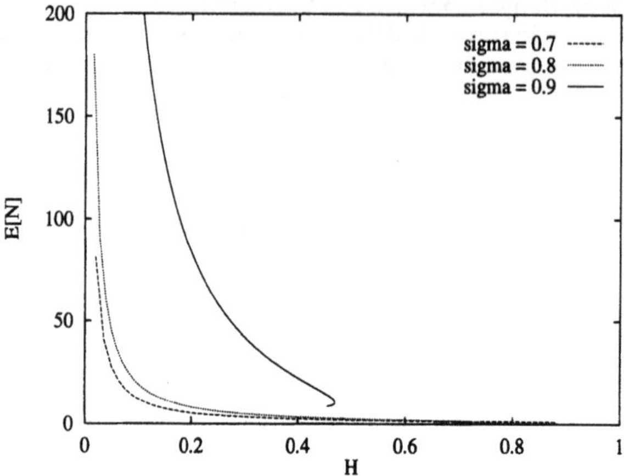

Figure 3 Average Queue Size vs. Entropy

The second moment of the queue size is determined by $E(N^2) = U''(1) + U'(1)$, and is given below.

$$\begin{aligned} E(N^2) &= \frac{\sigma m[6(\sigma m)^2 + (m^2 + 3m - 16)\sigma m + 2m^2 - 9m + 13]\gamma^2}{6(1-\gamma)^2(1-\sigma m)^2} \\ &+ \frac{\sigma m[-12(\sigma m)^2 + (-2m^2 + 6m + 20)\sigma m + 8m^2 - 18m - 2]\gamma}{6(1-\gamma)^2(1-\sigma m)^2} \\ &+ \frac{\sigma m[6(\sigma m)^2 + (m^2 - 9m - 4)\sigma m + 2m^2 + 3m + 1]}{6(1-\gamma)^2(1-\sigma m)^2} \end{aligned} \tag{14}$$

In Figure 4 we trace the variance $V(N) = E(N^2) - E(N)^2$ in terms of the entropy via a parametric plot. We observe similar behavior here as in the previous plot; namely that $V(N)$ is a monotonically decreasing function of H in the region of interest $\gamma > 0$.

We have found that for a fixed load, ON-OFF input streams with higher entropy find lower average queue sizes and see smaller queue size variation. Note that the ON-OFF model does **not** include the case of deterministic periodic streams since there is no setting for α and β that can generate such streams (other than $X_i = 0 \;\; \forall i$ or $X_i = m \;\; \forall i$). Since $H = 0$ in our model corresponds to a stream of either all 0's or either all m's, the above claims - which hold for ON-OFF traffic - cannot be trivially compared to other results for deterministic periodic streams. We believe that such relationships between entropy and queueing, as depicted in Figures 3 and 4, may be generalized to similar models that exhibit a variable amount of randomness such as Interrupted Poisson Process (IPP).

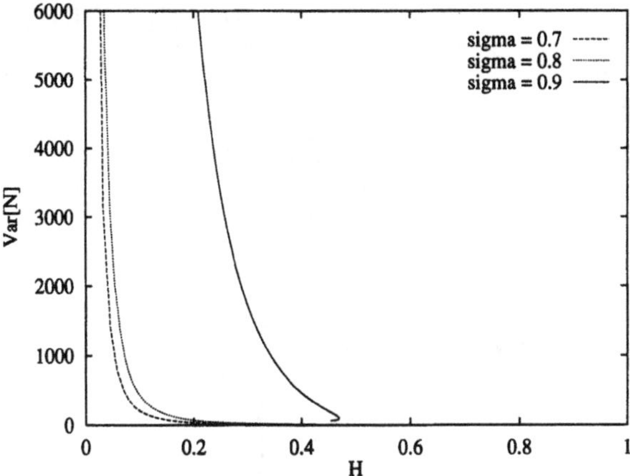

Figure 4 Variance of Queue Size vs. Entropy

As mentioned in the Introduction, we can interpret the *smoothness* of a traffic stream to be related to the amount of buffer space needed to handle a specific stream for a given load. One stream is then considered smoother than another if it requires less buffer space. This suggests a first application of the entropy metric, namely as a measure of smoothness of a traffic stream. Alternatively, we can say that the goal of a smoothing technique should be to increase the entropy of the traffic stream.

5 ENTROPY BOOSTER

Attractive smoothing techniques are those which reduce average queue size and queue variance, because reducing the average queue size leads to a reduction in average delays, and reducing the queue variance leads to a reduction in jitter. Lowering average delays and jitter are desirable from the point of view of network performance. We can see from Figures 3 and 4, that increasing the entropy, for a given stream with a given load, would lead to improved network performance. We introduce the following definition of smoothness.

DEFINITION: Consider two traffic streams X and Y whose entropies are given by $H(X)$ and $H(Y)$ respectively. If $H(Y) > H(X)$, then stream Y is considered *smoother* than stream X.

We therefore design a smoothing technique whose goal is to boost the entropy of a stream, and we call this an *Entropy Booster*. Our entropy booster is depicted in Figure 5. The idea is to achieve $H' > H$ which in turn leads to $E(Q2) < E(Q1)$. One entropy boosting method is is to insert "holes" into the stream at random. (There are a number

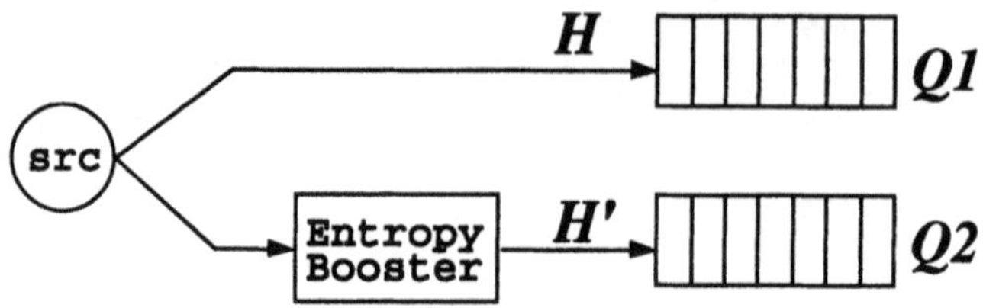

Figure 5 Randomized Smoother

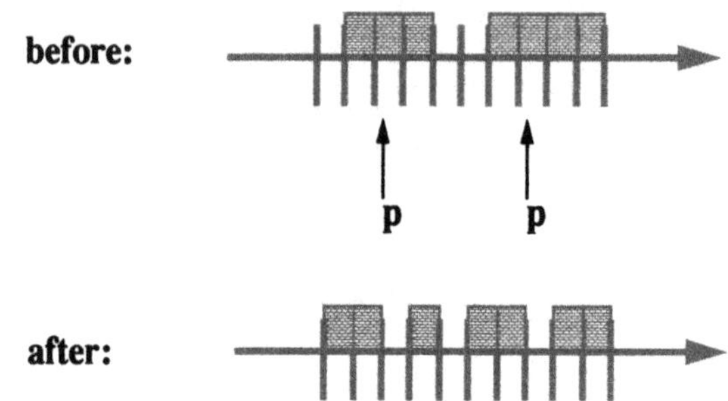

Figure 6 Entropy Boosting Scheme.

of ways to boost the entropy.)

> <u>ENTROPY BOOSTING SCHEME</u>
> At each slot time, flip a coin with bias p.
> IF the coin yields a head AND there is a cell in the slot,
>   Insert a hole into the stream.
> ELSE do nothing.

By "inserting a hole" we mean that if there is a cell in the slot, then the cell should be held back (i.e., buffered) by one slot time, so that the slot in question is left empty. This method is depicted in Figure 6. This procedure of randomly introducing spaces between consecutive cells breaks up correlations and scatters the cells, thus increasing the stream's entropy. We call the coin bias p the boost parameter. (For simplification in implementation, one need not actually flip the coin at *every* time slot, but rather once every n slots where n is chosen to suit the nature of the traffic being carried.)

To check the validity of the entropy booster scheme, we carried out preliminary simulation experiments. The testing environment emulated the scenario shown in Figure 5. We computed the entropy at the entrance to the queues using the entropy measurement algorithm described in [Taft-Plotkin, 1994].

Figure 7 shows an ON-OFF stream with a load of 0.5 both before and after it has been smoothed. We varied the average burst size (and corresponding idle period size) while

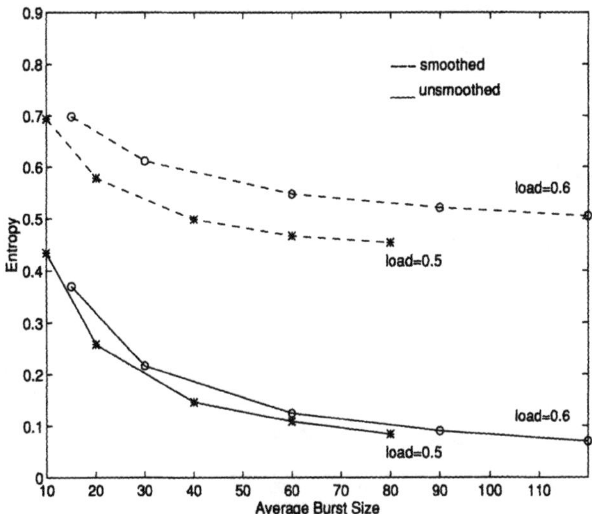

Figure 7 Change in entropy after smoothing.

keeping the load constant. The entropy boost parameter used was 0.2. The example of a load with 0.6 is also given. We see that, for a given burst size, the entropy booster is quite efficient in raising the entropy of a stream. This Figure should be directly compare to Figure 8 in which the exact same situations are considered, only here we plot the corresponding queue size. For both loading cases, the smoothed streams result in much smaller average queue sizes. These tests indicate the correct behavior of the proposed entropy boosting scheme.

It is interesting to note that this approach to traffic smoothing is seemingly in contrast to some of the implicit assumptions in the literature that a smoother stream is one which has been more precisely shaped by a leaky bucket. Future studies which directly compare the entropy-boosting smoother to a token bucket smoother could help to elucidate the relative advantages and disadvantages of each approach. It may be possible to combine the best of both philosophies in the same network.

6 EQUIVALENT BUFFER

Since we saw in Sections 4 and **??** that smooth streams have less buffering needs than non-smooth ones, we would therefore expect that smooth streams also have smaller equivalent buffers than nonsmooth ones. This motivates us to study the precise relationship between the equivalent capacity and the entropy of an ON-OFF traffic stream. In [Guerin, 1991] the authors determine the equivalent bandwidth for a single ON-OFF source using a fluid-flow model in which both the ON and OFF periods are exponentially distributed. Since

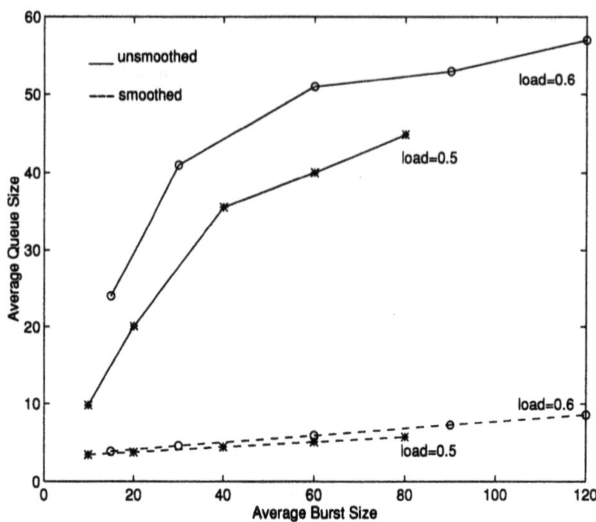

Figure 8 Expected queue size for smoothed and unsmoothed streams.

our traffic model is simply the discrete time version of theirs, we can apply their results. They determine the equation for the overflow probability ϵ to be

$$\epsilon = \nu \exp\left(-\frac{B(c-\sigma m)(1-\gamma)}{c(m-c)}\right) \tag{15}$$

where c is the queue service rate, B is the buffer needed to provide an overflow probability of ϵ, and the constant ν is given by

$$\nu = \frac{(c-\sigma m)+\epsilon\sigma(m-c)}{(1-\sigma)c} \tag{16}$$

We solve equation (15) for the equivalent buffer B which yields

$$B = \frac{ac(m-c)}{(c-\sigma m)(1-\gamma)} = \frac{B_0}{1-\gamma} \tag{17}$$

where the constant a stands for the quantity $\ln(\nu/\epsilon)$. The equivalent buffer B is expressed in number of cells. Let B_0 represent the minimum value of the equivalent buffer in the region of interest. B reaches this value when γ is null, i.e., in the case of a completely uncorrelated stream.

By eliminating the quantity $(1-\gamma)$ between equations (9) and (17), we can rewrite the entropy as a function of the equivalent buffer like this:

$$\begin{aligned}\mathcal{H} &= -\sigma\left\{(1-\sigma)\frac{B_0}{B}\log\left((1-\sigma)\frac{B_0}{B}\right)+\left(1-(1-\sigma)\frac{B_0}{B}\right)\log\left(1-(1-\sigma)\frac{B_0}{B}\right)\right\}\\ &\quad -(1-\sigma)\left\{\sigma\frac{B_0}{B}\log\left(\sigma\frac{B_0}{B}\right)+\left(1-\sigma\frac{B_0}{B}\right)\log\left(1-\sigma\frac{B_0}{B}\right)\right\}\end{aligned} \tag{18}$$

This formula is not easy to handle as it is. So we use a development to the first order in $1/B$, which yields the following approximation

$$\mathcal{H} \approx 2\sigma(1-\sigma)\frac{B_0}{B}\log\left(\frac{e}{\sqrt{\sigma(1-\sigma)}}\frac{B}{B_0}\right) \tag{19}$$

where e is the base of natural logarithms. The residual error that we make when approximating $\mathcal{H}$ by the preceding quantity is in $O(1/B^2)$, that is to say extremely small as soon as B is large.

The form of this approximation motivates us to plot $1/\mathcal{H}$ with respect to B, using the exact value given by equation (18). This plot is given in Figure 9 for different values of the utilization factor σ. We used the numerical values $\epsilon = 10^{-9}$ and $c = 1$. Since the inequalities $\sigma m \leq c \leq m$ must be satisfied for the system to be stable, we chose $m = 1.105$. This leads to $B_0 \approx 9.47$.

We see that these curves are very close to straight lines, which means that $1/\mathcal{H}$ is almost linear with respect to B. In fact we found that this quasi-linearity property holds on any interval $[B_0, B_m]$, with a slope that depends on the upper bound B_m. Due to the properties of the function $B/\log B$, the curves exhibit the same general shape for any value of B_m. This property can be interpreted as follows: once an upper bound (or order of magnitude) B_m is given for B, B is essentially proportional to $1/\mathcal{H}$, i.e. $B \approx \mu/\mathcal{H}$, where the proportionality coefficient μ is given by

$$\mu = 2\sigma(1-\sigma)B_0\log\left(\frac{e}{\sqrt{\sigma(1-\sigma)}}\frac{B_m}{B_0}\right) \tag{20}$$

The simplicity of this relationship implies that, in the case of Markovian ON-OFF processes, the measure of the entropy of a cell stream can be used directly to estimate the equivalent buffer of the stream. Since the utilization factor σ is used in determining the proportionality coefficient, we see that it is the pair (σ, H) which is a useful traffic descriptor. This result is interesting since it gives us a first idea of the meaning of the entropy of a cell stream from the point of view of network performance. Moreover, if we are able to extend this result to more general processes, this may open interesting perspectives for entropy as a traffic descriptor in ATM networks.

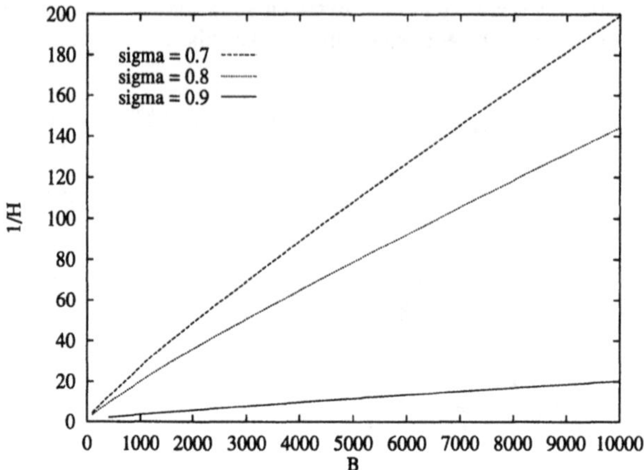

Figure 9 Inverse of Entropy vs. Equivalent Buffer

7 CONCLUSIONS

In this paper we have explored the properties of a promising new traffic descriptor for ATM networks. The entropy is a measure of the disorganization among cells and can be used to estimate the degree of scattering in cell streams. The traffic descriptor (σ, H) appears to be a promising traffic descriptor since the two quantities are both measurable and meaningful. First, we have demonstrated that the relationship of the entropy of an ON-OFF input source to both the average and variance of the corresponding queue can be found. In particular, we showed that the average and variance of the queue are monotonically decreasing with increasing entropy. Second, our study of equivalent buffer revealed a remarkable property that the inverse of the entropy behaves essentially according to the function $B/\log B$. This straightforward relationship between the entropy and the network performance parameter, equivalent buffer, renders the entropy an attractive traffic descriptor.

We introduced a definition of smoothness of a traffic stream which is measured in terms of its entropy. We then presented a sample smoother whose goal is to boost the entropy of the traffic stream. A few simple tests were conducted to demonstrate that this entropy boosting smoother can be quite efficient in reducing delays. Further testing of this traffic smoothing approach under more general traffic conditions is needed. Future research could explore the suitability of the entropy metric as an estimator of the effectiveness of traffic shaping mechanisms. Since the entropy measure is independent of the traffic shaping mechanism itself, it could be used to compare shaping schemes with one another. In particular, it would be interesting to compare the buffered leaky bucket mechanism with the entropy boosting mechanism since these mechanisms are based on opposing philosophies.

Another direction is to examine the usefulness of the entropy metric is measuring dependence across streams which can arise, for example, in multimedia traffic flows. As another example, it has been shown in [Floyd, 1993] that independent periodic processes (e.g. in the TCP protocol, in periodic routing protocols, and others) can become inadvertently synchronized and hence dependent. Such behavior could potentially be measured using entropy estimation techniques. They showed that this dependence can degrade network performance. They also adopted the approach of introducing randomness into the network timing in order to avoid performance degradation.

This work indicates that it would be useful to continue to examine the relationship between entropy and queueing. The extended application of this descriptor to a broader class of sources, such as superposed ON-OFF processes, should be studied. Future work could also examine other traffic models with more complex correlational structures than those reflected in traditional Markov chain models.

APPENDIX

We show here how the k-th order autocorrelation coefficient $\gamma(k)$ given in equation (2) was derived. Let A_n denote the number of cells generated in slot n by an ON-OFF process. Thus $A_n \in \{0, m\}$. The k-th autocorrelation coefficient is defined by

$$\gamma(k) = \frac{E(A_{n+k}A_n) - E(A_{n+k})E(A_n)}{\sqrt{Var(A_{n+k})Var(A_n)}} = \frac{E(A_{n+k}A_n) - E(A_n)^2}{Var(A_n)} \tag{21}$$

The second equality is true due to stationarity. For the term $E(A_{n+k}A_n)$ we have

$$E(A_{n+k}A_n) = \sum_{x\in\{0,m\}} \sum_{y\in\{0,m\}} xy\Pr(A_{n+k} = x, A_n = y) = m^2\Pr(A_{n+k} = m/A_n = m)\sigma \tag{22}$$

where the second equality is true since most terms in the summation are zero. To find $\Pr(A_{n+k} = m/A_n = m)$ let

$$\begin{aligned} s(k) &= \Pr(A_{n+k} = m/A_n = m) \\ q(k) &= \Pr(A_{n+k} = 0/A_n = m) \end{aligned} \tag{23}$$

We can define the following set of recursive equations.

$$\begin{aligned} s(k+1) &= \alpha s(k) + (1-\beta)q(k) \\ q(k+1) &= (1-\alpha)s(k) + \beta q(k) \end{aligned} \tag{24}$$

Solving these equations for $s(k)$, with the initial conditions $s(1) = \alpha$ and $q(1) = 1 - \alpha$, and substituting back into the definition for $\gamma(k)$ yields $\gamma(k) = (\alpha + \beta - 1)^k$.

REFERENCES

Bruneel, H. and Kim, B. (1994) *Communications Systems Including ATM.* Kluwer Academic Publishers.

CCITT Recommendation I.371 (1992) *Traffic control and congestion control in B-ISDN.* Geneva, Switzerland. June.

Cover, T. and Thomas, J. (1991) *Elements of Information Theory.* Wiley.

Eckberg, A. (1992) B-ISDN/ATM Traffic and Congestion Control. *IEEE Network Magazine*, September, 6(5):28–37.

Floyd, S. and Jacobson, V. (1993) The Synchronization of Periodic Routing Messages. *ACM SigComm Proceedings*, September.

Guerin, R. and Ahmadi, H. and Naghshineh, M. (1991) Equivalent Capacity and Its Application to Bandwidth Allocation in High-Speed Networks. *IEEE Journal On Selected Areas in Communications*, September.

Leland, W. and Taqqu, M. and Willinger, W.and Wilson, D. (1993) On the Self-Similar Nature of Ethernet Traffic. *ACM SigComm Proceedings*, September.

Low, S. and Varaiya, P. (1993) Burstiness Bounds for Some Burst Reducing Servers. *IEEE InfoCom'93 Proceedings*, March.

Plotkin, N. and Varaiya, P. (1994) The Entropy of Traffic Streams in ATM Virtual Circuits. *IEEE InfoCom'94 Proceedings*, June.

Takagi, H. (1993) *Queueing Analysis Volume 3: Discrete-Time Systems.* Elsevier Science.

Taft-Plotkin, N. (1994) *High-Speed Network Traffic: Characterization and Control.* PhD Thesis, University of California, Berkeley.

Nina Plotkin received the B.S.E. degree in computer science from the University of Pennsylvania, USA, in 1985, and the M.S. and Ph.D. degrees in electrical engineering from the University of California at Berkeley in 1991 and 1994, respectively. From 1985 to 1988 she worked at the Raytheon Company in Massachusetts working on hardware-software interfaces. Dr. Plotkin is currently working in the Telecommunications Theory and Technology Group at SRI International in California. Her research interests include call admission control, traffic characterization, congestion control, ATM networks and neural networks.

Christian Roche received the Engineer's degree from the Ecole Polytechnique, France, in 1990, and the M.S. and Ph.D. degrees from the University of Paris VI in 1992 and 1995, respectively. Dr. Roche currently has a postdoctoral position in the computer science department at the University of California at Los Angeles. He is a recipient of the INRIA postdoctoral studies abroad fellowship award. His research interests include data traffic management, high-speed networks and feedback congestion control.

PART SIX

Traffic Management 1

16

Analytic Models for Separable Statistical Multiplexing

Keith W. Ross[1] *and Véronique Vèque*[2]
(1) University of Pennsylvania and (2) Université de Paris-Sud
(1) Department of Systems, University of Pennsylvania, Philadelphia, PA 19104, USA. Telephone: 215-898-6069. Fax: 215-573-2065.
email: ross@eniac.seas.upenn.edu
(2) Laboratoire de Recherche en Informatique - CNRS, Université Paris-Sud, 91405 ORSAY CEDEX, FRANCE.
Telephone: 33-1-69416702. Fax: 33-1-69416586. email: vv@lri.fr

Abstract

We investigate a multiplexing scheme for ATM that statistically multiplexes VCs of the same service, but does not statistically multiplex across services. The scheme is implemented by allocating bandwidth to each service. In the static version, the allocations are fixed ; in the dynamic version, the allocations depend on the numbers of VCs in progress. Under minimal assumptions, we show that the distribution of the VC configuration has a product form. We use the product-form result to construct an efficient convolution algorithm to calculate VC blocking probabilities. We give a numerical example that demonstrates the rapidity of the algorithm and the potential efficiency of separable statistical multiplexing.

Keywords

Admission control, loss networks, performance evaluation, statistical multiplexing.

1 INTRODUCTION

It has long been known that statistical multiplexing of cell streams of the same service type can be highly cost efficient. This is true for delay-sensitive as well as delay-insensitive services. For example, statistical multiplexing of packet streams emanating from voice sources has long been used by telephone companies to increase efficiency, particularly on overseas links (Sriram, 1993).

On the otherhand, statistical multiplexing of VCs across services rarely gives significant gains in performance when services have greatly different QoS (Quality of Service) requirements or greatly different cell generation properties (Gallassi,1990), (Takagi, 1991), (Bonomi, 1993). Indeed, if services with greatly different QoS requirements are statistically multiplexed, then an overall QoS must realize the most stringent QoS requirement ; thus some services enjoy an overly generous QoS, leading to inefficient use of resources. Similarly, if services with substantially different traffic characteristics are multiplexed, then the cell loss probabilities for the various sources can differ by more than one order of magnitude ; thus the network has to be engineered for a QoS requirement that may be overly stringent for a large fraction of the traffic.

A more serious problem is that with statistical multiplexing across services it is difficult to determine the acceptance region for admission control. The analytic models of cell loss for multiplexers which integrate multiservice VCs are not always accurate, and they typically rely on dubious assumptions. Determining the acceptance region with discrete-event simulation is also difficult, since the QoS requirements must be verified at each boundary point of the multidimensional acceptance region, and because the cell loss probabilities are minuscule.

In this paper we investigate a multiplexing scheme for ATM that statistically multiplexes VCs of the same service, but does not statistically multiplex across services. We refer to this scheme as separable statistical multiplexing. In many scenarios this scheme is almost as efficient as statistical multiplexing across and within services. Moreover, determining the acceptance region for separable statistical multiplexing is substantially easier, whether by analytic models or by discrete-event simulation.

Although separable statistical multiplexing has been proposed by many authors, under different names, analytic models to evaluate its VC-level performance are not available in the literature to the best of our knowledge. Explicitly taking into account cell-level QoS requirements of the heterogeneous services, we develop an analytic model for estimating VC blocking probability for separable statistical multiplexing. We make only two assumptions in our model: (1) VC establishment requests arrive according to Poisson processes ; (2) If a VC establishment request finds insufficient resources available, it is blocked and lost. We make no assumptions about the distribution of VC holding times, nor about the cell generation processes of the heterogeneous sources. Our analytic model leads to an efficient convolution algorithm to calculate VC blocking probabilities.

In Section 2 we define separable statistical multiplexing. In Section 3 we develop an efficient convolution algorithm to calculate VC blocking probability. In Section 4 we present some examples and numerical results.

2 SEPARABLE STATISTICAL MULTIPLEXING

Types of services

Going by different names, separable statistical multiplexing has been proposed for ATM by many authors (for example, Gallassi et al (1990), Sriram (1993), Bonomi et al(1993)). We describe this scheme with the aid of Figure 1. In Figure 1 there is a multiplexer that schedules for transmission on the link the cells that are queued in the buffers. Each buffer aggregates the cell streams from one or more VCs. In Figure 1, the first buffer collects cells from VCs emanating from voice sources ; the second from Continuous Bit Rate (CBR) video sources ; the third from Variable Bit Rate (VBR) video sources ; the fourth from LAN-LAN interconnection sources ; and the fifth from delay insensitive sources, such as low-speed data, bulk data, and video delivery. Thus we have classified the VCs into four real-time services and one non-real-time service. The analytic model that we describe in the next section is independent of this classification, however. The number n_k next to the kth real-time service denotes the number of VCs of this service that are currently in progress. Except for the delay-insensitive services, we assume that the VCs belonging to the same service have identical cell generation statistics. Thus, if the multiplexer were to support two different types of video VBR – say, VHS and HDTV quality – then two services would have to be distinguished for video VBR.

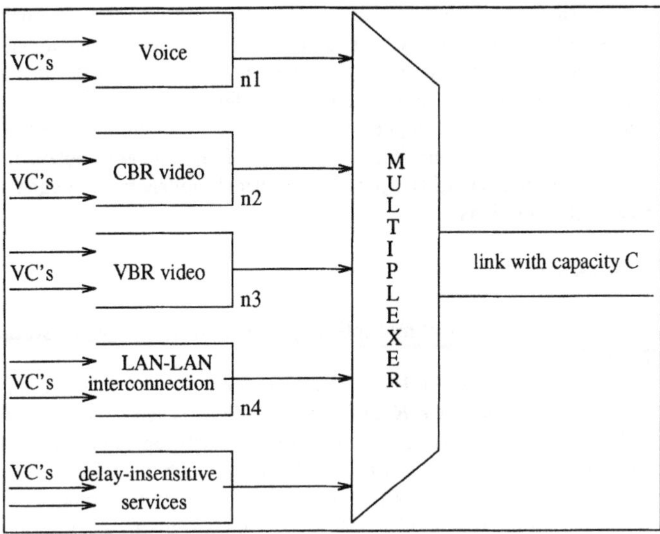

Figure 1 An ATM multiplexer integrating multiple services.

We assume throughout this paper that the buffer capacity allocated to each service is fixed. If a cell of a specific service arrives to find its buffer full, it is lost. Priority schemes for which a high-priority cell pushes out of the buffer a low-priority cell from the same service can also modeled. We neglect such priorities, however, in order not to obscure our main points about admission control.

Each service has a QoS requirement, which might be defined in terms of cell loss, cell delay, cell jitter, or a combination of these measures. The multiplexer must serve each buffer with sufficient frequency in order for the QoS requirements to be met for all VCs in progress. Obviously, the frequency with which the kth buffer must be serviced increases with n_k, the number of service-k VCs in progress.

Equivalent capacity

Before defining separable multiplexing, we digress and consider a link multiplexing n permanent service-k VCs, but no VCs from services other than k. This multiplexer's buffer is the kth buffer in the original multiplexer. Denote $\beta_k(n)$ for the minimum amount of link capacity needed in order for the QoS requirements to be met for the n service-k VCs. We call this function the **service-k capacity function**. Since $\beta_k(\cdot)$ is a function of a single parameter n, it should not be difficult to determine. For a CBR service and for peak-rate multiplexing the capacity function takes the form $\beta_k(n) = b_k n$, where b_k is the bit rate of a single VC. For bursty sources, the capacity function will reflect the economies of scale associated with statistical multiplexing : as n increases, the capacity function will increase, but its slope will decrease.

In particular, since this system involves only one service, it is substantially easier to analyze with discrete-event simulation than analytically a multiplexer which integrates multiple service types and allows statistical multiplexing across services. Furthermore, numerous simple, analytic models are available in the literature for approximating the capacity function for multiplexers with homogeneous sources. These models determine the **equivalent capacity** needed by n bursty homogeneous sources for a given QoS. (There are also in the literature some analytic models for heterogeneous sources, but they are not always accurate and depend on questionable assumptions.) Throughout the remainder of this paper we assume that the capacity functions are known. In Section 4, as an example we shall use one of the popular anaytical models for homogeneous sources to construct capacity functions.

Static Partitions

We now define separable statistical multiplexing. There are two versions: static partitions and dynamic partitions.

Consider again the multiplexer in Figure 1 with link capacity C. It is convenient to generalize the model so that there are K buffers for K delay-sensitive services and another buffer (labeled 0) for all the delay-insensitive services. Partition the capacity C into allocations $C_0, \ldots, C_K$ such that $C_0 + \cdots + C_K = C$. The $K+1$ buffers are served by the link in a weighted round-robin fashion, with the weights being proportional to the capacity allocations. For example, if $K = 2$, $C = 150$, $C_0 = 10$, $C_1 = 40$, $C_2 = 100$, then in a cycle of fifteen cells, the first buffer is served one time, the second four times, and the third ten times. If during the cycle the multiplexer finds one of the buffers empty, it instead serves the 0th buffer (delay-insensitive services). There are several specific algorithms in the literature for weighted round-robin scheduling ; for example see the fluid algorithm of Parekh and Gallager (1992,1993-1) or the dynamic-time-slice algorithm of Sriram (1993). Instead of this schedule, we could also use the Generalized Processor Sharing (GPS) scheduling (Parekh,1993-2) which is generally implemented in ATM multiplexers. In fact, the scheduling scheme has no consequence on our call admission technique.

Separable statistical multiplexing with a static partition admits a newly arriving delay-sensitive service-k VC if and only if $\beta_k(n_k + 1) \leq C_k$ when n_k service-k VCs are already in progress. Thus this scheme statistically multiplexes VCs within the same service k, but does not allow service-k VCs to interfere with service-j VCs for all $j \neq k$. Note that this scheme coupled with the round-robin service mechanism essentially guarantees that the QoS requirements are met for all VC configurations. We write "essentially" because the cells from the kth service are not served at a constant rate of C_k, as is required in the definition of $\beta_k(n)$. Instead, due to the round-robin discipline, these cells are served at rate C in batches ; but the average service rate is C_k and the fluctuation should be negligible if the granularity of the round robin discipline is sufficient.

Dynamic Partitions

Since VC arrivals are random, there will be time periods when the number of VC establishment requests for a particular service are unusually large. With static partitions, the VC blocking for this service might be excessive during these periods. The following multiplexing scheme alleviates this problem by dynamically allocating bandwidth to the services. It is similar to the scheme proposed by Gallassi et al (1990) and to the scheme proposed by Sriram (1993). Let β_0 be a number less than C.

We again assume that the buffers are served by the link in a weighted round-robin fashion, but now with the weights being proportional to $\beta_0, \beta_1(n_1), \ldots, \beta_K(n_K)$. For example, suppose $K = 2$, $n_1 = 4$, $n_2 = 6$, $\beta_1(4) = 50$ Mbps, $\beta_2(6) = 80$ Mbps, and $\beta_0 = 10 Mbps$. Then in a cycle of 15 cells, the first buffer is served 5 times, the second eight times, and the third (for time-insensitive services) is served two times (once for its allocation and once because there is a free slot in the cycle). Again, if during a cycle the multiplexer finds one of the buffers empty, then it instead serves the buffer for delay-insenstive services. Thus the round-robin weights dynamically change, but on the relatively slow time scale of VC arrivals and departures.

Separable statistical multiplexing with dynamic partitions admits a newly arriving service-k VC, $k = 1, \ldots, K$, if and only if

$$\beta_1(n_1) + \cdots + \beta_k(n_k + 1) + \cdots + \beta_k(n_K) \leq C - \beta_0. \tag{1}$$

This scheme again statistical multiplexes the VCs of the same service, but it does not limit a service to a fixed bandwidth allocation. Indeed, any one delay-sensitive service can consume up to $C - \beta_0$ of the bandwidth over a period of time. This scheme coupled with a dynamic round-robin service mechanism essentially guarantees that the QoS requirements are met for all VCs.

3 PERFORMANCE EVALUATION

In order to simplify the discussion, we henceforth assume that all services are delay-sensitive. Thus there is no longer a buffer delay-insensitive traffic in our model. We also assume that service-k VC establishment requests arrive according to a Poisson process with rate λ_k. The holding time of a service-k may have an arbitrary distribution ; denote $1/\mu_k$ for its mean. Also let $\rho_k := \lambda_k/\mu_k$.

We can easily analyze VC blocking for static partitions. The maximum number of service-k VCs that can be present in this system is $\lfloor \beta_k^{-1}(C_k) \rfloor$. Since there is no interaction between services, the probability of blocking a service-k VC is given by the Erlang loss formula with offered load ρ_k and capacity $\lfloor \beta_k^{-1}(C_k) \rfloor$.

Each partition $(C_1, \ldots, C_K)$ defines one static partition policy. If we define a revenue rate r_k for each service k, we can employ dynamic programming to find the optimal separable multiplexing policy with static partitions ; see Ross (1995).

For the remainder of this paper we focus on separable statistical multiplexing for dynamic partitions. The set of all possible VC configurations for this scheme is

$$\Lambda^s := \{\mathbf{n} : \beta_1(n_1) + \cdots + \beta_K(n_K) \leq C\}$$

where $\mathbf{n} := (n_1, \ldots, n_K)$ is a VC configuration. Of course Λ^s is a subset of Λ, the set of all possible VC configurations that meet the QoS requirements (including those resulting from statistical multiplexing across services). Nevertheless, Λ^s may closely approximate Λ for certain scenarios, in which case little is lost by disallowing statistical multiplexing across services.

We now present a methodology for calculating VC blocking probabilities for separable statistical multiplexing with dynamic partitions. Let $\pi(\mathbf{n})$, $\mathbf{n} \in \Lambda^s$, be the equilibrium probability of being in VC configuration $\mathbf{n}$.

Theorem 1 *The equilibrium probability that the VC configuration is* $\mathbf{n}$ *has the following product form:*

$$\pi(\mathbf{n}) = \frac{1}{G} \prod_{k=1}^{K} \frac{\rho_k^{n_k}}{n_k!}, \quad \mathbf{n} \in \Lambda^s \tag{2}$$

where

$$G := \sum_{\mathbf{n} \in \Lambda^s} \prod_{k=1}^{K} \frac{\rho_k^{n_k}}{n_k!}. \tag{3}$$

Proof. First assume that the holding times are exponentially distributed and that $C = \infty$. Then the stochastic process corresponding to n_k is a birth-death process with equilibrium probability

$$\pi(n_k) = \frac{\rho_k^{n_k}}{n_k!} e^{-\rho_k}. \tag{4}$$

Furthermore, the K birth-death processes are independent, and hence the joint stochastic process corresponding to n is reversible. Imposing a finite value for C corresponds to truncating the state space of the joint stochastic process. The resulting truncated process has the equilibrium probabilities given above (Kelly, 1979). Finally, it follows from standard arguments that this result is insensitive to the holding time distributions (Kelly, 1979). □

The set of VC configurations for which a newly arriving service-l VC is accepted is

$$\Lambda_l^s := \{\mathbf{n} : \beta_1(n_1) + \cdots + \beta_l(n_l + 1) + \cdots + \beta_K(n_K) \leq C\}.$$

Therefore, from Theorem 1, the probability of blocking a newly arriving service-l VC is

$$B_l = 1 - \sum_{\mathbf{n}\in\Lambda_l^s} \pi(\mathbf{n}) = 1 - \frac{\sum_{\mathbf{n}\in\Lambda_l^s}\prod_{k=1}^K \rho_k^{n_k}/n_k!}{\sum_{\mathbf{n}\in\Lambda^s}\prod_{k=1}^K \rho_k^{n_k}/n_k!}. \quad (5)$$

Thus, to obtain the probability that a service-l VC is blocked, it suffices to calculate the sums in (**??**). One possible approach is to use Monte Carlo summation for loss networks (Ross and al, 1992) (Ross, 1995). Another way for calculating blocking probabilities is to use a recursive algorithm as developed by Kaufman (1981) but it only works when the $\beta_k(n)$ functions are linear ; it not the case here (see figures 2 and 3 as examples of $\beta_k(n)$ functions). Below we give alternative approach based on a convolution algorithm.

Henceforth assume that $\beta_k(n)$ is integer valued. Consider calculating the sum in denominator of (**??**):

$$G := \sum_{\mathbf{n}\in\Lambda^s}\prod_{k=1}^K \frac{\rho_k^{n_k}}{n_k!}. \quad (6)$$

Note that

$$\begin{aligned} G &= a\sum_{\mathbf{n}\in\Lambda^s}\prod_{k=1}^K e^{-\rho_k}\frac{\rho_k^{n_k}}{n_k!} \\ &= a\sum_{\mathbf{n}\in\Lambda^s} P(Y_1 = n_1,\ldots,Y_K = n_K) \\ &= aP(\beta_1(Y_1)+\cdots+\beta_K(Y_K)\le C) \\ &= a\sum_{c=0}^{C} P(\beta_1(Y_1)+\cdots+\beta_K(Y_K) = c) \end{aligned}$$

where

$$a = e^{\rho_1+\cdots+\rho_K}. \quad (7)$$

and the Y_k's are independent random variables, with Y_k having the Poisson density

$$P(Y_k = n) = \frac{e^{-\rho_k}\rho_k^n}{n!} \qquad n = 0,1,2,\ldots \quad (8)$$

Let

$$g_k(c) = P(\beta_k(Y_k) = c), \quad c = 0,1,\ldots,C \quad (9)$$

and

$$\mathbf{g}_k = [g_k(0), g_k(1),\ldots,g_k(C)]. \quad (10)$$

Then

$$G = a \sum_{c=0}^{C} (\mathbf{g}_1 \otimes \cdots \otimes \mathbf{g}_K)(c), \tag{11}$$

where $\otimes$ denotes the convolution operator, that is,

$$(\mathbf{g}_1 \otimes \mathbf{g}_2)(c) = \sum_{d=0}^{c} g_1(d) g_2(c-d). \tag{12}$$

Since $\beta_k(\cdot)$ is (almost certainly) an increasing function, it should not be difficult to obtain the $\mathbf{g}_k$'s. The $K-1$ convolutions in (**??**) can be done in a total of $O(KC^2)$ time. (This complexity depends on the granularity of the units for C.) Calculating the numerator in (**??**) can be done in the same manner by replacing $\beta_l(n)$ by $\beta_l(n+1)$ for all n. The techniques in Section 3.5 of Ross (1995) can accelerate the calculation of the K blocking probabilities, $B_1, \ldots, B_K$.

We conclude this section by mentioning some generalizations and extentions. First, the assumption of Poisson arrivals can be relaxed — the same convolution algorithm can be used for arrival rates of the form $\lambda_k(n_k)$ and, in particular, for finite-population arrivals. Second, since derivatives of blocking probabilities can also be represented in terms of normalization constants, the above convolution algorithm can also be used to obtain these performance measures. Third, our model for separable multiplexing can be used to obtain the optimal admission control policy subject to the constraint that the statistical multiplexing is separable ; see Ross (1995).

4 NUMERICAL EXAMPLE

As we mentioned earlier, the capacity functions, $\beta_k(\cdot)$'s can be obtained with discrete-event simulation or approximated analytically. We now outline one analytical approach, due to Guérin et al (1991). For $k = 1, \ldots, K$, assume the following QoS requirement for a service-k VC: No more than the fraction ϵ_k of the VC's cells may be lost.

Digress again and consider a multiplexer supporting n permanent service-k VCs. Assume that each VC alternates between *On Periods* and *Off Periods.* The VC generates cells at the peak rate during an On Period ; it generates no cells during an Off Period. Let b denote the peak rate (in the same units as C) during an On Period. Assume that the lengths of these periods are independent and exponentially distributed. Denote Δ for the average On Period (in seconds). Denote u for the utilization of a VC, that is, the average On Period divided by the sum of the average On Period and the average Off period. Let Q be the capacity of the input buffer and ϵ be the QoS requirement. Guérin et al approximate the capacity function as follows :

$$\beta_k(n) = \min\{\beta_k^{(1)}(n), \beta_k^{(2)}(n)\}, \tag{13}$$

where

$$\beta_k^{(1)}(n) = n[\frac{\ln(\epsilon)\Delta(u-1)b - Q + \sqrt{[\ln(\epsilon)\Delta(u-1)b - Q]^2 + 4Q\ln(\epsilon)\Delta u(u-1)b}}{2\ln(\epsilon)\Delta(u-1)}] \quad (14)$$

and

$$\beta_k^{(2)}(n) = nbu + b\sqrt{nu(1-u)}\sqrt{-2\ln(\epsilon) - \ln(2\pi)}. \quad (15)$$

Clearly this method for estimating $\beta_k(n)$ is quite simple. Note that b, ϵ, Δ, and u are different for different services.

Our numerical example is for a multiplexer of capacity $C = 150$ Mbps, integrating three delay-insensitive services ($K = 3$). We set the buffer capacity, Q, equal to 6 Mbits for each service. We have used the following parameters for the three services as defined in Table 1.

Table 1 Parameters for Multiplexer with Three Services

Class k	Peak Rate b_k	Burst Length Δ_k	Utilization u_k	QoS ϵ_k
1	1Mbps	100msec	0.4	10^{-5}
2	10Mbps	100msec	0.2	10^{-4}
3	5Mbps	100msec	0.5	10^{-6}

We use the above procedure to determine the capacity functions for the three services. We have rounded up all the $\beta_k(n)$'s to the nearest integer. Figure 2 compares the used capacity for service 3 and three allocation schemes : mean rate, peak rate and equivalent capacity. We see that curves for mean rate and peak rate are linear because each time a new connection arrives the capacity increases by 2.5 Mbit for mean rate and by 5 Mbit for peak rate. At the opposite, the $\beta_3(n)$ is not linear. The mean rate function gives always the minimal capacity but does not guarantee the QoS. At the opposite, the peak rate function is maximal.

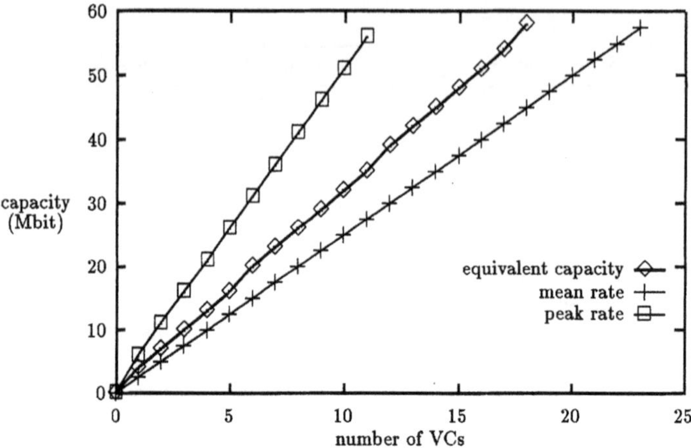

Figure 2 Capacity for peak rate, mean rate and equivalent capacity allocation schemes versus number of VCs (service 3).

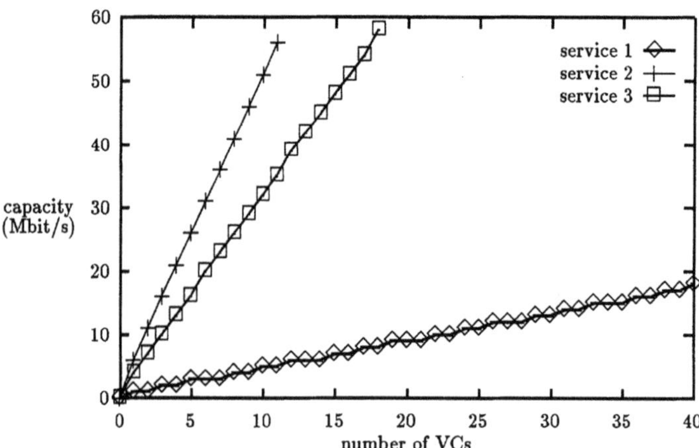

Figure 3 Equivalent capacity for the three types of services depending on the number of VCs.

Figure 3 presents the $\beta_k(n)$'s for the three services versus n. They logically increases with n and depends on their peak rate.

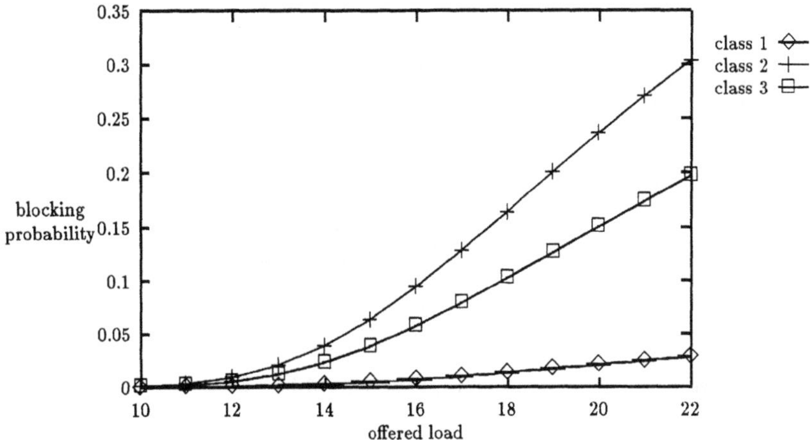

Figure 4 Blocking probabilities for the three classes versus offered load.

Figure 4 presents the blocking probabilities for separable statistical multiplexing obtained from the convolution algorithm. In this and the subsequent figures we set $\rho_1 = \rho_2 = \rho_3$ and plot blocking probabilities as a function of ρ_1. The amount of time required by the convolution algorithm for a specific value of ρ_1 is less than a second on a SPARC 2 workstation.

Figure 4 shows that blocking probabilities depend mainly on the peak rate b_k as b_1 is upper to b_3 which is upper to b_2. They depend also on mean rate $b_k u_k$ of the service, and to a lesser extent on the QoS parameter because probabilities for class 3 are close to those of class 2. As expected, service-1 VCs have the lowest blocking probability because of their low peak and average cell generation rates. It is interesting to note that although class-2 has a lower average rate and a less stringent QoS requirement than class-3, it has a higher VC blocking probability. This is due to its high peak rate, which renders its cell stream very bursty. We also note that VC blocking probabilities greatly vary from service to service.

Figures 5 to 7 compare the performance of separable statistical multiplexing to peak-rate multiplexing. There is one figure for each service. The curves for peak rates are obtained by setting $\beta_k(n) = b_k n$ for all services.

As expected, these figures show that the blocking probabilities for separable statistical multiplexing is less than that for peak-rate multiplexing. What may be surprising is how dramatic this difference in performance can be. For example, with $\rho_1 = 12$, the blocking probabilities for all three services with separable multiplexing is less than 1% ; this blocking probability is roughly 4%, 33%, and 19% for the three services with peak-rate multiplexing. The curves for statistical multiplexing *with* statistical multiplexing across

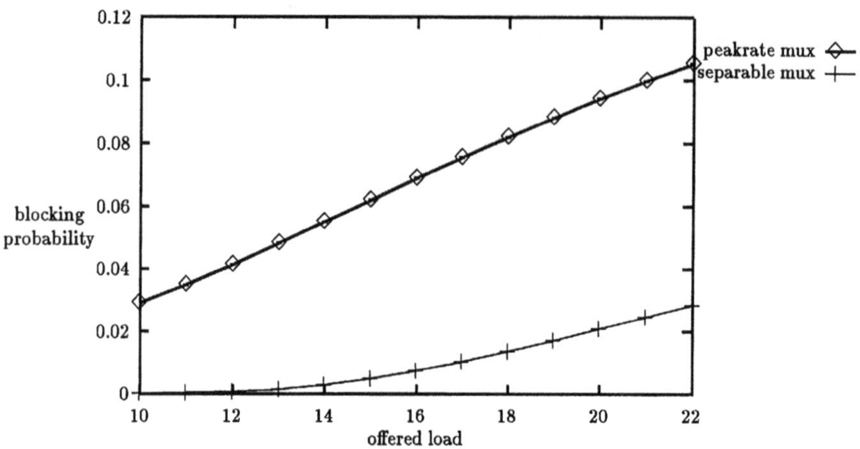

Figure 5 Blocking probabilities versus offered load for service-1 VCs.

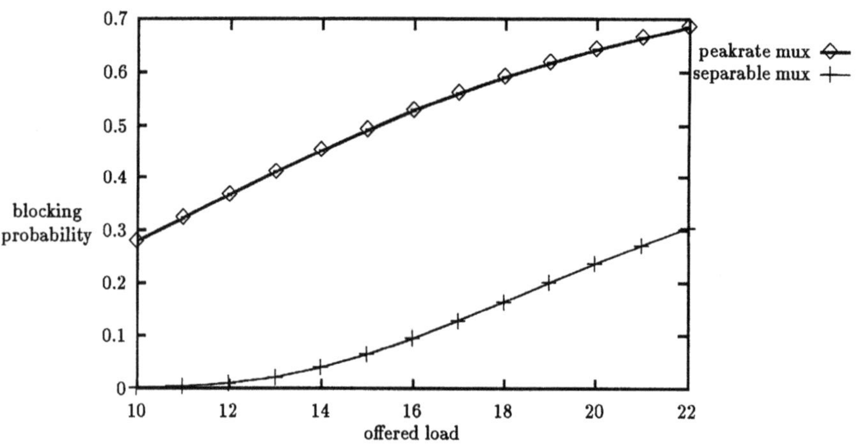

Figure 6 Blocking probabilities versus offered load for service-2 VCs.

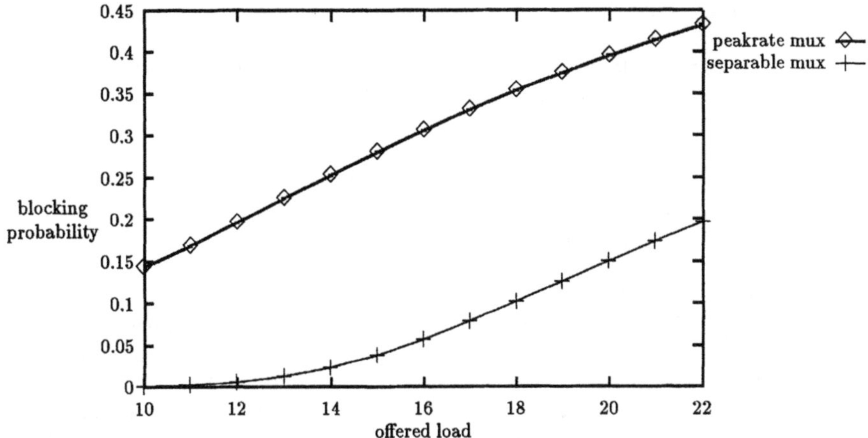

Figure 7 Blocking probabilities versus offered load for service-3 VCs.

classes would lie somewhere below the curves for separable statistical multiplexing. We conjecture that they would be not far below.

5 CONCLUSION

We have developed an efficient convolution algorithm to estimate VC blocking probabilities for separable statistical multiplexing. The numerical results show that separable statistical multiplexing can give substantial gains in performance over peak-rate multiplexing.

There are several related problems that merit attention. (1) A detailed study comparing the blocking probabilities for separable statistical multiplexing with "maximal multiplexing", that is, multiplexing across and within services. Estimating blocking with maximal multiplexing would require discrete-event simulation at the cell. (2) For separable statistical multiplexing, a cell-layer simulation should verify that the QoS requirements are indeed met with the weighted round-robin scheduling disciplines. (3) A theory for separable statistical multiplexing for *networks* should be developed (see Ross, 1995).

REFERENCES

F. Bonomi, S. Montagna, and P. Paglino.(1993) A further look at statistical multiplexing in ATM networks. *Computer Networks and ISDN Systems*, **26**, 119–38.

G. Gallassi, G. Rigolio, and L. Verri.(1990) Resource management and dimensioning in ATM networks. *IEEE Network Magazine*, **05**, 8–17.

R. Guérin, H. Ahmadi, and M. Naghshineh. (1991) Equivalent capacity and its application to bandwidth allocation in high-speed networks. *IEEE Journal on Selected Areas in Communications*, **9**, 968–81.

J.S. Kaufman. (1981) Blocking in a shared resource environment. *IEEE Trans. on Comm.*,**COM-29**, 1474–81.

F.P. Kelly. (1979) *Reversibility and Stochastic Networks.* Wiley, Chichester.

A.K. Parekh and R.G. Gallager. (1992) A generalized processor sharing approach to flow control in integrated services networks. In *Proceedings of IEEE INFOCOM'92.*

A.K. Parekh and R.G. Gallager. (1993) A generalized processor sharing approach to flow control in integrated services networks. In *Proceedings of IEEE INFOCOM'93.*

A.K. Parekh and R.G. Gallager. (1993) A generalized processor sharing approach to flow control in integrated services network: the single node case. *IEEE/ACM Transaction on Networking*, **1**, 344–57.

K. W. Ross. (1995) *Multiservice Loss Models for Broadband Telecommunication Networks.* Springer-Verlag, London.

K.W. Ross and J. Wang. (1992) Monte Carlo summation applied to product-form loss networks. *Probability in the Engineering and Informational Sciences*, 323–48.

K. Sriram. (1993) Methodologies for bandwidth allocation, transmission scheduling, and congestion avoidance in broadband ATM networks. *Computer Networks and ISDN Systems*, **26**, 43–60.

Y. Takagi, S. Hino, and T. Takahashi. (1991) Priority assignment control of ATM line buffers with multiple QOS classes. *IEEE Journal on Selected Areas in Communications*, **9**, 1078–92.

17

On the Effective Bandwidth of Arbitrary on/off Sources *

K. M. Elsayed[†]
BNR INC.
P.O.Box 833871
Richardson, TX 75083-3871, U.S.A.
khalede@bnr.ca

H. G. Perros
Department of Computer Science
North Carolina State University
Raleigh, NC 27695-8206, U.S.A.
hp@csc.ncsu.edu

Abstract

The effective bandwidth approximation is an attractive mechanism for performing call admission control in ATM networks. We present a methodology for evaluating the effective bandwidth of arbitrary on/off sources. An arbitrary on/off source is described by a stochastic process which alternates between on and off periods that have arbitrary probability density functions. We present two approximate methods for evaluating the effective bandwidth of an arbitrary on/off source. The first method is based on moment matching of the on and off periods. The second method is called the bandwidth matching procedure and is considerably more efficient than the moment matching procedure.

Keywords
Effective bandwidth, call admission control, arbitrary on/off sources

1 Introduction

In this paper, we devise a methodology for calculating the effective bandwidth of an arbitrary on/off source. The effective bandwidth is a mechanism for carrying out the call admission control process in the high speed networks environment. The methodology of the effective bandwidth is in general applicable to the case of Markov modulated sources. Arbitrary on/off source are a special case of semi-Markov processes and are in general non-Markovian. Using an elaborate mapping, we find a Markov chain which can be used to characterize an arbitrary on/off source. We then apply the effective bandwidth method to the resulting Markov source.

*Supported in part by BellSouth, GTE Corporation, and NSF and DARPA under cooperative agreement NCR-8919038 with the Corporation for National Research Initiatives and in part by a gift from BNR INC.

[0]Work was done when K. Elsayed was with Dept. of Computer Science, N. Carolina State Univ.

For a Markov modulated source, the computation of the effective bandwidth involves finding the spectral radius or sometimes the largest positive eigenvalue of a matrix which is a function of the modulating Markov chain, the arrival rates at the states of the Markov chain, the multiplexer buffer size, and the required cell loss probability. This is usually a computationally intensive process. In order to be able to perform call admission decisions using the effective bandwidth scheme in real-time, we propose a heuristic matching procedure in which the original source is mapped to an *equivalent* binary Markov modulated on/off source. The equivalent binary Markov modulated on/off source, for which the effective bandwidth has a simple characterization, can then be used in place of the original source in the call admission control process. As evidenced by the numerical examples reported in the paper, the accuracy of the approximation is valid for practical purposes.

Guibert [8] obtains results for the overflow probability for heterogeneous fluid queues with arbitrary on/off sources input. The effective bandwidth of an arbitrary on/off fluid source is obtained as a side result. However, the expression for the effective bandwidth obtained depends only on the first two moments of the on and off periods. The squared coefficient of variation of the on and off periods contributes equally to the effective bandwidth of the source which contradicts the well known fat that the variation in the on period has more impact on the effective bandwidth than the off period [3, 2].

This paper is organized as follows. In section 2 we discuss the concept of effective bandwidth for Markov modulated sources. In section 3, we show how to approximately calculate the effective bandwidth of an arbitrary on/off source. We refer to this approximation as the Markovian-transformation method. We present two less computationally-intensive approximate methods for characterizing the effective bandwidth of an arbitrary on/off source in section 4. In section 5 we compare the evaluation of the effective bandwidth using the Markovian-transformation with the two other proposed approximations. Conclusions are given in section 6.

2 Effective Bandwidth of Markov Modulated Sources

The primary role of a network congestion control procedure is to protect the network and the user in order to achieve network performance objectives and optimize the usage of network resources. In ATM-based B-ISDN, congestion control should support a set of ATM quality of service classes sufficient for all foreseeable B-ISDN services. Call admission control (CAC) is one of the primary mechanisms for preventive congestion control in an ATM network. CAC is one particular type of many possible resource allocation mechanisms performed by the network provider. In an ATM network, resource allocation can be identified on three different levels: call, burst, and cell levels [10].

In the context of ATM networks, the CAC process is described as follows. During the call setup phase, users declare and/or negotiate with the network their connection characteristics and their required quality of service. Some of the parameters that may be used to specify

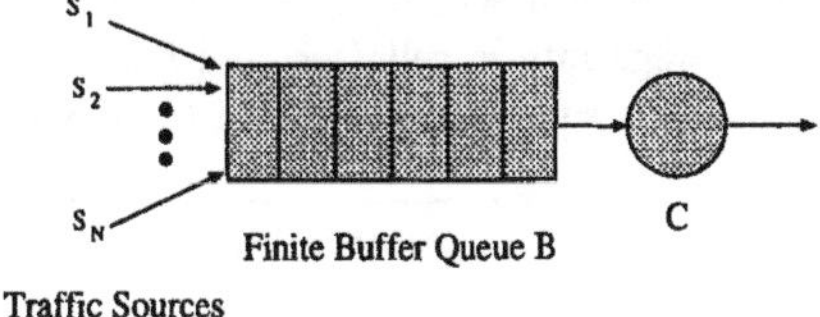

Figure 1: The multiplexer model

a call characteristics are sustainable (average) bit rate, peak bit rate, cell delay variation tolerance, and maximum burst length, as recommended by the ATM Forum [5].

In a loss network, where the probability of lost cells is the primary grade of service, the effective bandwidth scheme emerged as an attractive scheme for performing call admission control. For a finite buffer multiplexer with buffer size B and output link capacity C as shown in Figure 1, the effective bandwidth of a call is defined as the link capacity needed in order to keep the call's cell loss probability below a specific value ϵ, assuming that the given call is the sole user of the multiplexer.

Guérin, Ahmadi and Naghshineh [7], and also Gibbens and Hunt [6], provide an efficient method to evaluate the effective bandwidth of a single call and the aggregate bandwidth of multiplexed calls. The arrival process is modeled as a binary Markov modulated on/off fluid source. The results of Anick et al. [1] were used to calculate the effective bandwidth.

Elwalid and Mitra [3] generalize these results to general Markovian traffic sources for both fluid and point processes models. The effective bandwidth is shown to be the maximal real eigenvalue of a matrix directly obtained from the source characteristics and admission criterion. The approximation is mostly valid in the asymptotic regime as buffer size approaches infinity and the cell loss probability tends to zero.

Kesidis, Walrand, and hang [12] use large deviations theory to find an approximation of the effective bandwidth for a variety of traffic sources. We use their results to handle the case of discrete-time Markov modulated sources. The main result for discrete-time sources can be summarized as follows. Consider a traffic source modeled as an L-state source $(\boldsymbol{P}, \vec{\boldsymbol{\lambda}})$ where $\boldsymbol{P}$ is the probability transition matrix of the modulating Markov chain and $\vec{\boldsymbol{\lambda}} = (\lambda_0, \lambda_1, \cdots, \lambda_{L-1})$ is the vector of peak arrival rates at the various states. Let $\bar{\lambda}$ be the mean rate and $\hat{\lambda}$ be the maximum rate. The buffer is served by a channel of capacity C ($C = 1$ after normalization). Define $G(B) = Pr[x \geq B]$, where x is the stationary buffer content. Then the grade of service is $G(B) \leq \epsilon$. The effective bandwidth e of a source $(\boldsymbol{P}, \vec{\boldsymbol{\lambda}})$ was shown to be [12]:

$$e = log(\Omega\{exp(\delta\boldsymbol{\Lambda})\boldsymbol{P}\})/\delta \tag{1}$$

where $\boldsymbol{\Lambda} = diag(\vec{\boldsymbol{\lambda}})$ and $\delta = -\frac{log(\epsilon)}{B}$, and $\Omega\{M\}$ is the spectral radius of matrix M. The value of e satisfies the relation: $\bar{\lambda} \leq e \leq \hat{\lambda}$. When N sources are multiplexed the total effective bandwidth is approximated by $\sum_{i=1}^{N} e_i$.

For a 2-state Markovian source described by the triplet (R, r, b), where R is the peak rate in cells/sec, r is the average rate in cells/sec, and b is the mean burst length in cells, the effective bandwidth, e, of the source can be shown to be equal to:

$$e = log(\frac{1}{2}\left[P_{11} + exp(\delta R)P_{22} + \sqrt{(P_{11} + exp(\delta R)P_{22})^2 + 4exp(\delta R)(1 - P_{11} - P_{22})}\right])/\delta \quad (2)$$

where $P_{11} = ((R-r)b - r)/(R-r)b$ and $P_{22} = 1 - 1/b$.

Other related work is Kelly [11] and Whitt [15].

3 Effective Bandwidth of an Arbitrary on/off Source

Consider a traffic source which alternates between on and off periods where the lengths of the on and off periods have an arbitrary probability distribution. The distributions of the off and the on periods are given by:

$$\begin{array}{ll} f^{off}(i) = Pr[t^{off} = i \text{ slots}] = \alpha_i & 1 \leq i \leq K \\ f^{on}(i) = Pr[t^{on} = i \text{ slots}] = \beta_i & 1 \leq i \leq L \end{array} \quad (3)$$

where t^{off} (t^{on}) is a random variable indicating the length of an off (on) period. Arrivals occur periodically every T slots during the on period.

To describe an arbitrary on/off source as a Markov modulated source, we define the following quantities:

$$\begin{array}{ll} a_i \triangleq Pr[t^{off} > i | t^{off} > i-1] = 1 - \alpha_i/(1 - \sum_1^{i-1} \alpha_j), & 1 \leq i \leq K \\ b_i \triangleq Pr[t^{on} > i | t^{on} > i-1] = 1 - \beta_i/(1 - \sum_1^{i-1} \beta_j), & 1 \leq i \leq L \end{array}$$

We can then represent the off and on periods as a Markov chain with L and K states respectively. The states are represented by the pair (i, s), $0 \leq s \leq 1$. States with $s = 0$ represent the off period, where $1 \leq i \leq K$, and states with $s = 1$ represent the on period, where $1 \leq i \leq L$. The variable i in the state descriptor (i, s) indicates that the Markov chain has reached slot i in the current period. When the source is in state (i, s), the only two possible transitions are:

1. The source moves to state $(i + 1, s)$ with probability a_i respectively b_i if $s = 0$ respectively $s = 1$. This means that the source stays for at least one more slot in its current state.
2. The source moves to state $(1, \bar{s})$, where $\bar{s} \neq s$ is the other type of period, with probability $1 - a_i$ respectively $1 - b_i$ if $s = 0$ respectively $s = 1$. This represents the end of the current period after i slots and the start of the other type of period in the following slot.

The discrete-time Markov chain representing the source is then given by:

$$
P = \left[\begin{array}{cccc|cccc}
0 & a_1 & & & 1-a_1 & 0 & \cdots & 0 \\
 & 0 & a_2 & & 1-a_2 & 0 & \cdots & 0 \\
 & & \ddots & \ddots & \vdots & \vdots & \ddots & \vdots \\
 & & 0 & a_{K-1} & 1-a_{K-1} & 0 & \cdots & 0 \\
 & & & 0 & 1 & 0 & \cdots & 0 \\
\hline
1-b_1 & 0 & \cdots & 0 & 0 & b_1 & & \\
1-b_2 & 0 & \cdots & 0 & & 0 & b_2 & \\
\vdots & \vdots & \ddots & \vdots & & & \ddots & \ddots \\
1-b_{L-1} & 0 & \cdots & 0 & & & & b_{L-1} \\
1 & 0 & \cdots & 0 & & & & 0
\end{array}\right]
$$

Now let us define $O_{i \times j}$ is matrix of i rows and j columns with entries equal to zero, $\Lambda = \begin{bmatrix} O_{K\times K} & O_{K\times L} \\ O_{L\times K} & RI_{L\times L} \end{bmatrix}$, $R = 1/T$, and $\delta = -log(\epsilon)/B$. The effective bandwidth of the source is *approximated* by

$$log(\Omega\{exp(\delta\Lambda)P\})/\delta$$

where $\Omega\{M\}$ is the spectral radius of matrix M.

3.1 The Eigenvalue Problem

As noted above, we need to solve for the spectral radius of the matrix $exp(\delta\Lambda)P$ This matrix is very sparse, since only $2(K+L-1)$ of the whole $(K+L)^2$ matrix elements are non-zero. This suggests that using sparse matrix techniques can be useful here, specially when K and/or L are large. However, for our purposes, we preferred to use the inverse-iteration method which proved very successful in solving for the effective bandwidth. In order to use the inverse-iteration method, we need to invert the matrix $exp(\delta\Lambda)P$. The inversion process requires the storage of the whole matrix.

Other methods of interest, which make use of the sparsity of the matrix, include the Krylov subspace method [14] which does not require the storage of the whole matrix. This method evaluates the characteristic equation of a given matrix. (The roots of the characteristic equation are the eigenvalues of the matrix.) A possible solution of the problem would then be to solve for the root of the characteristic equation in the period $[r, R]$, where r is the average rate of the source. It has been reported in Faddeeva [4] that the accuracy of the method is not satisfactory and poor-conditioning can occur easily. This is due to the fact that the coefficients of the characteristic equation, which are quantities of different order of magnitude, are solved by the same system of linear equations. A good candidate for solving this eigenvalue problem is the simultaneous iteration method for obtaining the set of eigenvalues of largest absolute magnitude as described by Stewart and Jennings [13]. This method is particularly suitable for large sparse matrices.

The power method, a method for finding the eigenvalue with the largest magnitude which is known to be very suitable for sparse-matrix calculations, can not be used here efficiently. This is because of the highly periodic nature of the matrix $\boldsymbol{P}$.

4 Approximations for the Effective Bandwidth of an Arbitrary on/off Source

The disadvantage of the methodology presented in the previous section is clearly the computational and storage complexity. In this section, we propose two approximation methods for characterizing the effective bandwidth of an arbitrary on/off source. The first method is based on *moment matching* of the statistics of the off and on periods. The second method finds a binary Markov modulated on/off source which, for some carefully chosen values of the buffer size and required cell loss probability, has the same peak rate, average rate, and effective bandwidth as the arbitrary on/off source. The resulting binary Markov modulated on/off source is then used in place of the original source for performing call admission control. We call this method the *bandwidth matching* method.

4.1 The Moment Matching Procedure

From the distributions of the off and on periods $f^{off}(i)$ and $f^{on}(i)$, we evaluate the mean and squared coefficient of variation of each period. We can then use standard techniques for fitting phase-type distributions to given moments. The arrival rate during the on period is set equal to $R = 1/T$. In this case, an on/off source with phase-type distribution for the on and off periods is used to substitute the original source. The effective bandwidth of the Markov modulated phase-type on/off source is calculated as we have discussed in section 3.

Alternatively, we can use the means of the on and off periods and the peak rate of the source to define a binary Markov modulated on/off source. The on and off periods have an exponential distribution with mean equal to the mean on and off periods of the original arbitrary on/off source. This source is referred to as the *plain* binary Markov modulated on/off source when it is used to substitute for the original source.

4.2 The Bandwidth Matching Procedure

Consider a binary Markov modulated on/off source specified by the triplet (R, r, b). The effective bandwidth of such a source can be obtained using equation 2 for a given value of the required cell loss probability ϵ and the buffer size B. The proposed bandwidth matching procedure is based on finding a binary Markov modulated on/off source which has the same effective bandwidth as that of the arbitrary on/off source, for a given ϵ and B. The peak rate and average rate of the binary source are set equal to those of the original source. We argue that the two sources have approximately the same effective bandwidth over a range of values of ϵ and B. The binary Markov modulated on/off source is used instead of the complex original source for call admission control decisions at different points of the ATM network. Our results strongly support the validity of this approximation.

The matching procedure is outlined as follows:

- Find the effective bandwidth e of the arbitrary on/off source for a chosen value of B and ϵ using the method developed in section 3. Our experience indicates that choosing $B = \lceil b = \text{mean length of the on period} \rceil$ provides a good approximation (which can be seen from the results in section 5.4).

- Find the mean length of the on period b' of the binary Markov modulated on/off source as follows:

$$\begin{aligned} b' &= \frac{1}{1-Q_{22}}, \\ Q_{22} &= [\eta^2 - (1-\omega)\eta - exp(\delta R)\omega] \,/\, [(\omega + exp(\delta R))\eta - (\omega+1)exp(\delta R)], \\ \omega &= \frac{r}{R-r}, \\ \eta &= exp(e\delta), \text{ and} \\ \delta &= -log(\epsilon)/B \end{aligned} \tag{4}$$

  The above relation is obtained from equation 2 relating the effective bandwidth of a binary Markov modulated on/off source to the parameters (R, r, b'), the required cell loss probability ϵ, and the buffer size B.

- Set the peak rate and average rate of the binary Markov modulated on/off source equal to those of the arbitrary on/off source.

The accuracy of the bandwidth matching procedure is better than the moment matching procedure (using one or two moments fitting). This will be demonstrated in section 5.

A related methodology was suggested by Gün [9] for approximating a complex source by a simple binary on/off source. In this method, however, it is necessary to obtain the complete queue length distribution through the numerical solution of the multiplexer with the complex source at the input. Explicit expressions for the queue length distribution of the multiplexer with the binary Markov modulated source are known from the work of Anick, Mitra, and Sondhi [1]. Gün used these expression to find the parameters of a binary Markov modulated on/off source that would make the queue length distribution of the original source and that of the binary Markov modulated source equal at a given value of the queue length. This method can only be applied to Markov modulated sources. Our method is more general in this regards as it can handle non-Markovian sources and avoids solving for the entire queue length distribution.

Table 1: The distribution of the off and on periods

Length	Probability
1	0.1
50	0.05
100	0.05
150	0.2
200	0.6

(a) Off Period

Length	Probability
1	0.6
100	0.4

(b) On Period

5 Results

In this section, we provide some numerical examples with a view to validating the methodologies of sections 3 and 4.

5.1 Example 1

For our first example, we choose an arbitrary on/off source with the distribution of the on and off periods shown in Tables 2(a) and 2(b). We obtain the effective bandwidth using four different approximations: the Markovian-transformation procedure of section 3, the bandwidth matching binary on/off source, the phase-type on/off source obtained by matching the first two-moments of the off and on periods, and the plain binary on/off source obtained by matching the means of the off and on periods. The bandwidth matching is carried out at a cell loss probability equal to 10^{-6} and a buffer size equal to $\lceil b \rceil = \lceil 40.6 \rceil = 41$. As it can be seen from Figure 2, the bandwidth matching provides the most accurate approximation over the range of the required cell loss probability and buffer sizes. It is noted here that as the buffer size increases, all the methods converge to an effective bandwidth equal to the average rate.

5.2 Example 2

We now consider an arbitrary on/off source where the distribution of the off period is a uniform distribution in the period [1,130]. The on period has a probability of 0.71 of being of length 1 and a probability of 0.29 of being of length 40.

We obtain the effective bandwidth using four different approximations: the Markovian-transformation procedure of section 3, the bandwidth matching binary on/off source, the phase-type on/off source obtained by matching the first two-moments of the off and on

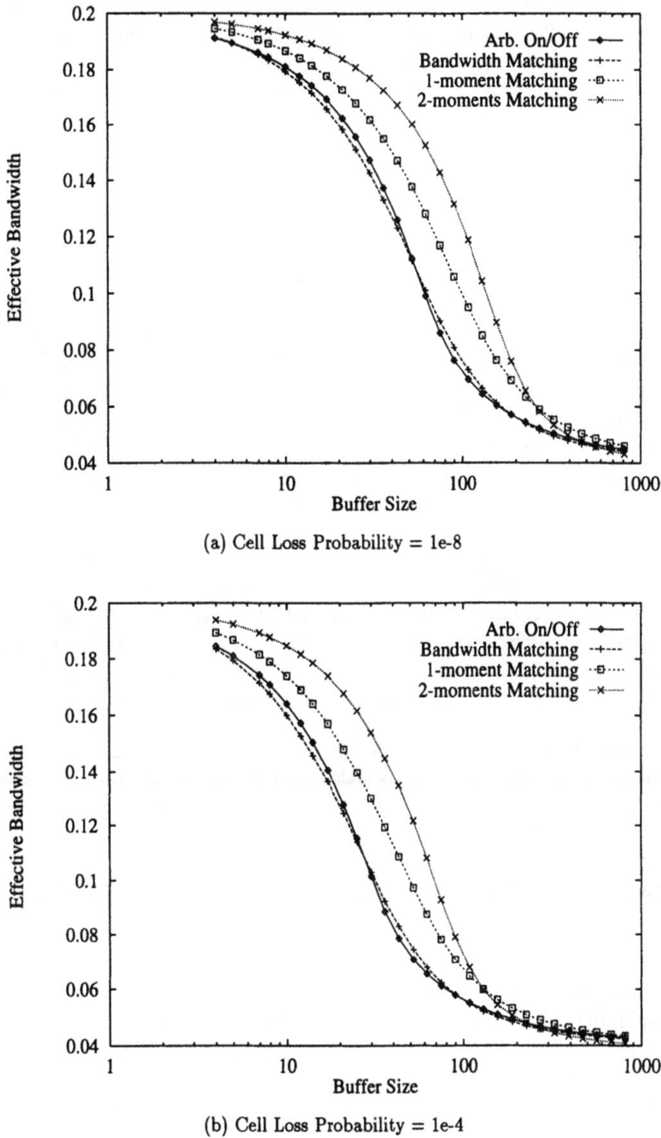

Figure 2: Comparison of various effective bandwidth calculation schemes

periods, and the plain binary on/off source obtained by matching the means of the off and on periods. The bandwidth matching is carried out at a buffer size equal to $\lceil b \rceil = \lceil 12.31 \rceil = 13$ and a cell loss probability equal to 10^{-6}. The results are plotted in Figure 3. Again, we note here that the bandwidth matching procedure provides the best overall accuracy. As observed in the other examples, as the buffer size grows, the effective bandwidth estimate of all the methods tends to the average rate of the source.

5.3 Example 3

We now use an on/off source with a phase-type distribution for the off and on periods. The mean and squared coefficient of variation of the off and on periods are given by the pairs (150.0,5.0) and (100.0,10.0) respectively. The peak rate during the on period is equal to 0.025. The mean and squared coefficient of variation are used to fit a hyper-exponential distribution of the off and on periods. In Figure 4, we plot the effective bandwidth as obtained by the effective bandwidth method for the phase-type on/off source, the bandwidth matching binary on/off source, and the plain binary on/off source in Figure 4. The bandwidth matching was done at a buffer size equal to $\lceil b \rceil = 100$ and a cell loss probability $\epsilon = 10^{-6}$.

In this example too, it can be seen that the bandwidth matching procedure provides an accurate estimate of the effective bandwidth. This is not the case with the plain binary on/off source whose estimate of the effective bandwidth is far lower than its actual value for a wide range of buffer sizes. If we use the plain binary source in the call admission control decision process, the required cell loss probability may not be achievable since we are admitting a larger number of sources than what can be handled.

Similarly, we tested the case when the on and off periods have a coefficient of variation which is smaller than one. The bandwidth matching procedure provided good accuracy whereas the plain binary on/off source provided an upper bound of the effective bandwidth.

5.4 Selection of Buffer Size for the Bandwidth Matching Procedure

The selection of the buffer size, at which the effective bandwidth of the arbitrary on/off source is evaluated, affects the accuracy of the bandwidth matching procedure. We use the on/off sources of examples 1, 2, and 3 to demonstrate this relationship. For the sources of examples 1 and 2, we obtain the effective bandwidth by the Markovian transformation method. For the source of example 3, we obtain its effective bandwidth by directly applying the effective bandwidth method since the source is already Markovian. The calculations of the effective bandwidth is compared with three different binary Markovian on/off sources obtained by bandwidth-matching. These three sources are obtained by applying the bandwidth-matching procedure at three different values of the buffer size. The three different values of the buffer size are designated as low, medium, and large.

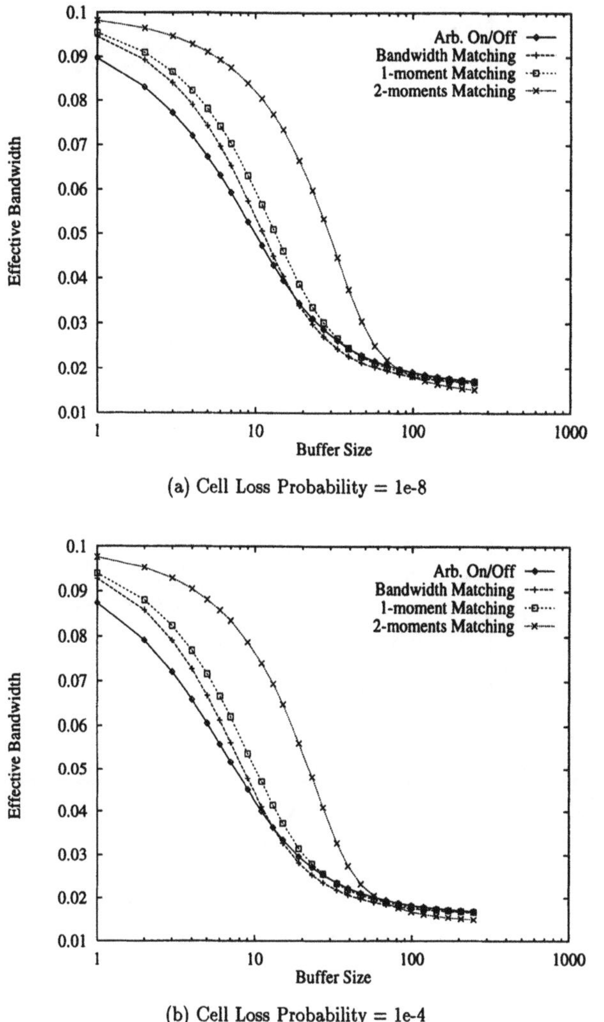

Figure 3: Comparison of various effective bandwidth calculation schemes – off period: uniform(1,130) – on period: P[1] = 0.71, P[40] = 0.29

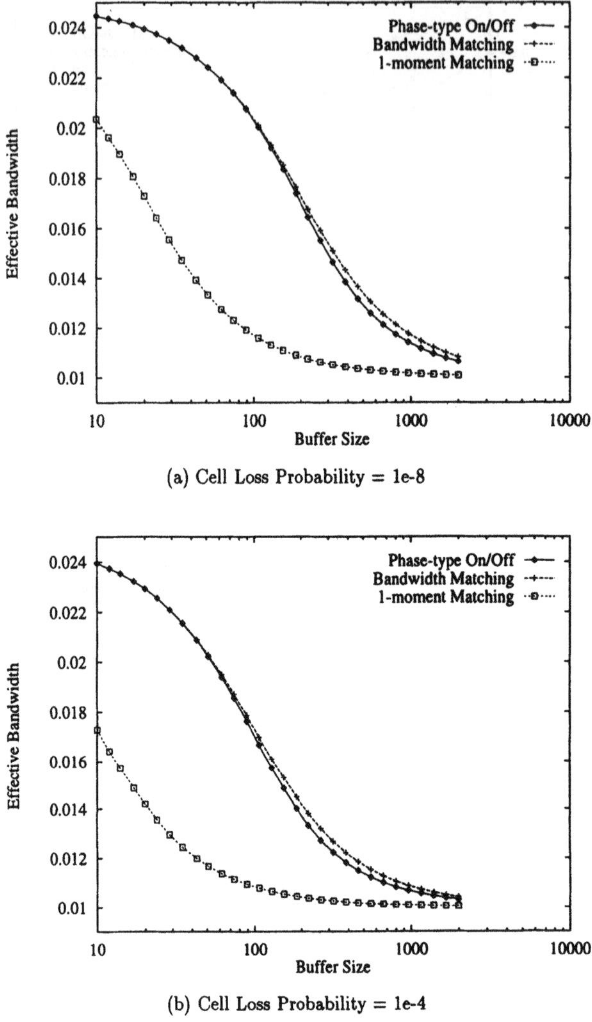

(a) Cell Loss Probability = 1e-8

(b) Cell Loss Probability = 1e-4

Figure 4: Comparison of various effective bandwidth calculation schemes – the source has a hyper-exponential on and off periods

and 7 equi-distant points for the required cell loss probability in the range $[10^{-9}, 10^{-3}]$. The results are shown in Tables 3(a), 3(b), and 3(c) for the sources of examples 1, 2, and 3 respectively. In the tables the column BWM buffer size indicates the buffer size at which the bandwidth matching procedure is performed. Matching at the buffer size value equal to $\lceil$mean length of the on period$\rceil$ provides the best overall performance of the three alternatives.

6 Conclusions

We have introduced the Markovian-transformation methodology for characterizing the effective bandwidth of an arbitrary on/off source. We proposed the bandwidth matching procedure as an effective mechanism for performing the effective bandwidth calculation in real-time. The bandwidth matching procedure provides good accuracy for the set of sources we tested. An interesting extension would be to investigate if it is feasible to provide upper and/or lower bounds for the cell loss probability in a statistical multiplexer with arbitrary on/off sources input using the bandwidth-matching procedure.

References

[1] D. Anick, D. Mitra, and M. M. Sondhi. Stochastic Theory of a Data-Handling System with Multiple Sources. *Bell Sys. Tech. J.*, 61:1871–1894, 1982.

[2] K. Elsayed. *Performance Analysis of Statistical Multiplexing and Call Admission Control in High-Speed Networks.* PhD thesis, North Carolina State University, 1995.

[3] A. I. Elwalid and D. Mitra. Effective Bandwidth of General Markovian Traffic Sources and Admission Control of High Speed Networks. *IEEE Transactions on Networking*, 1:329–343, 1993.

[4] D. K. Faddeev and V. N. Faddeeva. *Computational Methods of Linear Algebra.* W. H. Freeman and Company, 1963.

[5] ATM Forum. ATM User-Network Interface Specification, Version 2.2, June 1991.

[6] R. J. Gibbens and P. J. Hunt. Effective Bandwidths for the Multi-type UAS Channel. *Queueing Systems*, 9:17–28, 1991.

[7] R. Guérin, H. Ahmadi, and M. Naghshineh. Equivalent Capacity and Its Application to Bandwidth Allocation in High-Speed Networks. *IEEE Journal on Selected Areas in Communications*, 9:968–981, 1991.

Table 2: Effect of performing the bandwidth matching procedure at different values of the buffer size

BWM Buffer Size	RMS Error	Average Error
4	0.30%	0.64%
41	0.24%	-0.60%
812	0.31%	-2.57%

(a) Example 1

BWM Buffer Size	RMS Error	Average Error
1	1.36%	-15.61%
13	0.46%	-0.21%
248	0.52%	1.76%

(b) Example 2

BWM Buffer Size	RMS Error	Average Error
10	0.19%	2.14%
100	0.14%	1.44%
2000	0.11%	0.99%

(c) Example 3

The low buffer value is equal to maximum(1, $\lceil$ mean length of the on period/10 $\rceil$), the medium value is equal to $\lceil$mean length of the on period$\rceil$, and the large value is equal to $\lceil 20 \times$ mean length of the on period$\rceil$.

We calculate the root mean square relative error and average relative error between the three approximating sources and the arbitrary on/off source. The root mean square relative error and average relative error are defined as follows. Let N be the total number of readings, x be the actual value of a reading, and $\hat{x}$ be an estimated value for x. The root mean square relative error is equal to

$$\sqrt{\sum_{i=1}^{N}\left(\frac{\hat{x}_i - x_i}{x_i} \times 100\right)^2 / N},$$

and the average relative error is equal to

$$\sum_{i=1}^{N}\left(\frac{\hat{x}_i - x_i}{x_i} \times 100\right) / N.$$

The calculations are performed at logarithmically equidistant values of buffer sizes in the range [maximum$(1, \lceil$mean length of the on period/10$\rceil)$, $\lceil 20\times$mean length of the on period$\rceil$]

[8] J. Guibert. Overflow Probability Upper Bound for Heterogeneous Fluid Queues handling on-off Sources. In *Proceedings of 14th International Teletraffic Congress (ITC)*, pages 65–74, 1994.

[9] L. Gün. An Approximation Method for Capturing Complex Traffic Behavior in High Speed Network. *Performance Evaluation*, 19:5–23, 1994.

[10] J. Hui. Resource Allocation for Broadband Networks. *IEEE Journal on Selected Areas in Communications*, 6:1598–1608, 1988.

[11] F. P. Kelly. Effective Bandwidth at Multi-Class Queues. *Queueing Systems*, 9:5–16, 1991.

[12] G. Kesidis, J. Walrand, and C.-S. Chang. Effective Bandwidths for Multiclass Markov Fluids and Other ATM Sources. *IEEE Transactions on Networking*, 1(4):424–428, August 1993.

[13] W. J. Stewart and A. Jennings. A Simulatenous Iteration Algorithm for Real Matrices. *ACM Transactions on Mathematical Software*, 7:184–198, 1981.

[14] F. Stummel and K. Haines. *Introduction to Numerical Analysis*. Scottish Academic Press, Edinburgh, 1980.

[15] W. Whitt. Tail Probabilities with Statistical Multiplexing and Effective Bandwidth for Multi-Class Queues. *Telecommunications Systems*, 2:71–107, 1993.

18

The impact of the reactive functions on the LAN interconnection by a Frame-Relay Net

T. Atmaca
Institut National des Télécommunications
9, rue Charles Fourier, 91011 Evry Cedex, France
email: tulin@etna.int-evry.fr

T. Czachórski
Polish Academy of Sciences, Institute of Theoretical and Applied Computer Science
44-100 Gliwice, ul. Bałtycka 5, Poland (at sabbatical stay at INT, Evry, during the preparation of this article)
email: tadek@atos.iitis.gliwice.pl

Abstract

This article proposes the use of a diffusion approximation to model the effect of preventive and reactive functions implemented in a Frame-Relay network interconnecting LAN networks. These functions are based on two mechanisms: discard eligibility (DE) and explicit congestion notification (ECN). Diffusion models applied in this paper allow us to define how the input stream issued by LANs is changed as a result of the DE mechanism and to estimate the losses due to queue overflow when either a push-out or a threshold policy is used to manage the queue. The transient solution to the diffusion equation serves to estimate the dynamics of the evolution of queues during the changes of input flow issued by LANs and moderated by the ECN mechanism. It is the basis of a closed-loop, control theory model describing the influence of the ECN mechanism on FR network performance.

Keywords

Performance evaluation, Frame-Relay network, diffusion approximation.

1 INTRODUCTION

Frame Relay (FR) is a new fast packet technique that has been defined as a packet mode bearer service for the ISDN (CCITT,1991). It is widely seen as a solution for LAN-to-LAN interconnection. FR promises to combine circuit and packet switching and to achieve better response times than existing packet-switch networks can provide. It has been conceived to satisfy the requirements of emerging high speed data applications, to minimize transit delay or maximize throughput.

Like most high speed networks, FR networks require effective congestion control mechanisms to cope with unanticipated network component failures and overloads. Consequently, effective and efficient alternative congestion controls are particularly important in the architectural design of ISDN FR networks. In fast packet networks, alternative techniques based on congestion notification have been proposed to operate at the source level. These techniques use *Discard Eligibility* (DE) and *Explicit Congestion Notification* (ECN) mechanismes. They include the preventive and reactive control mechanismes.

First, each time permission is asked by a source to send additional traffic (a stream of frames), *Connection Admission Control (CAC)* procedures estimate whether this additional traffic will not deteriorate the quality of service, measured in frame loss probability due to overflow of buffers in transmission nodes. If the loss probability of frames does not overpass a predefined level, the new traffic is allowed but the agreed parameters are monitored constantly. The admission control mechanism makes use of four parameters:
• T_c – *Commited Rate Measurement Interval*, the reference interval for B_c and B_e which are defined below,
• B_c – *Commited Burst Size*: maximum amount of data which can be sent with high priority during the interval T_c,
•B_e – *Excess Burst Size*: maximum amount of data which can be sent in addition to B_c with low priority during the interval T_c,
• *CIR* – *Commited Information Rate*: average transmission rate guaranteed, $CIR = \frac{B_c}{T_c}$.
T_c is the basic interval of control performed at the *interface* between LAN and FR networks: as long as the volume of traffic received in the interval T_c remains below the level B_c, all frames are marked as "high priority" (DE bit set to 0). When the threshold B_c is reached, frames are marked as "low priority" (DE bit set to 1). All frames received in excess of $B_c + B_e$ are discarded. Frames with low priority are first to be discarded in FR nodes when the danger of congestion arises.

Apart from the *preventive functions* sketched above, *reactive functions* watch if the current situation in the network does not indicate congestion and, if necessary, inform the source that it should limit its activity. It is done with the use of ECN messages which are sent forward by network nodes (FECN bit in frames is set to 1) and/or sent backward (BECN bit is set and a special control packet is sent back to the source) to notify the receiver and/or the sender that the node is entering a rejection rate (e.g. the loss of frames exceeds a certain threshold or the queue length is greater than a determined value). According to CCITT recommendations (CCITT,1991) this function is optional and the details are not specified. The source will subsequently reduce its traffic input rate to the network. If control packets are not received during a certain time interval, then the source can start to increase its traffic rate.

In this paper, we consider a model of LAN interconnection through a Frame Relay network. The model is shown in Figure 1; Figure 2 represents the queueing model. The

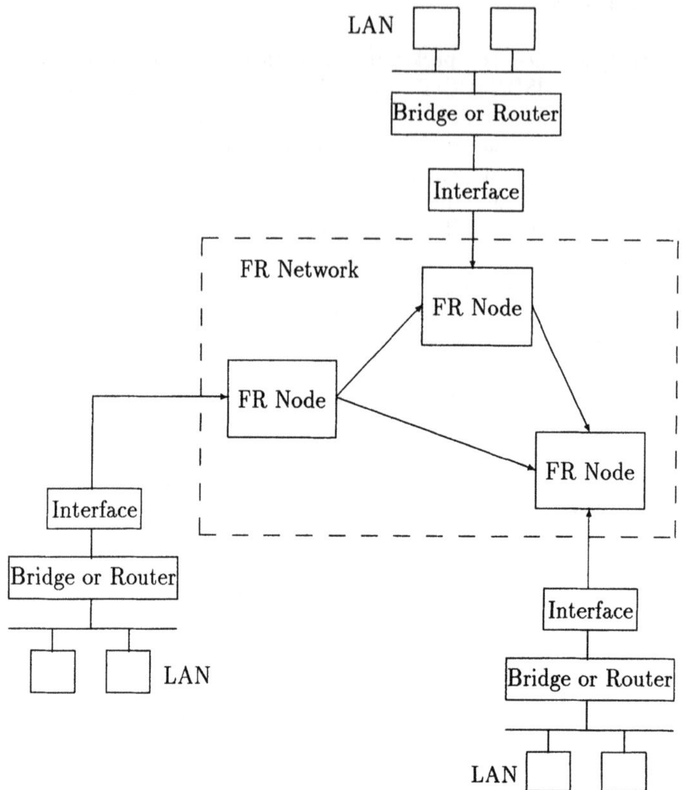

Figure 1 Model of LANs interconnected by a FR network.

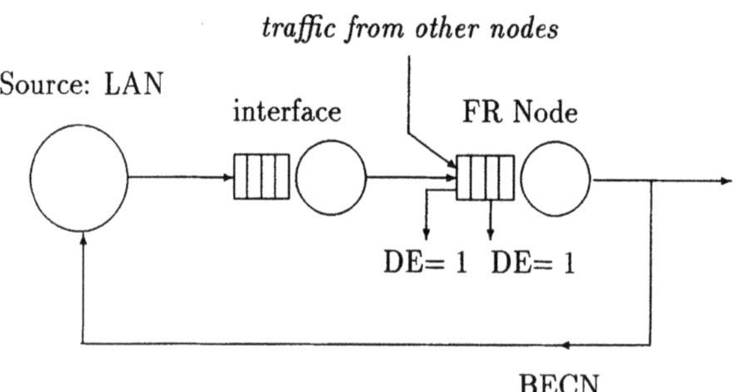

Figure 2 Queueing model of FR network.

network nodes can perform both access and transit functions and are interconnected with transmission links. The interface between the LAN and FR network is used to perform the CAC procedures and to mark the frames with two priority levels with the use of the DE bit. The network nodes have a single finite buffer shared by all the connections using that link. The buffers are managed according to FIFO and, in the case of congestion, they discard low priority the frames.

To date, most studies of the impact of control functions on the performance of a network have used to simulation, (Castelli,1993, Fratta,1993). In this paper, we propose an analytical queueing model based, a on diffusion approximation. We assume that a FR node sends a ECN signal to the source. This feedback signal is a request to lower the LAN activity and it is transmitted to the LAN with a certain delay.

The article is organised as follows. Section 2 presents a model of the admission control performed by the interface and studies the influence of this procedure on the parameters of the stream of high and low priority frames. Section 3 introduces the model of a FR transmission node with push-out and threshold policies. Section 4 presents a transient analysis of the behaviour of a FR node. With these results, we are able to propose in Section 5 a closed-loop model including the ECN mechanism. Section 6 presents our conclusion.

Diffusion approximation which is used here developes models proposed in (Gelenbe,75,76): the solution of diffusion equation

$$\frac{\partial f(x,t;x_0)}{\partial t} = \frac{\alpha}{2}\frac{\partial^2 f(x,t;x_0)}{\partial x^2} - \beta\frac{\partial f(x,t;x_0)}{\partial x} \qquad (1)$$

with properly chosen boundary conditions (they have the form of instantaneous return processes) and parameters α, β yields the density function $f(x)$ which approximates the queue distribution: $f(n,t;n_0) \approx p(n,t;n_0)$, in steady state $f(n) \approx p(n)$.

2 INTERFACE NODE MODEL

In our model, the input streams from LANs are characterized by two parameters, namely the mean and variance of interarrival times. Suppose that the frame arrival rate is λ and the squared coefficient of the interarrival time distribution $A(x)$ is C_A^2. The values of these parameteres are modified by the admission policies performed by the interface described in the introduction. In the interface, the stream of frames is observed in an interval T_c. This implies a type of Jumping Window process. The interval is divided into three periods: T_1, T_2 and T_3. Their length is random and depends on the traffic intensity. During T_1, B_c volume of information is guaranteed to transmit within T_c and the frames are marked as high priority. During T_2, the volume of information in excess of B_c (until B_e) is transmitted but the frames are marked with low priority. If the B_e level is reached, the period T_3 starts and lasts until the end of T_c. During T_3, the arriving frames are discarded.

The queueing model determines the parameters of streams for two classes of customers leaving the interface and representing high and low priority frames. Denote their rates by $\lambda^{(1)}$ and $\lambda^{(2)}$, respectively. During the period T_1, $\lambda^{(1)} = \lambda$ and outside it $\lambda^{(1)} = 0$; during T_2, $\lambda^{(2)} = \lambda$ and $\lambda^{(2)} = 0$, otherwise. Let $h_1(x)$, $h_2(x)$ and $h_3(x)$ denote the densities of the duration of the three mentioned periods within T_c. In terms of the diffusion model $h_1(x)$ is the density of the first passage time of the diffusion process between $x = 0$ and

$x = B_c$; $h_2(x)$ is the density of the first passage time of the process between $x = B_c$ and $x = B_c + B_e$. Naturally, T_2 occurs only if the duration of T_1 is inferior to T_c. Note that we replace the volume of information by the number of frames.

In a $G/G/1$ or $G/G/1/N$ diffusion model, diffusion process $X(t)$ represents the number of custommers $N(t)$ present in the system, hence coefficients of the diffusion equation take in account the arrivals and departures of customers: $\beta = \lambda - \mu$, $\alpha = C_B^2\mu$, where $1/\mu$ and C_B^2 are the mean and squared coefficient of variation of the service time distribution $B(x)$.

Here, the diffusion process represents the number of arrivals to the interface, hence we choose the diffusion parameters $\beta = \lambda$, $\alpha = \sigma_A^2\lambda^3 = C_A^2\lambda$ and we model the arrival process as a diffusion process initiated at $t = 0$ at the point $x_0 = 0$ and having an absorbing barrier at $x = B_c$. Once the process reaches $x = B_c$, it remains there. It does not seem necessary to bound the process at $x = 0$, due to the constant growing of the process ($\beta > 0$). Hence we omit it for simplicity. The pdf of such a process is (Cox,1965)

$$p(x,t) = \frac{1}{\sqrt{2\alpha\Pi t}}\left[\exp\left\{-\frac{(x-\beta t)^2}{2\alpha t}\right\} - \exp\left\{\frac{2\beta N}{\alpha} - \frac{(x-2N-\beta t)^2}{2\alpha t}\right\}\right]. \tag{2}$$

The density of the first passage time from $x_0 = 0$ to the barrier at $x = B_c$ is

$$h_1(t) = -\frac{d}{dt}\int_{-\infty}^{B_c} p(x,t)dx = \frac{B_c}{\sqrt{2\alpha\Pi t^3}}\exp\left[-\frac{(B_c-\beta t)^2}{2\alpha t}\right] \tag{3}$$

Similarily, the density $h_2(t)$ of the second period is expressed as

$$h_2(t) = \frac{B_e}{\sqrt{2\alpha\Pi t^3}}\exp\left[-\frac{(B_e-\beta t)^2}{2\alpha t}\right]. \tag{4}$$

The n-th moments of the duration of T_1 and T_2 are

$$E[T_1^n] = \int_0^{T_c} h_1(x)x^n dx + \left[1 - \int_0^{T_c} h_1(x)dx\right] T_c^n \tag{5}$$

$$E[T_2^n] = \int_0^{T_c} h_1(x)\left\{\int_0^{T_c-x} \xi^n h_2(\xi)d\xi + \left[1 - \int_0^{T_c-x} h_2(\xi)d\xi\right](T_c-x)^n\right\}dx. \tag{6}$$

The sum of three periods equals T_c. Hence $h_1(x) * h_2(x) * h_3(x) = \delta(x - T_c)$, or

$$\bar{h}_3(s) = \frac{e^{-T_c s}}{\bar{h}_1(s)\bar{h}_2(s)} = e^{-T_c s - \frac{B_c+B_e}{\alpha}(\beta-\sqrt{\beta^2-2\alpha s})}$$

and the departure rates of high and low priority frames are $\lambda^{(1)} = \frac{E[T_1]}{T_c}\lambda$, $\lambda^{(2)} = \frac{E[T_2]}{T_c}\lambda$. An estimate of the variation of the output streams of both classes is

$$C_A^{(1)^2} \approx C_A^2\left(\frac{E[T_1]}{T_c}\right)^2, \qquad C_A^{(2)^2} \approx C_A^2\left(\frac{E[T_1]}{T_c}\right)^2. \tag{7}$$

3 FR TRANSMISSION NODE MODEL

We consider two types of queue policy at a FR node: push-out and threshold policies. The FR node is represented by a multiclass $G/G/1/N$ diffusion model (Gelenbe,1975,1976) adapted to include either push-out or threshold policy of selection of frames to be discarded.

3.1 $G/G/1/N$ diffusion model with push-out policy

The model was proposed and validated in (Czachórski,1992) for the case of ATM networks. We recall here its principles making minor but necessary changes regarding FR networks.

As long as the number of customers n in the $G/G/1/N$ queue with push-out policy is less than N, it acts as a standard $G/G/1/N$ queue with two classes of customers. During non-saturation periods, we obtain from the $G/G/1/N$ model the function $f(x)$ and probabilities p_0, p_N that the process is at lower or upper boundary, hence the distribution $p(n)$ of n customers of both classes taken together. The conditional distribution $p(n|n < N)$, which corresponds to the non-saturation period, can be easily obtained.

The process enters the saturation period with probability $p(N)$. The conditional distribution of the number of class-k customers calculated without replacements is as follows:

$$p^{(k)}(n^{(k)}|N) = \binom{N}{n^{(k)}} \left(\frac{\lambda^{(k)}}{\lambda}\right)^{n^{(k)}} \left(1 - \frac{\lambda^{(k)}}{\lambda}\right)^{N-n^{(k)}}, \qquad k = 1,2\,. \tag{8}$$

Now we study the policy of replacement. If we approximate the stream of class-1 customers during a saturation period by a Poisson process with parameter $\lambda^{(1)}$, the probability of n arrivals of class-1 customers within a single saturation period is

$$p_{\text{arriv}}(n) = \int_0^\infty \frac{(\lambda^{(1)}x)^n}{n!} \exp^{-\lambda^{(1)}x}\, b(x)dx\,, \qquad n = 0,1,\ldots\,.$$

where $b(x)$ is the density of service time distribution and the probability $p_{\text{rep}}(n)$ of n replacements is

$$p_{\text{rep}}(n) = p_{\text{arriv}}(n) \sum_{n_2=n+1}^{N} p^{(2)}(n^{(2)}|N) + p^{(2)}(n^{(2)}|N) \sum_{i=n}^{\infty} p_{\text{arriv}}(i).$$

The first sum corresponds to situations where there are n arrivals and at least $n+1$ class-2 customers which could be replaced. The second sum corresponds to situations where there are n class-2 customers and at least n class-1 arrivals.

In the case of a non-Poisson input stream, the pdf $a(x)$ or distribution function $A(x)$ of interarrival times is required to determine the probability of n arrivals:

$$p_{\text{arriv}}(n) = \int_0^\infty b(x) \int_0^x a^{*n}(t)[1 - A(x-t)]dt\, dx,$$

where $*n$ denotes the n-fold convolution.

Due to the policy of replacements, the effective throughputs are

$$\lambda_{\text{eff}}^{(1)} = \lambda^{(1)}(1 - p_N) + p_N \lambda^{(1)}\varepsilon, \qquad \lambda_{\text{eff}}^{(2)} = \lambda^{(2)}(1 - p_N) - p_N \lambda^{(1)}\varepsilon, \tag{9}$$

where ε is the probability that a class-1 customer arriving at a saturation period may replace a class-2 customer, that is the ratio of mean number of replacements to mean number of arrivals in this period:

$$\varepsilon = \frac{\sum_{k=1}^{N} p^{(2)}(k|N)[\sum_{i=0}^{k-1} i p_{\text{arriv}}(i) + k\sum_{i=k}^{\infty} p_{\text{arriv}}(i)]}{\sum_{k=1}^{\infty} k\, p_{\text{arriv}}(k)} . \tag{10}$$

Taking these throughputs into account, we recalculate new $f(x)$, p_0, p_N in the $G/G/1/N$ model and we iterate until convergence is achieved. The loss ratios of class-1 and class-2 customers are

$$L^{(1)} = \frac{\lambda^{(1)} - \lambda^{(1)}_{\text{eff}}}{\lambda^{(1)}} = p_N(1-\varepsilon) , \qquad L^{(2)} = \frac{\lambda^{(2)} - \lambda^{(2)}_{\text{eff}}}{\lambda^{(2)}} = p_N\Big(1 + \frac{\lambda^{(1)}}{\lambda^{(2)}}\varepsilon\Big) . \tag{11}$$

In the case where the input varies with the time, the above steady-state model uses transient solution of $G/G/1/N$ queue presented in the next section. In order to correct the values of $\lambda^{(1)}_{\text{eff}}$ and $\lambda^{(1)}_{\text{eff}}$, the algorithm reflecting the push-out mechanism should be restarted every fixed time-interval chosen sufficiently small with respect to the time-scale of changes of input parameters.

Figures 3 – 4 present some numerical results obtained with the model described above. They refer to the loss ratios $L^{(1)}$ and $L^{(2)}$ of priority and non-priority frames. The service time is exponential and its mean value is equal to the time unit; the input stream is not Poisson, $C_A^2 = 5$. The loss ratio is much higher than that in the case of $C_A^2 = 1$. Note that the estimation of such small losses, of the order 10^{-15} would be difficult to obtain with other methods, e.g. simulation.

In (Gravey,1991) the authors studied a $M_1 + M_2/G/1/N$ queue with non-preemptive HOL and pushout priorities. The behaviour of the system is similar to the one of our system in term of loss probabilities for pushout cells.

3.2 The $G/G/1/N$ diffusion model with threshold

If the number $N(t)$ of frames at the moment t is less than a defined threshold N_1, both classes of frames are queued and served on a FIFO. When the number of frames is equal or greater than N_1, only priority frames are admitted and ordinary ones are lost. Hence, the arrival stream depends on the state of the queue and the diffusion parameters α, β should reflect this fact and depend on the value of x: $\alpha(x)$, $\beta(x)$. We assume that these parameters are piecewise constant. A natural choice is as follows:

$$\beta(x) = \begin{cases} \beta_1 = \lambda^{(1)} + \lambda^{(2)} - \mu & \text{for} \quad 0 < x \le N_1 , \\ \beta_2 = \lambda^{(1)} - \mu & \text{for} \quad N_1 < x < N \end{cases} \tag{12}$$

and

$$\alpha(x) = \begin{cases} \alpha_1 = \lambda^{(1)} C_A^{(1)^2} + \lambda^{(2)} C_A^{(2)^2} + \mu C_B^2 & \text{for} \quad 0 < x \le N_1 , \\ \alpha_2 = \lambda^{(1)} C_A^{(1)^2} + \mu C_B^2 & \text{for} \quad N_1 < x < N. \end{cases} \tag{13}$$

Let $f_1(x)$ and $f_2(x)$ denote the pdf function of the diffusion process in intervals $x \in (0, N_1]$ and $x \in [N_1, N)$. We suppose that:

- $\lim_{x\to 0} f_1(x,t;x_0) = \lim_{x\to N} f_2(x,t;x_0) = 0$,

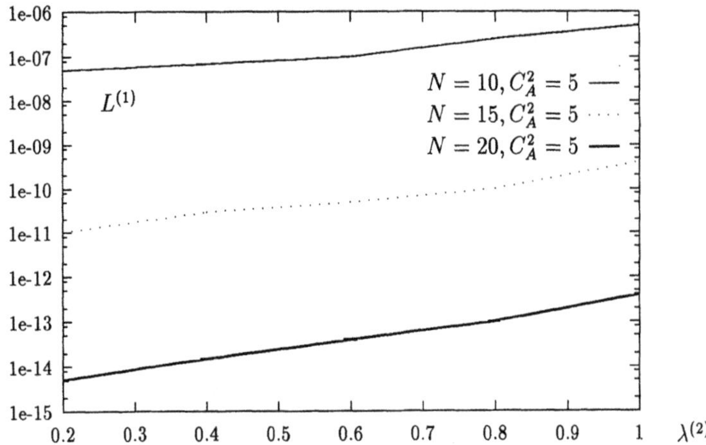

Figure 3 Loss ratio $L^{(1)}$ as a function of $\lambda^{(2)}$, ($\lambda^{(1)}$=0.2); buffer length $N = 10, 15, 20$; $C_A^2 = 5$; push-out mechanism.

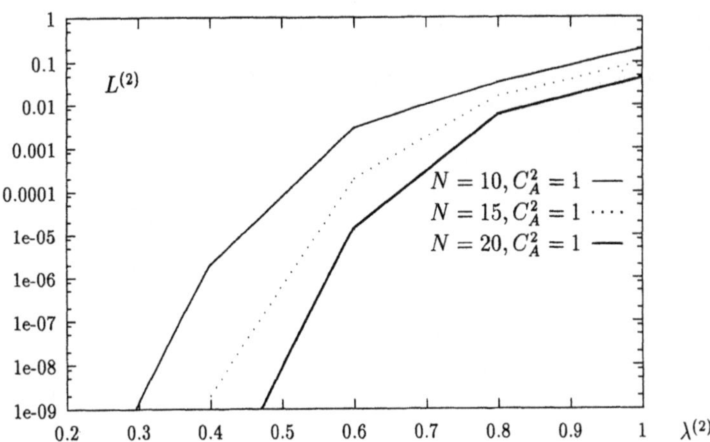

Figure 4 Loss ratio $L^{(2)}$ as a function of $\lambda^{(2)}$, ($\lambda^{(1)}$=0.2); buffer length $N = 10, 15, 20$; $C_A^2 = 5$; push-out mechanism.

- $f_1(x)$ and $f_2(x)$ functions have the same value at the point N_1: $f_1(N_1) = f_2(N_1)$,
- there is no probability mass flow within the interval $x \in (1, N-1)$:

$$\frac{\alpha_n}{2}\frac{d\,f_n(x)}{d\,x} - \beta_n f_n(x) = 0\ , \qquad x \in (1, N-1),\ \ n = 1,2,$$

and we obtain the solution of the diffusion equations:

$$f_1(x) \ = \ \begin{cases} \dfrac{\lambda_0 p_0}{-\beta_1}(1 - e^{z_1 x}) & \text{for } \ 0 < x \le 1\ , \\ \dfrac{\lambda_0 p_0}{-\beta_1}(1 - e^{z_1})e^{z_1(x-1)} & \text{for } \ 1 \le x \le N_1\ , \end{cases} \tag{14}$$

$$f_2(x) \ = \ \begin{cases} f_1(N_1)e^{z_2(x-N_1)} & \text{for } \ N_1 \le x \le N-1\ , \\ \dfrac{\mu p_N}{-\beta_2}\left[1 - e^{z_2(x-N)}\right] & \text{for } \ N-1 \le x < N\ , \end{cases} \tag{15}$$

where $z_n = \dfrac{2\beta_n}{\alpha_n}$, $n = 1,2$. Probabilities p_0, p_N are obtained with the use of the normalization condition. The loss ratio $L^{(1)}$ is expressed by the probability p_N , the loss ratio $L^{(2)}$ is determined by the probability $P[x > N_1] = \int_{N_1}^{N} f_2(x)dx + p_N$. We omit here numerical examples but they are easily obtained from Eqs. (15, 14).

4 TRANSIENT BEHAVIOUR OF THE NODE

In the transient solution of the $G/G/1/N$ model, we consider a diffusion process with two absorbing barriers at $x = 0$ and $x = N$, starting at $t = 0$ from $x = x_0$. Its probability density function $\phi(x, t; x_0)$ has the following form (Cox,1965)

$$\phi(x,t;x_0) = \begin{cases} \delta(x - x_0) & \text{for } \ t = 0 \\ \dfrac{1}{\sqrt{2\Pi\alpha t}}\displaystyle\sum_{n=-\infty}^{\infty}\left\{\exp\left[\frac{\beta x'_n}{\alpha} - \frac{(x - x_0 - x'_n - \beta t)^2}{2\alpha t}\right]\right. & \\ \qquad\qquad \left. - \exp\left[\dfrac{\beta x''_n}{\alpha} - \dfrac{(x - x_0 - x''_n - \beta t)^2}{2\alpha t}\right]\right\} & \text{for } \ t > 0\ , \end{cases} \tag{16}$$

where $x'_n = 2nN$, $x''_n = -2x_0 - x'_n$.
If the initial condition is defined by a function $\psi(x)$, $x \in (0, N)$, $\lim_{x\to 0}\psi(x) = \lim_{x\to N}\psi(x) = 0$, then the pdf of the process has the form $\phi(x,t;\psi) = \int_0^N \phi(x,t;\xi)\psi(\xi)d\xi$.

The probability density function $f(x, t; \psi)$ of the diffusion process with barriers and jumps from $x = 0$ to $x = 1$ and from $x = N$ to $x = N - 1$ is composed of the function $\phi(x, t; \psi)$ which represents the influence of the initial conditions and of a spectrum of functions $\phi(x, t - \tau; 1)$, $\phi(x, t - \tau; N - 1)$ which are pd functions of diffusion processes with absorbing barriers at $x = 0$ and $x = N$, starting at time $\tau < t$ at points $x = 1$ and $x = N - 1$ with densities $g_1(\tau)$ and $g_{N-1}(\tau)$, cf. (Czachòrski,1993):

$$f(x,t;\psi) = \phi(x,t;\psi) + \int_0^t g_1(\tau)\phi(x,t-\tau;1)d\tau + \int_0^t g_{N-1}(\tau)\phi(x,t-\tau;N-1)d\tau\ . \tag{17}$$

Densities $\gamma_0(t)$, $\gamma_N(t)$ of probability that at time t the process enters to $x = 0$ or $x = N$ are

$$\begin{aligned}
\gamma_0(t) &= p_0(0)\delta(t) + [1 - p_0(0) - p_N(0)]\gamma_{\psi,0}(t) + \int_0^t g_1(\tau)\gamma_{1,0}(t-\tau)d\tau \\
&\quad + \int_0^t g_{N-1}(\tau)\gamma_{N-1,0}(t-\tau)d\tau \,, \\
\gamma_N(t) &= p_N(0)\delta(t) + [1 - p_0(0) - p_N(0)]\gamma_{\psi,N}(t) + \int_0^t g_1(\tau)\gamma_{1,N}(t-\tau)d\tau \\
&\quad + \int_0^t g_{N-1}(\tau)\gamma_{N-1,N}(t-\tau)d\tau \,, \qquad (18)
\end{aligned}$$

where $\gamma_{1,0}(t)$, $\gamma_{1,N}(t)$, $\gamma_{N-1,0}(t)$, $\gamma_{N-1,N}(t)$ are densities of the first passage time between corresponding points. The functions $\gamma_{\psi,0}(t)$, $\gamma_{\psi,N}(t)$ denote densities of probabilities that the initial process, starting at $t = 0$ at the point ξ with density $\psi(\xi)$ will end at time t by entering respectively $x = 0$ or $x = N$.

We may express $g_1(t)$ and $g_N(t)$ with the use of functions $\gamma_0(t)$ and $\gamma_N(t)$:

$$g_1(\tau) = \int_0^\tau \gamma_0(t)l_0(\tau - t)dt \,, \qquad g_{N-1}(\tau) = \int_0^\tau \gamma_N(t)l_N(\tau - t)dt \,, \qquad (19)$$

where $l_0(x)$, $l_N(x)$ are the densities of sojourn times in $x = 0$ and $x = N$; note that the distributions of these times are not restricted to exponential ones. Laplace transforms of Eqs. (18,19) give us $\bar{g}_1(s)$ and $\bar{g}_{N-1}(s)$. The Laplace transform of the density function $f(x,t;\psi)$ is obtained as

$$\bar{f}(x,s;\psi) = \bar{\phi}(x,s;\psi) + \bar{g}_1(s)\,\bar{\phi}(x,s;1) + \bar{g}_{N-1}(s)\,\bar{\phi}(x,s;N-1) \,. \qquad (20)$$

Probabilities that at the moment t the process has the value $x = 0$ or $x = N$ are

$$\bar{p}_0(s) = \frac{1}{s}\left[\bar{\gamma}_0(s) - \bar{g}_1(s)\right], \qquad \bar{p}_N(s) = \frac{1}{s}\left[\bar{\gamma}_N(s) - \bar{g}_{N-1}(s)\right]. \qquad (21)$$

Numerical inversions of these functions have been carried out using Stehfest's algorithm (Stehfest,1970).

Figs. 5 and 6 present the dynamics of mean queue changes studied by the model of this section. In the case of Figure 5, at $t = 0$ frames begin to arrive with constant intensity λ to a $G/G/1/N$ queue, empty previously. The mean queue $E[N(t)]$ begins to increase and finally reaches its steady state value $E[N(\infty)]$. Figure 6 presents the opposite case: at time $t = 0$ a $G/G/1/N$ queue is in steady state and has the mean value $E[N(0)]$. There are no arrivals to the queue after $t = 0$ and its mean value decreases. The shapes of the curves are very similar to the exponential functions $1 - e^{-t/T}$ and $e^{-t/T}$. The time-constant T defines the length of the transient period. We see how the speed of changes depends on the value of $C^2 = C_A^2 = C_B^2$. Another factor that strongly influences the length of transient period is the utilisation ϱ. Figure 7 presents how the value of T depends on C^2 (linear dependance) and on ϱ (nonlinear dependance) during the growth of the queue. The character of dependance $T(C_A^2, C_B^2, \varrho)$ is the same when the decrease of the queue is considered, but the values of T are different (smaller), hence we distinguish T_{load} and T_{unload}.

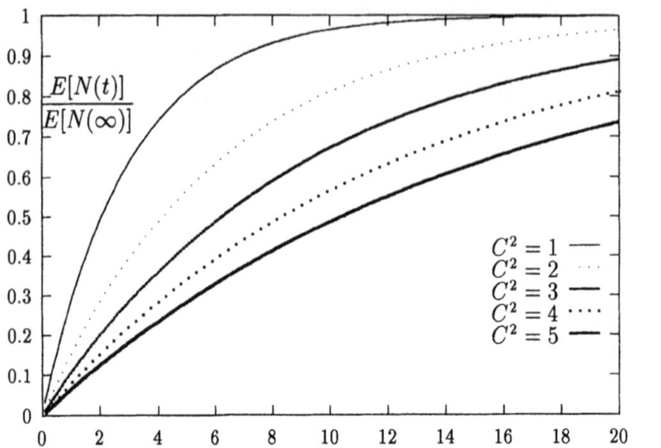

Figure 5 Loading of a queue.

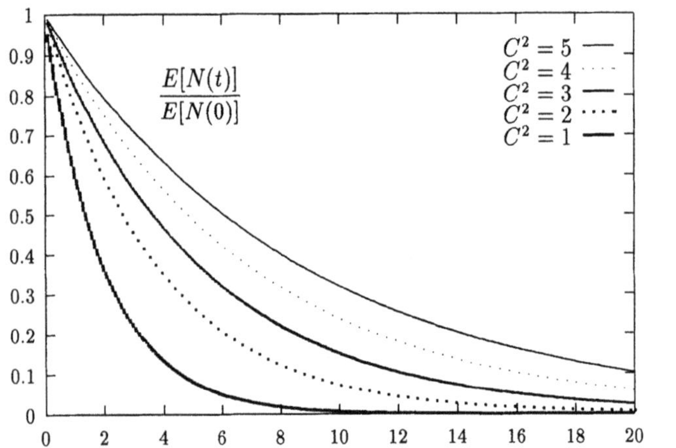

Figure 6 Unloading of a queue.

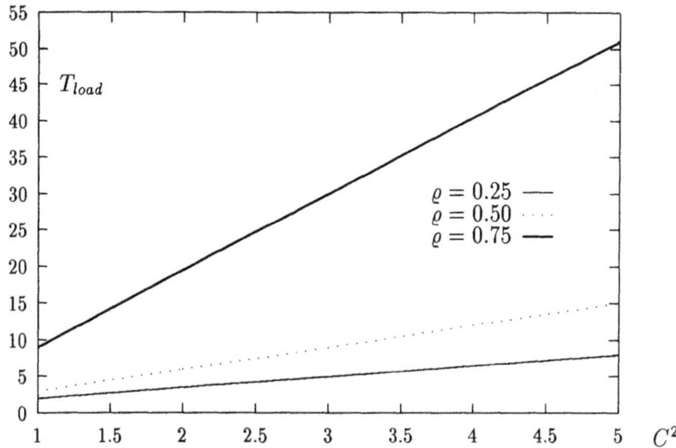

Figure 7 Constant T_{load} during growing of the queue as a function C^2.

The above transient solution of a $G/G/1/N$ model assumes that the parameters of the model are constant. In a network of queues however the output flows of stations change continuously, hence the input parameters of each station are also changing during a transient period. We are obliged to discretize this changes and keep the parameters constant within relatively small time-interval Δt. The transient solution at the end of each interval Δt allows one to determine $\varrho(t)$ and then $\lambda(t)$ and C_D^2. This solution serves also as the initial condition for the solution in the next interval: for the n-th interval, $t \in [(n-1)\Delta t, n\Delta t$, the density function of the diffusion process of station i is $f(x,t;\psi_n(x))$, where $\psi_n(x) = f[x, t = (n-1)\Delta t; \psi_{n-1}(x)]$. This method has been already successfuly applied in the analysis of a single ATM node and of the virtual path composed of such nodes (Czachórski 1992,1994).

5 CLOSED-LOOP MODEL

The conclusion of the study performed in the previous section may be as follows. We can approximately analyse the dynamics of a node seeing it as a first order (inertia) system. We apply here a notion frequently used in control theory; inertia is a system which transforms an input signal having the form of the unit step function $\mathbf{1}(t)$ ($\mathbf{1}(t) = 0$ for $t < 0$, $\mathbf{1}(t) = 1$ for $t \geq 0$) into the function $k(1 - e^{-t/T})$ where constant k denotes amplification of the system and constant T characterizes the speed of the output changes. It means that if we consider the time-varying intensity $\lambda(t)$ of traffic as an input signal and the mean queue $E[N(t)]$ as an output signal, the dynamics of the output given by a node is similar to the answer of a first order system. The same applies to the loss ratio $L^{(1)}(t)$ or $L^{(2)}(t)$ considered as the output signal.

If $\bar{X}_i(s)$ is the Laplace transform of the input signal (traffic intensity changing in time) at the node i of a network and $\bar{Y}_i(s)$ is the Laplace transform of the output signal (mean

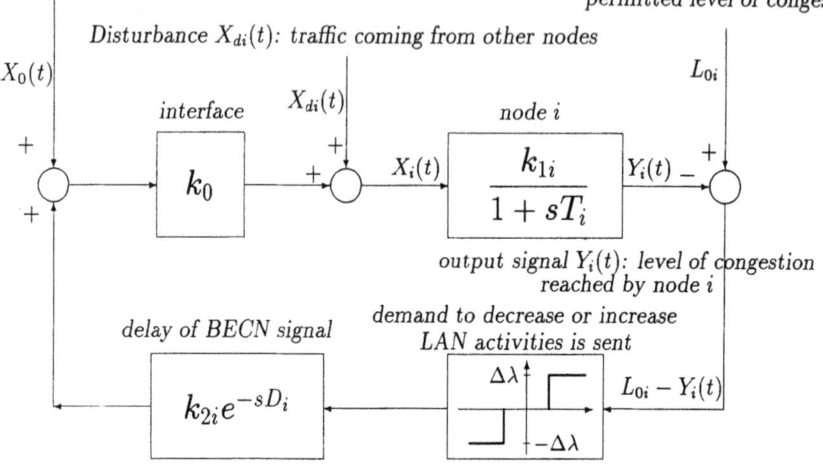

Figure 8 Block diagram of feedback control system using BECN bit.

queue length or the loss ratio changing in time), the transition function of the system is $\bar{K}_i(s) = \frac{k_i}{1+sT_i}$, i.e. the output signal is given by

$$\bar{Y}_i(s) = \bar{K}_i(s)\bar{X}_i(s) = \frac{k_i}{1+sT_i}\bar{X}_i(s). \tag{22}$$

Parameters k_{1i} and T_i depend on the utilisation of the queue i and variances of interarrival and service time distributions: $k_{1i} = k_{1i}(\varrho_i, C^2_{A_i}, C^2_{B_i})$, $T_i = T_i(\varrho_i, C^2_{A_i}, C^2_{B_i})$ in the linear and nonlinear manner presented in the previous section. Naturally, the outputs $E[N(t)]$, $L^{(1)}(t)$ and $L^{(2)}(t)$ demand different values for the coefficients k_{1i}.

The interface station changes only the parameters of the input stream, hence its transition function in the model is a constant k_0.

The warning about congestion which is sent to the source with the use of the BECN is received with a fixed delay D_i. This can be modelled by a transition function $k_{2i}e^{-sD_i}$.

The whole model is represented in Figure 8. A simple example of the time-evolution of signals is given in Figure 9. At $t = 0$, the level of traffic increases and the output of the node increases until it reaches a given level Y_u. At this moment, a message demanding the diminuation of traffic by the given value $\Delta\lambda$ is issued; it reaches its destination after time D_i; meanwhile the output still rises. Then the input is limited and the output starts to decrease: periodic oscillations are seen in the figure. This is only a simple example; once the transfer function is determined, we can consider any form of input signal.

6 CONCLUSIONS

Diffusion approximations seem to be a proper tool to model the behaviour of high speed networks where we can observe a large number of entities in the network. It allows us (a)

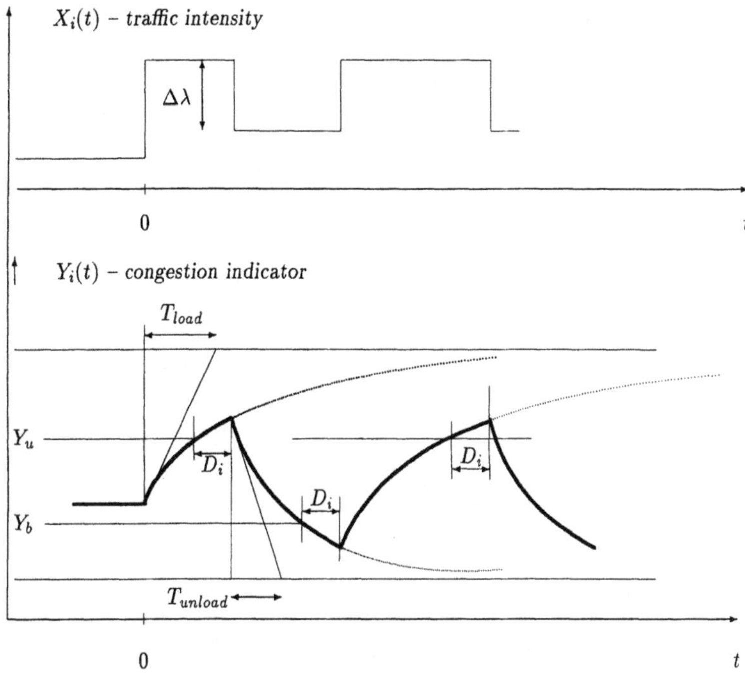

Figure 9 An example of evolution of traffic and congestion in the feedback control system using BECN bit.

to treat transient states and (b) regard second moments of traffic.

FR networks adopt the Explicit Congestion Notification (ECN) mechanism to implement flow control at the source level and to avoid the congestion in the network. In this paper, we have presented an analytical model to study the impact of the reactive functions on the system performance. We have analysed the FR node under two policies: push-out and threshold. The ECN mechanism is modelled as a closed-loop scheme using signal theory.

We have also presented a transient behaviour study of the FR node in order to estimate the queue length changes in time. This model serves us to construct the feed-back scheme which predicts the effect of the ECN mechanism on the network performance.

Preventive control is used at the interface between the LAN and FR networks to make sure that the transmission resources can support a new connection and whether the agreed traffic parameters are not violated. At the interface, the frames receive high or low priority, or are discarded, depending whether the transmitted amounts of data during an interval T_c exceed defined levels.

Numerical results show that the presented method can easily be used to study cases where the loss ratios of frames are small, e.g. less than 10^{-12} for higher priority frames and less than 10^{-7} for lower priority frames.

7 REFERENCES

Browning, D.W. (1994) Flow Control in High Speed Communication Networks. *IEEE Transactions on Communications*, **42**, No. 7.

Castelli, P. (1993) Frame Relay over ATM: Traffic Control Aspects, in *Proceedings of the International Teletraffic Congress, seminaire Teletraffic Analysis Methods for Current and Future Telecom Network*, Bangalore, India.

CCITT (1992) Recommandation Q.922, *ISDN Data link layer specification for frame mode bearer services*, Recommandation I.370, *Congestion management for the ISDN Frame Ralying Bearer Service*, CCITT/Com XVIII; Q.922, CCITT/Com XI, Geneva.

Cox, D.R. and Miller, H.D. (1965) *The Theory of Stochastic Processes.* Methuen, London.

Czachórski, T., Fourneau, J.M. and Pekergin, F. (1992) Diffusion model of the push-out buffer management policy, in *Proceedings of IEEE INFOCOM '92, Conference on Computer Communications*, Florence.

Czachórski, T. (1993) A Method of Solving the Diffusion Equation with Instantaneous Return Process Acting as Boundary Conditions. *Bulletin of the Polish Academy of Sciences, Technical Sciences* **41**, 421-427.

Czachórski T., Fourneau, J.M. and Pekergin, F. (1994) Diffusion Models to Study Nonstationary Traffic and Cell Loss in ATM Networks, in *Proceedings of ACM 2nd Workshop on ATM Networks*, Bradford.

Doshi, B.T. and Nguyen, H.Q. (1988) Congestion Control in ISDN Frame Relay Networks *A.T.& T. Technical Journal*, November-December.

Fratta L., Musumeci, L., Passalia, A. and Rigolio, G. (1993) Performance of Frame-Relay Services on ATM Networks, in *Proceeding of the Fifth International Conference on Data Communication Systems and Their Performance*, Raleigh.

Gelenbe, E. (1975) On Approximate Computer Systems Models. *Journal of ACM*, **22**, No. 2.

Gelenbe, E. and Pujolle, G. (1976) The Behaviour of a Single Queue in a General Queueing Network. *Acta Informatica*, **7**, 123-136.

Gravey, A., Hébuterne, G. (1991) Mixing Time and Loss Priorities in a Single Server Queue, in *Teletraffic and Datatraffic ITC-13* (ed. A. Jensen and V.B. Iversen), North-Holland, Amsterdam.

Stehfest, H. (1970) Algorithm 368: Numeric Inversion of Laplace Transform. *Communications of ACM*, **13**, No. 1, 47-49.

PART SEVEN

Traffic Management 2

19

Dimensioning the Continuous State Leaky Bucket for Geometric Arrivals

Alain Dupuis		Gérard Hébuterne
FRANCE TELECOM - CNET		INT
LAA/RSL/ATM		Dépt RST
2 avenue Pierre Marzin		9 rue Charles Fourier
F-22307 LANNION Cedex		91011 Evry Cedex
Alain.Dupuis@lannion.cnet.fr	e-mail	Gerard.Hebuterne@int-evry.fr
(33) 96.05.28.07	Tel	(33) 1.60.76.45.83
(33) 96.05.11.98	Fax	(33) 1.60.77.60.82

Abstract : In the ATM network, conformance of the connection with the negotiated traffic contract is monitored by means of the "Continuous State Leaky Bucket", or "Virtual Scheduling Algorithm" (VSA). It allows to verify both T (peak emission interval) and τ (cell delay variation) of the connection.

Usual models of the VSA make use of a ./D/1/N queue. We show that in the general case where T and τ take arbitrary (non integer) values, the exact model is given by the queue with bounded waiting time. In the simple case of Bernoulli arrivals this allows dimensioning the VSA for any value of the parameters. An asymptotic formula is given, and an independence property is stressed, easing the dimensioning.

1 Introduction

In the ATM network, conformance of the actual cell flow to the negotiated traffic parameters has to be verified. For peak-rate allocated connections, the *Virtual Scheduling Algorithm* (VSA), or equivalently the *Continuous State Leaky Bucket* is the reference algorithm for cell conformance at the UNI or NNI. It is defined by the two parameters T (Peak Emission Interval) and τ (Cell Delay Variation Tolerance), cf. [3].

The VSA is traditionally modelled as a finite-capacity queue, with total size $1 + \tau/T$. However, this holds only in case where the ratio τ/T takes an integer value. A general model has to be given for arbitrary values of τ/T. Another problem arises due to arbitrary value of T, when measured in cell transmission time. Since arrivals occur on a synchronous basis, a discrete-time model would be appropriate. However, T may not be integer, implying that services may begin somewhere between slots.

In the following, we first show that the general form of the VSA (i.e. with arbitrary values for T and τ) can be exactly modelled as a queue with Bounded Waiting Time. This improves previous models considering the VSA as a finite-capacity queue (see e.g. [2]). In the case of Bernoulli input, we write down the equations in a form allowing an exact, numerical solution for rational T's (i.e. T of the form r/s, with r, s integers).

The numerical results show an interesting independence property, allowing to express τ/T as a function of a single parameter. This gives a practical way to dimension the system, that is to find the bound τ on the CDV Tolerance which achieves a given Cell Loss Probability.

2 The VSA as a Queue with Bounded Waiting Time

Consider a queueing system with *limited virtual offered waiting time* (or, equivalently, with *impatient customers*). Namely, each arriving customer is characterized by an amount of time it accepts to wait before beginning being served. If the unfinished work at its arrival (the *virtual waiting time*) is larger than this delay, then the customers gives up immediately (or equivalently, it enters the queue and gives up as soon as this delay is exceeded).

Such a system has already received attention (see e.g. references [8] [9]). In what follows, one is restricted to the case where all customers have the same patience time and require the same amount of service : this corresponds to the system analysed in [9]. One may, algorithmically, define the system using the following rules:

- Let $T_i = T$ be the service time of the i-th accepted customer (a customer is accepted if his patience is long enough; otherwise it is said to be rejected).
- Let τ be the (common) patience time of customers.
- Let LT denote the last time a customer has been accepted.
- Let X denote the value of the virtual offered waiting time immediately after a customer is accepted.
- Let X'_t denote the value of the unfinished work at time t (or virtual offered waiting time).
- The queue with impatient customers works according to the following rules :
  Upon an arrival at time t, the customer estimates the value of the unfinished work :

$$X'_t = \max\{0, X - (t - \mathrm{LT})\}$$

  - If $X'_t > \tau$, then the customer is rejected.
  - If $X'_t \leq \tau$, the customer is accepted, and both X and LT are updated :

$$X = X'_t + T \quad \text{and} \quad LT = t$$

These rules are the same as the *Continuous State Leaky Bucket* as defined in reference [3], and which is equivalent to the *Virtual Scheduling Algorithm* ([3], [1]).

Since all customers have the same service time T, the integer number $\lceil \frac{X'}{T} \rceil$ represents the number of customers in the system at the arrival epoch ($\lceil x \rceil$ is the smallest integer number greater than or equal to x : $\lceil 5 \rceil = 5$, $\lceil 5.1 \rceil = 6$). The fractional part of the ratio accounts for the customer being served.

Now, let us assume that $\tau/T + 1 = N$ (integer). In this case, the test "$X' > \tau$?" is equivalent to "$n > N$?". In other words, for τ/T integer, that is in the configuration of the classical *Leaky Bucket* [15], the system is equivalent to the finite-capacity queue X/D/1/N [2].

3 Analysis of the VSA with Discrete-Time Arrivals

3.1 The General Case

Let δ be the slot duration. The input flow is of discrete nature, cells arriving at epochs of the form $k\delta$. The VSA is characterized by parameters (T, τ). These values are arbitrary (integer or real). Note that T may not be a multiple of δ, so that services may begin at arbitrary epochs and the system is generally not a discrete-time queue. To simplify notations, we assume that $\delta = 1$ (equivalently, all times are measured in units of δ).

Let W_n denote the virtual offered waiting time at the slot number n (unfinished work just prior to a possible arrival in the slot). It obeys the following recurence equations :

- If an arrival occurs in slot n and is accepted :

$$W_{n+1} = W_n + T - 1 \tag{3.1}$$

- If no arrival occurs in slot n, or if the arrival is rejected :

$$W_{n+1} = \max\{0, W_n - 1\} \tag{3.2}$$

The domain in which $Wsubn$ varies is bounded : $0 \leq W_n \leq T + \tau - 1$. Moreover, W_n takes only values of the general form $(kT - j)$ with $k \geq 0$ and $j \geq 0$. For instance, let us assume that $T = 1.4$. In this case, $W_n \in \{0, 0.2, 0.4, 0.6, 0.8, 1., 1.2, \ldots\}$. However, if T is irrational W_n takes an infinite number of values in the interval $[0, T + \tau - 1]$.

3.2 The Case with Bernoulli Arrivals

In the following, we assume a Bernoulli arrival process with parameter p : in each slot, a cell arrives with probability p, independently of what happened in previous slots.

Let $\mathcal{W}_n(t) = P\{W_n \leq t\}$ be the Probability Distribution Function (PDF) of the virtual waiting time. According to eqn. (3.1) and (3.2), one has :

$$\left.\begin{array}{rcll} \mathcal{W}_n(0) & = & (1-p)\mathcal{W}_{n-1}(1) & \\ \mathcal{W}_n(t) & = & (1-p)\mathcal{W}_{n-1}(t+1) & \text{if} \quad t < T - 1 \\ \mathcal{W}_n(t) & = & (1-p)\mathcal{W}_{n-1}(t+1) + p\mathcal{W}_{n-1}(t - T + 1) & \text{if} \quad T - 1 \leq t < \tau - 1 \end{array}\right\} \tag{3.3}$$

As usual, one is interested in the limiting distribution (if it exists) as $n \to \infty$. The limit always exists (see [8]), and let $\mathcal{W}(t) = \lim_{n\to\infty} P\{W_n \leq t\}$. One has :

$$\left.\begin{array}{rcll} \mathcal{W}(t) & = & (1-p)\mathcal{W}(t+1) & \text{if} \quad 0 \leq t < T - 1 \\ & = & (1-p)\mathcal{W}(t+1) + p\mathcal{W}(t - T + 1) & T - 1 \leq t < \tau - 1 \\ & = & \mathcal{W}(t+1) - p\mathcal{W}(\tau) + p\mathcal{W}(t - T + 1) & \tau - 1 \leq t < T + \tau - 1 \\ \mathcal{W}(\tau + T - 1) & = & 1 & \end{array}\right\} \tag{3.4}$$

4 Calculation of the Rejection Probability

4.1 Asymptotic Case of Poisson Arrivals

In the case where the parameter p of the Bernoulli process decreases, while T increases so that the product $\rho = pT$ remains constant, it is known that the system goes to a limit given by the $M/D/1$ queue (with impatient customers).

This system has been studied in depth, see e.g. [8], [9]. We refer here to the results given in [9] :

Let $a = \frac{\tau}{T}$ and $n = 1 + \lfloor a \rfloor$ (that is, n is such that $n \leq 1 + \tau/T < n + 1$). Let $\rho(= pT)$ be the load offered to the $M/D/1$ system. Then, the rejection probability π is given by :

$$\begin{array}{rcl} \pi & = & 1 - \dfrac{1 - Q}{\rho} \\ \text{with} \quad \dfrac{1}{Q} & = & 1 + \rho e^{\rho a} \displaystyle\sum_{j \leq n-1} \frac{e^{-j\rho}}{j!} (-1)^j [\rho(a - j)]^j \end{array} \tag{4.5}$$

For low rejection probabilities the above relation gives poor results, and it is worth transforming it using a combinatorial identity (see e.g. [12]) :

$$\frac{e^{a\rho}}{1-\rho} = \sum_{j=0}^{\infty} \frac{(\rho e^{-\rho})^j}{j!}(j+a)^j \tag{4.6}$$

The transformation yields an infinite series of positive terms, instead of a finite (*but of alternating signs*) series.

4.2 Solution when T is a rational number

As already mentioned, if T is an irrational number, the W_n's can take all real values in $[0, T+\tau-1]$. On the other hand, in the case where T is a rational number, say $T = r/s$, possible values of the virtual waiting time W_n are of the form $x = k/s$ with $k \in \{0, 1, \cdots, \lfloor \tau s \rfloor + r - s\}$. That is, the PDF given by eqn. (3.4) only varies by jumps at k/s. Transitions between these points are given by Table 1.

$\mathbf{sW_n}$	$\mathbf{sW_{n+1}}$	**Probability**	**Condition**
k	$k+r-s$	p	$k \leq \lfloor \tau s \rfloor$
	$\max(0, k-s)$	$1-p$	$k \leq \lfloor \tau s \rfloor$
	$\max(0, k-s)$	1	$k > \lfloor \tau s \rfloor$

Table 1: Transition Table - $W_n = k/s \longrightarrow W_{n+1}$

The transition matrix is very sparse as it can be seen in Figure 1 which gives the affectation algorithm of the non zero values.

```
for k = 0 ⟶ ⌊τs⌋ do
    a_{k,max(0,k-s)} = 1 - p
    a_{k,k+r-s} = p
end do
for k = ⌊τs⌋ + 1 ⟶ ⌊τs⌋ + r - s do
    a_{k,max(0,k-s)} = 1
end do
```

Figure 1: Transition Matrix construction.

The resolution proceeds by computing the pdf $P(W = k/s)$ from the transition matrix, instead of using eqns. (3.4). For some special cases a direct resolution is available, see Appendix. In any case, the GASTA property[7] is used in order to derive the Loss Probability from state probabilities:

$$P_L = \sum_{k=\lfloor \tau s \rfloor+1}^{\lfloor \tau s \rfloor+r-s} \Pr(s.W = k)$$

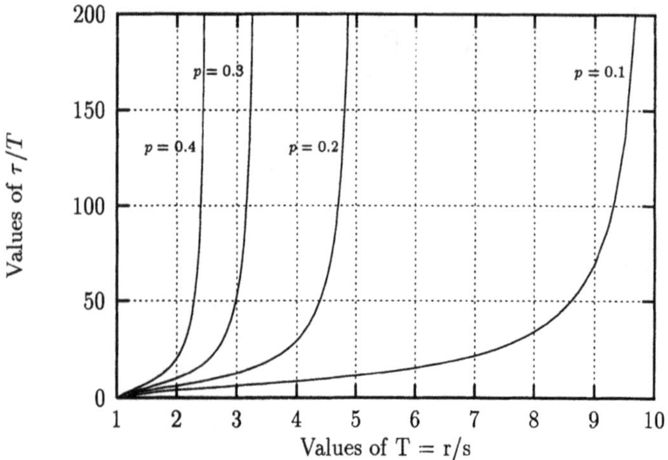

Figure 2: τ/T vs T for different values of p and for $P_L \leq 10^{-8}$.

5 Numerical results

Calculations have been made – using a stochastic matrices reduction method (see, for example [5]) – for the general case

$$T > 1 \quad \text{and} \quad \begin{cases} T \in \mathbb{N} & \text{and} \quad \tau \geq 1 \\ \text{or} \\ T \notin \mathbb{N} & \text{and} \quad \tau \geq T - \lfloor T \rfloor \end{cases}$$

Figure 2 shows the evolution of τ/T vs T for different values of the arrival probability ($p = 0.1,\ 0.2,\ 0.3$ et 0.4) in order to obtain a maximum Cell Loss Probability less or equal to 10^{-8}. There is, as expected, a vertical asymptote for $T = 1/p$ which corresponds to the load $\rho = 1$.

Note that the points cannot be chosen arbitrarily in the general case. Assume for instance $T = 1.2$, which is naturally represented as $6/5$ (that is, $r = 6, s = 5$). The state space is composed of values $k \times 0.2$, for $0 \leq k \leq \tau + 0.2$. As a consequence, the loss probability is the same for values of τ in the interval $[k \times 0.2, (k+1) \times 0.2[$.

The choice of a rational representation for T is of importance. For an arbitrary, real value of T, it is always possible to find a pair (r, s) such as $T \sim r/s$, the ratio approximating T as closely as needed. On the other hand, since the matrix size is $1 + \tau s + r - s = 1 + s(\tau + T - 1)$, s has to be chosen as low as possible.

6 An approximation formulae

This approach has been described in [10]. It is based on estimates for the tail behaviour of the virtual waiting time in the GI/G/1 queue obtained by Kingman [11] and extended by Ross [13] :

Let A be the intercell distribution for the GI arrival process and v be defined by the equation :

$$E\left[e^{-vA}\right] = e^{-vT} \tag{6.7}$$

The following upper bound applies for the tail of the virtual waiting time distribution (result given by Kingman) :

$$\Pr(W > \tau) \leq e^{-v\tau}$$

Therefore, in order to ensure a proportion of non-conforming cells smaller than 10^{-r}, it is sufficient to choose τ as

$$\tau = \frac{r \ln 10}{v}$$

The value of v is obtained by solving equation (6.7), which requires the transform for the distribution of A.

For the Bernoulli input considered in this paper, the transform is $H(z) = \dfrac{p.z}{1-(1-p).z}$ so that, finally,

$$\frac{\tau}{T} = \frac{-\gamma}{\ln\left[H(e^{-\frac{\gamma}{\tau}})\right]} \tag{6.8}$$

Figure 3 shows the accuracy of the approximation.

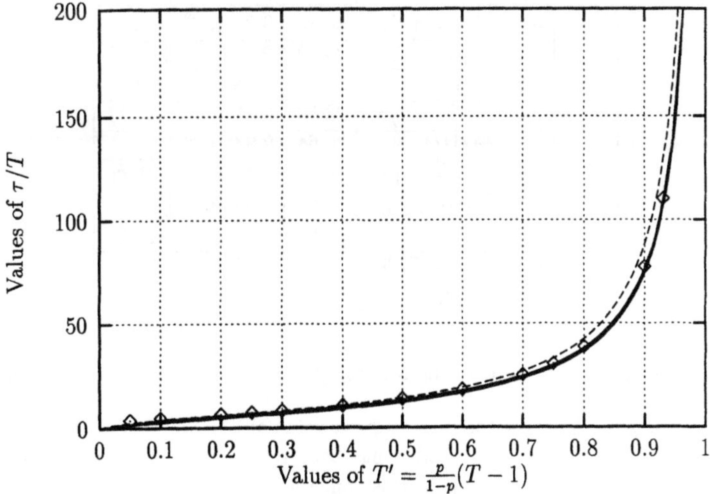

Figure 3: τ/T vs $T' = \frac{p}{1-p}(T-1)$ for different values of p and for $P_L \leq 10^{-8}$. The squares above correspond to the M/D/1 values and the dotted line to the approximation (6.8).

7 An Independence Property. Application to VSA Dimensioning

For T in the interval $]1, \frac{1}{p}]$, an interesting property appears thanks to the transformation $T \longrightarrow T' = \frac{p}{1-p}(T-1)$ (so that $T' \in]0,1]$). The result (see Figure 3) gives the curves almost perfectly superimposed.

The property is illustrated for some T' in Table 2. The rows "M/D/1" refer to the asymptotic case where $p \to \infty$, with $\rho = T'$.

T'	p	T = r/s	τ/T	Loss	τ_a/T
0.25	0.1	3.25=13/4	6.77	8.82 10^{-9}	7.17
	0.2	2.0=2/1	6.00	8.94 10^{-9}	6.64
	0.3	1.583=19/12	5.89	8.58 10^{-9}	6.23
	0.4	1.375=11/8	5.55	8.28 10^{-9}	5.89
		M/D/1	7.44	9.99 10^{-9}	
0.50	0.1	5.5=55/10	13.27	9.55 10^{-9}	14.19
	0.2	3.0=3/1	12.67	9.76 10^{-9}	13.78
	0.3	2.167=13/6	12.38	9.97 10^{-9}	13.40
	0.4	1.75=7/4	12.00	9.69 10^{-9}	13.07
		M/D/1	13.78	9.99 10^{-9}	
0.75	0.1	7.75=31/4	30.13	9.89 10^{-9}	33.12
	0.2	4.0=4/1	29.50	9.92 10^{-9}	32.77
	0.3	2.75=11/4	29.09	9.88 10^{-9}	32.43
	0.4	2.125=17/8	28.59	9.81 10^{-9}	32.11
		M/D/1	30.64	9.93 10^{-9}	
0.90	0.1	9.1=91/10	76.70	9.99 10^{-9}	88.60
	0.2	4.6=23/5	75.87	9.99 10^{-9}	88.28
	0.3	3.1=31/10	75.00	9.96 10^{-9}	87.97
	0.4	2.35=47/20	74.04	9.97 10^{-9}	87.65
		M/D/1	77.50	9.93 10^{-9}	

Table 2: A display of some of the results, illustrating the normalizing factor T'.

The results seem to show that (for a given value of T') τ/T is slightly decreasing as p increases. No satisfactory explanation has been given yet; note that Geometric burst arrival models would be appealing to this concern - since $\frac{p}{1-p}(T-\delta)$ is the remaining work at the end of a burst.

Anyway, this property can be used to dimension the VSA as follow (see Figure 4). For a given T - say $T = 4.5$ - an horizontal line crosses the staight line corresponding to a desired arrival probability - say $p = 0.2$. The abscissa obtained actually is $T' = \frac{p}{1-p}(T-\delta)$ so that the curve finally gives the requested value of $\frac{\tau}{T}$.

Table 2 also shows — as well as Figure 3 — the results obtained using approximation (6.8). That approximation gives greater values of τ/T that those computed exactly, but it remains quite good since the maximum increase is about 18.4 % for $T' = 0.9$ and $p = 0.4$. Nethertheless, its pessimistic character makes it practically usefull.

8 Conclusion

We have shown the equivalence between the general VSA (i.e. the VSA with arbitrary values of its parameters T and τ) and the discrete-time queue with impatient customers (queue with limited waiting time). This equivalence allows to write down the recurrence equation to which the waiting time distribution obeys. It must be noted that this property has been mentioned independently by [14]. In the case where the arrival process is of the Bernoulli type, the Markov chain analysis is worked out by representing the parameter T under the form r/s (r, s integers). Such a representation is always possible (with an error bounded by $1/s$). From a numerical viewpoint, the smaller s the better since the matrix size grows directly with s.

This allows to dimension the VSA, that is to calculate the bound on τ, such that for given p and T the loss probability is lower than the QoS requirement. The system exhibits a curious and interesting property : the normalized CDV Tolerance (ratio τ/T) does not depend on p and T but it depends only on the aggregated parameter $T' = \frac{p}{1-p}(T-\delta)$. This property allows an easy dimensioning procedure examplified by the abacus on Figure 4.

It remains to extend the analysis to more general input processes. Especially, it would be interesting to look for an analogous invariance property for other input processes.

References

[1] The ATM Forum : *UNI Specification.* Version 3.0, September 1993.

[2] P. Boyer, F. Guillemin, M. Servel, J.P. Coudreuse : *Spacing cells protects and enhances utilization of ATM networks links.* IEEE Networks Magazine, september 1992, pp 38-49.

[3] CCITT Recommendation I.371 : *Traffic Control and Congestion Control in B-ISDN.* Geneva, June 1992.

[4] P.E. Gounod : *Modèles de multiplexage de flux géométriques.* Note Technique CNET/LAA/EIA/15 , February 1994.

[5] W.K. Grassmann, M.I. Taksar, D.P. Heyman : *Regenerative Analysis ans Steady State Distributions.* Operations Research, 33, pp. 1107–1116, 1985.

[6] A. Gravey, J.R. Louvion, and P. Boyer : *On the Geo/D/1 and Geo/D/1/n Queues.* Performance Evaluation 11, pp. 117–125, 1990.

[7] F. Guillemin, J. Boyer, A. Dupuis : *Burstiness in Broadband Integrated Networks.* Performance Evaluation, pp. 163–176, September 1992.

[8] G. Hébuterne, F. Baccelli : *On Queues with Impatient Customers.* Performance'81, F.J. Kylstra ed., North Holland, 1981.

[9] P. Hokstad : *A Single Server Queue with Constant Service Time and Restricted Availability.* Management Science, Vol 25, No 2, February 1979.

[10] F. Kelly, P. Key : *Dimensioning Playout Buffers for an ATM Network.* 11th UK Teletraffic Symposium, Cambridge, 1994.

[11] J.F.C. Kingman : *Inequalities in the Theory of Queues.* Journal of the Royal Statistical Society, B32, pp. 102–110, 1970.

[12] J. Riordan : *Combinatorial Identities.* Krieger, 1979.

[13] S.M. Ross : *Bounds on the Delay Distribution in GI/G/1 Queues.* Journal of Applied Probabilities, 11, pp. 417–421, 1974.

[14] P. Tran-Gia : *Discrete-Time analysis Technique and Application to UPC Modelling in ATM Systems.* 8th Australian Teletraffic Research Seminar, Melbourne, December 1993.

[15] J. Turner : *New directions in communications (or which way in the information age ?).* Proceedings of the Zurich Seminar on Digital Communications, Zurich, March 1986, pp. 25–32.

Appendix: Resolution of eqn. (4) for special values

- $T \leq 1$:
  The waiting time is always zero and the distribution is simply $\Pr(W = 0) = 1$. This remains true, even if T is not a rational number. Note that τ does not matter and that the loss probability is always zero.

- $p = 1$:
  The terms $1 - p$ vanishe in the transition matrix and

$$P_{(sW=k)} = 0 \text{ for } 0 \leq k \leq \lfloor \tau s \rfloor - s$$

  The remainding equations are then

$$\begin{cases} P_{(sW=k)} = P_{(sW=k+s)} & \lfloor \tau s \rfloor - s < k \leq \lfloor \tau s \rfloor + r - 2.s \\ P_{(sW=k)} = P_{(sW=k-r+s)} & \lfloor \tau s \rfloor + r - 2.s < k \leq \lfloor \tau s \rfloor + r - s \end{cases}$$

  which solution is $\quad P_{(sW=k)} = 1/r \quad \lfloor \tau s \rfloor - s < k \leq \lfloor \tau s \rfloor + r - s$

  The Loss Probability is then $\quad P_L = \frac{r-s}{r} = 1 - \frac{1}{T} \quad$ which does not depend on τ so that dimensioning the VSA for an aimed Loss Probability leads to a unique value of T.

- **Case where a cell is accepted only if $W = 0$** :
  In that case, τ is small enougth to make all incoming cells refused, but if $W = 0$. The only reachable states are then $(W = 0)$ and $(W = T - k)$ for $1 \leq k \leq \lfloor T \rfloor$, linked by the relations $pP_{(W=0)} = P_{(W=T-\lfloor T \rfloor)} = \cdots = P_{(W=T-1)}$.
  Two alternatives are here to be considered depending on the fact that $T - \lfloor T \rfloor$ can be nul or not. Nethertheless, the loss probability is $P_L = 1 - P_{(W=0)}$ in both cases.

  1. $T \in \mathbb{N}$ **and** $\tau < 1$:
     Here, $s = 1$ and $T - \lfloor T \rfloor = 0$. The values taken by the virtual waiting time are integers and

$$\begin{cases} P_{(W=0)} = \dfrac{1}{1 + (T-1)p} \\ P_{(W=T-k)} = pP_{(W=0)} \quad 1 \leq k \leq T - 1 \end{cases}$$

  2. $T \notin \mathbb{N}$ **and** $\tau < T - \lfloor T \rfloor$:
     Here, $s > 1$, $T - \lfloor T \rfloor > 0$ and $\tau s < r \bmod s$ so that

$$\begin{cases} P_{(W=0)} = \dfrac{1}{1 + \lfloor T \rfloor p} \\ P_{(W=T-k)} = pP_{(W=0)} \quad 1 \leq k \leq \lfloor T \rfloor \end{cases}$$

  The cell loss probability is then a step function of T. Let n be a positive integer. If $T \in]n, n+1]$, the cell loss is constant in that interval (see Figure 5) :

$$P_L = \frac{np}{1 + np} \quad \text{with} \quad n \in \mathbb{N} \quad \text{such that} \quad T \in]n, n+1]$$

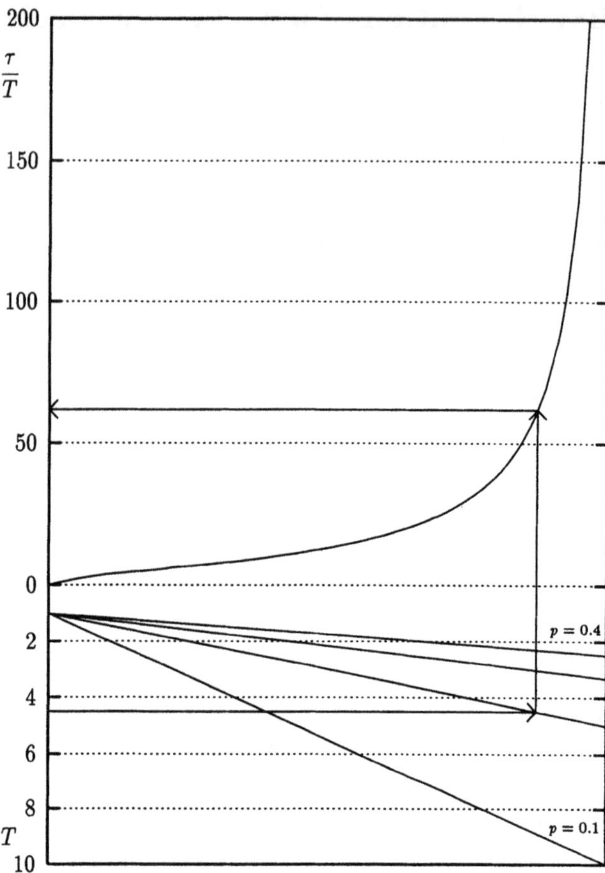

Figure 4: Abacus for finding the value of $\frac{\tau}{T}$ with given T and p to achieve a Cell Loss Probability less or equal to 10^{-8}.

Since P_L does not depend on τ, dimensioning the VSA for an aimed Loss Probability leads, as in the previous case, to a unique value of T. But here, τ has to be choosen properly, that is

$$\begin{cases} \tau < 1 & \text{if } T \in \mathbb{N} \\ \tau < T - \lfloor T \rfloor & \text{if } T \notin \mathbb{N} \end{cases}$$

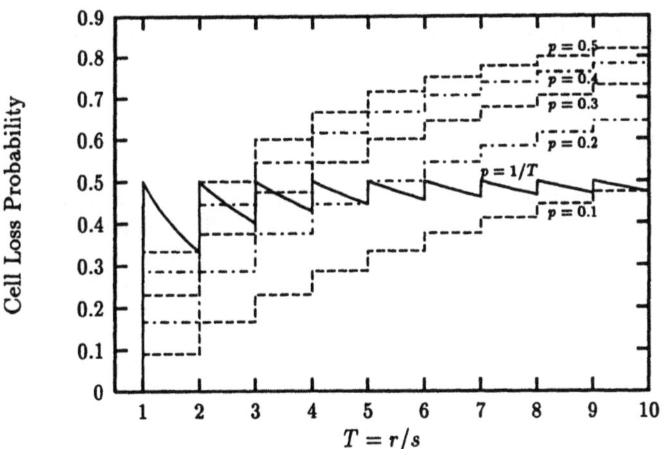

Figure 5: Cell Loss Probability vs T for $T \in \mathbb{N}$ and $\tau < 1$ or $0 < \tau < T - \lfloor T \rfloor$.

20

On-line Bandwidth and Buffer Allocation for ATM

E. Gelenbe and X. Mang
Department of Electrical Engineering, Duke University,
Durham, N.C. 27708-0291, USA
email: {erol,xmang}@ee.duke.edu

Abstract

We propose a novel on-line algorithm for shared Bandwidth and Buffer Allocation (BABA) in ATM networks. The objective of the BABA algorithm is to guarantee users' Quality of Service (QoS), while saving as much bandwidth and buffer space as possible to meet the needs of other potential network users. This algorithm proceeds incrementally on each link of a path, when a new user arrives to the network – or when a user terminates a connection. The algorithm uses gradient descent of a cost function which describes the "closest" available allocation for a given loss probability bound. BABA only requires simple algebraic operations, making it practical for fast on-line control. Numerical and simulation results show that BABA compares very favorably with currently proposed resource allocation policies.

Keywords

ATM networks, call admission control, buffer and bandwidth allocation, modeling and simulation

1 INTRODUCTION

ATM provides a universal bearer service for B-ISDN networks, which can carry all voice, data and video by the same cell transport arrangement. This technique allows complete flexibility in the choice of connection bit rates and enables the statistical multiplexing of variable bit rate traffic streams. It is well known that the traffic in B-ISDN will be bursty, and this can be lead to poor performance. However, if the burstiness is adequately reflected in network management, considerable economy of network resources can be achieved. In a bursty and dynamic traffic environment, all users will not send traffic at peak data rate at the same time. Therefore, one of the major challenges in traffic control is to achieve a statistical multiplexing gain while satisfying users' *Quality of Service (QoS)*.

An important functionality of traffic control in ATM is *Call Admission Control* (CAC). A connection can only be accepted if sufficient network resources are available to establish the connection end to end at its required quality of service. Also, the agreed QoS of pre-existing connections in the network must not be adversely influenced by the new connection. Thus a key issue in CAC is bandwidth allocation. Although this is not usually done, we also examine buffer allocation at nodes in conjunction with bandwidth allocation.

In this paper we propose an on-line method for Bandwidth and Buffer Allocation (BABA) in ATM switch nodes. This algorithm increases the buffer and/or bandwidth allocation on each link on the path that a new user u will take so as to satisfy the new user's QoS requirements, without adversely affecting the pre-existing users on each link. Similarly, when a user disconnects, the allocations will in general be adjusted. This allocation will be carried out using a gradient algorithm which seeks a new operating point to satisfy the resource requirements of the remaining users.

The algorithm we propose is simple and fast, and can be implemented in a distributed manner on each link. An evaluation of its effectiveness, and of the influence of source traffic parameters on network performance, is provided numerically and via extensive simulations. The efficiency of BABA as compared to well-known policies such as the "Peak Rate" and "Equivalent Bandwidth" allocation policies, is discussed.

1.1 Network Control

In ATM networks, cells from different sources are statistically multiplexed. Therefore, network resources such as buffers and transmission and switching facilities, will be shared dynamically. Statistical multiplexing will increase network efficiency if appropriate controls are applied. On the other hand, it also introduces a risk of overload due to traffic variations which cause network capacity to be exceeded. Overload is the main cause of cell loss and jitter. Therefore the number and nature of connections on each link must be limited so as to avoid link overload. On the other hand, the number of connections on each link should be increased so as to increase network utilization. Thus bandwidth allocation schemes which achieve a tradeoff between network utilization and performance have attracted considerable attention over the last decade.

Much work has been done on bandwidth allocation mechanisms based on the notions of *effective bandwidth* or *equivalent bandwidth,* which reflects the source's characteristics (including burstiness) and the QoS requirements. Related QoS computations are discussed in (Elwalid2 *et al.*, 1993) (Guerin *et al.* 1992), (Dziong *et al.*, 1993) (Kelly, 1991).

The *effective bandwidth* of a source is an explicitly identified, simply computed quantity. Though researchers offer different approaches to effective bandwidth, they all use the main property, which is that it is independent of traffic submitted by other sources to the multiplexer. This means that a source's effective bandwidth depends only on that source, and not on the system as a whole.

However, allocation schemes based on *effective bandwidth,* which do provide useful approximations and guidelines, either overestimate or underestimate the bandwidth which is actually needed because of insufficient consideration of other traffic sharing the same link, as indicated by many authors who propose this approach. Adaptive bandwidth allocation using various methods has been investigated extensively (Cheng, 1994) (Bolla *et al.*, 1993) (Tedijanto *et al.*, 1993) (Bolla *et al.*, 1990) (Xiao *et al.*, 1994) (Sriram, 1993). The numerical and simulation results in (Guerin *et al.*, 1992) show that for moderate and heavy traffic with On-Off sources, *equivalent bandwidth* may be represented by a Gaussian approximation. In (Tedijanto *et al.*, 1993) it is argued that providing control actions only at connection setup

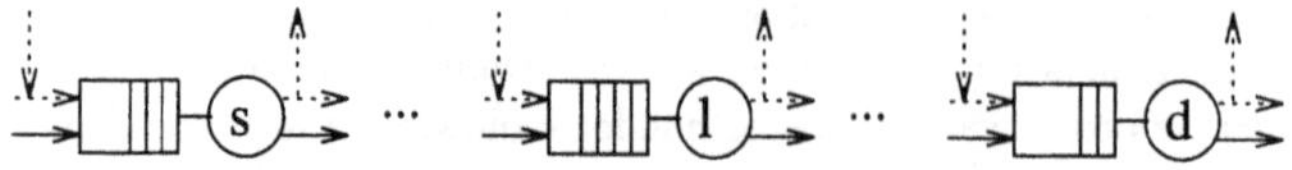

Figure 1 Each individual user views network as a tandem set of nodes.

is necessary but not sufficient for successful bandwidth management. Dynamic interaction between various controls during the establishment of a connection not only solves the shortcomings of static access control schemes in avoiding network congestion, but also leads to a more efficient and fair use of network resources.

In our research we take this dynamic nature into account by performing control actions both at call setup, at call disconnection, and also during the life time of connections.

1.2 The BABA Algorithm

Consider a network path which is schematically described in Figure 1. A user's connection from source node s to destination node d is composed of a tandem set of nodes connected by intermediate links along a selected path. On each link l along the selected route, the user will generally share bandwidth and buffer space with other users.

When a new incoming user u arrives at entry point s to the network and requests a connection to a destination d, the BABA algorithm is invoked. BABA will calculate the amount of bandwidth and the buffer size which will be allocated at each intermediate link of the selected route. If there is not enough bandwidth and buffer space to satisfy the new user's and the pre-existing users' QoS requirement, BABA will reject the new request for admission.

When a user terminates a connection, BABA will again be invoked to dynamically adjust the bandwidth and buffer space shared by all currently active users on corresponding links. One of the interesting aspects of this algorithm is that it will be executed independently for each link. Thus, BABA is also a *distributed control algorithm.*

1.3 Notation

The following notation will be used in this paper:

- M_l is the total number of current background users on the link l, before a decision for a new incoming u is taken.
- C_l is the total capacity of link l.
- C_o is the capacity allocated to current background users on the link l, before a decision for a new incoming user u is taken.
- B_l is the total buffer space on the l-th link.
- B_o is the total buffer space occupied by current background users on the link l, before a decision for a new incoming user u is taken.

- $L^l(b, C)$ denotes the cell probability estimate for link l which is a function of the total occupied buffer space b and total link capacity C, and of the aggregate traffic characteristics.
- P^*_{ln} is the upper bound to the cell loss probability on link l for all users including the new incoming one, as evaluated from the user's QoS requirements. Specifically, a worst-case value of P^*_{ln} would be the minimum of the tolerable cell loss probabilities of all users (including the new incoming user) on that link.

2 TRAFFIC REPRESENTATION USING DIFFUSION APPROXIMATIONS

Resource allocation studies for ATM networks are strongly influenced by considerations concerning the traffic which is expected to flow in B-ISDN systems. Bursty ATM traffic from a single source, can be characterized simply by a bit rate which changes randomly between different constant high and low rates. Thus ATM traffic is often simplified as a superposition of *On-Off sources.* Different mathematical models have been proposed to represent this kind of bursty traffic, such as Markov modulated arrival processes (Roberts *et al.*, 1991) (Yegenoglu *et al.*, 1994) (Friesen *et al.*, 1993) (Sole-Pareta *et al.*, 1994) (Chan *et al.*, 1994), fluid flow models (Elwalid *et al.*, 1991) (Baiocchi *et al.*, 1993) (Meempai *et al.*, 1993) (Wong *et al.*, 1993) (Elwalid *et al.*, 1992) (Guerin *et al.*, 1992) and diffusion models (Kobayashi *et al.*, 1993) (Kobayashi *et al.*, 1992) where the buffer content distribution is calculated by solving a partial differential "diffusion" equation.

In our study we use a diffusion approximation to derive cell loss probability estimates, based directly on the results in (Kobayashi *et al.*, 1993) (Kobayashi *et al.* 1992). However, as we will see below, the BABA algorithm can be used with any analytical representation which provides accurate estimates or bounds (such as – for instance – large deviation estimates) of cell loss as a function of traffic characteristics, bandwidth and buffer size.

There are several types of diffusion approximation models which differ according to the choice of boundary conditions. These boundaries relate to light traffic conditions (the *"boundary at 0"*), and to conditions which prevail when the buffers are full (the *"boundary at some value b"*). The simplest model uses reflecting boundaries, while a more sophisticated approach is based on the instantaneous return process (Gelenbe *et al.* 1980) (Medhi, 1991). The latter approach leads to better models of the queueing behavior of the system when the traffic is light and also when the effect of finite capacity is represented explicitly, while the former (Gelenbe, 1975) (Gelenbe *et al.* 1976) is used when the detailed behavior of the traffic close to the "boundaries" can be simplified.

We adopt a multi-dimensional diffusion model to characterize the collective behavior of users represented by "On-Off" sources (Kobayashi *et al.*, 1993) (Kobayashi *et al.* 1992). Let the source characteristics of user u, represented with an On-Off model, be given by:

- R_u the peak traffic rate during the "On" period;
- α_u^{-1} the average length of the "Off" period;
- β_u^{-1} the average length of the "On" period;
- the activity probability $a_u = \frac{\alpha_u}{\alpha_u + \beta_u}$.

Note that any three of the above four variables will suffice to characterize the source. The diffusion approximation model assumes that these sources may be represented by a semi-Markov model, so that times spent in each On and Off period can have a general distribution (rather than an exponential or related distribution).

The superposition of a large number of uncorrelated "On-Off" sources can thus be represented approximately by a diffusion process, which is used to estimate the cell loss probability $L^l(b,C)$ for the users at each link l (Kobayashi *et al.*, 1993) (Kobayashi *et al.* 1992). This cell loss expression is:

$$L^l(b,C) = \frac{\sigma_R}{(C-m_R)\sqrt{2\pi}} e^{-\frac{(C-m_R)^2}{2\sigma_R^2}} e^{zb} \tag{1}$$

where

$$z = -\frac{C-m_R}{\sum_{u=1}^{M_l} \frac{R_u^2\sigma_u^2}{\alpha_u+\beta_u}}, \quad \sigma_u^2 = \frac{\alpha_u\beta_u}{(\alpha_u+\beta_u)^2}, \quad \sigma_R^2 = \sum_{u=1}^{M_l} R_u^2\sigma_u^2, \quad m_R = \sum_{u=1}^{M_l} R_u a_u$$

Clearly $L^l(b,C)$ is a function of *total* bandwidth and buffer space, and of *all* users' characteristics at the multiplexer.

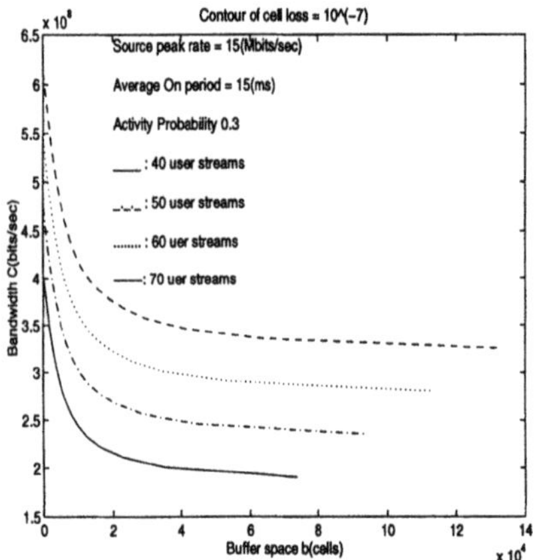

Figure 2 Admission region with different number of user streams.

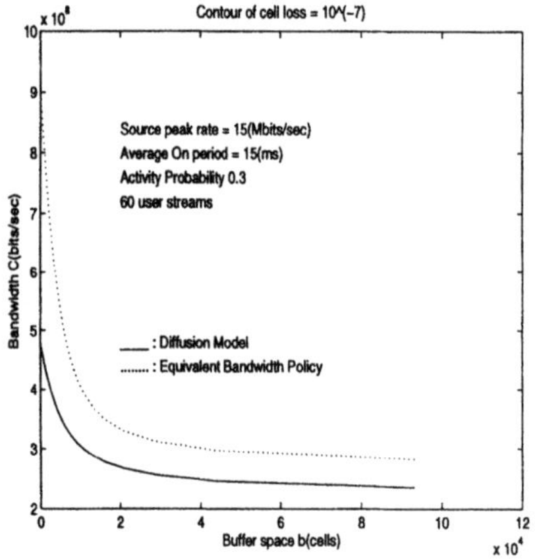

Figure 3 Comparison of admission region of diffusion model with equivalent bandwidth policy.

Figures 2 and 3 illustrate the use of this formula to determine the regions of satisfactory operation of a link. In Figure 2 we show how the number of simultaneous users affects the choice of buffer size and bandwidth, *i.e.* the pair (b,C), which needs to be allocated on a link so as to meet a cell loss QoS requirement. The acceptable cell loss rate is 10^{-7}. Each curve is the set of values of (b,C) which yield that cell loss rate according to the diffusion approximation, for a given number of simultaneous user streams. Each stream has the same traffic characteristics as described by the source peak rate, average "On" period, and activity probability given in the figure.

In Figure 3 we describe the values of buffer and bandwidth allocation which satisfy a given cell loss probability constraint. In this case, we choose the acceptable cell loss probability as being 10^{-7}. *The area above the solid line* represents the values of (b, C) which will yield a lower cell loss probability, when the diffusion approximation is used as an estimate for a set of 60 identical sources with parameters given on the figure. For the same set of sources but when the well-known *Equivalent Bandwidth Policy* (Guerin *et al.*, 1992) is used, the set of acceptable (b, C) pairs lie above the dotted line in the same figure. We see that the diffusion approximation formula tends to predict lower cell loss probability than the Equivalent Bandwidth approach.

2.1 Resource Allocation at Call Set-up Time Using BABA

We will now describe the manner in which BABA proceeds to decide whether to admit, or not to admit, a newly arriving user u. Then we will discuss what happens when an ongoing call is disconnected.

At any given instant of time *before* any decision concerning incoming user u is taken, the path from source s to destination d – which is assumed to contain L links – will be characterized by:

- The buffer size and channel capacity currently allocated to each link l on that path: (b_o^l, C_o^l), $l = 1, \ldots, L$,
- The acceptable maximum cell loss probability P_{lo}^* for each link in view of the current set of users,
- The current cell loss probability at each link $L(b_o^l, C_o^l)$ which is necessarily less than the corresponding maximum cell loss probability.

If user u is indeed admitted, we will denote the new values of these quantities by:

- (b_n^l, C_n^l), $l = 1, \ldots, L$,
- P_{ln}^*, $1, \ldots, L$, and
- $L(b_n^l, C_n^l)$, $l = 1, \ldots, L$.

User u will have an acceptable maximum total cell loss probability requirement of P^u. This implies it must have some maximum cell loss probability at each link l, which we denote by P_l^u, satisfying the following constraint:

$$P^u = \sum_{l \in path(s,d)} P_l^u$$

Before proceeding any further, we have to decide how the new allowable loss probability

P^*_{ln} on link l will be chosen. Let the allowable loss probability on the path from s to d, before the current allocation, be denoted by:

$$P^*_o = \sum_{l \in path(s,d)} P^*_{lo} \tag{2}$$

The BABA algorithm will first choose the new allowable loss probabilities as follows. First we will "spread out" user u's allowable loss probability P^u over the set of links in the path in a manner which is proportional to the current situation, to obtain the allowable link loss probability for the new user u:

$$P^u_l = P^u \frac{P^*_{lo}}{P^*_o} \tag{3}$$

Finally we will update the allowable loss probabilities on each link in the path (s, d) as follows so as to satisfy the QoS requirements of *all* the users, including the pre-existing users and the new user u whose admission is being considered:

$$P^*_{ln} = Min(P^*_{lo}, P^u_l) \tag{4}$$

The following inequalities summarize the constraints we need to satisfy as we consider the introduction of the new user u: the existing users' QoS must not be adversely perturbed by the new arrival, because we must satisfy user u's QoS requirements, and because we cannot exceed the available buffer space and bandwidth which can be allocated to link l:

- $P^*_{ln} \leq P^*_{lo}$,
- $P^u_l \leq P^*_{ln}$,
- $L(b^l_n, C^l_n) \leq P^*_{ln} \leq P^*_{lo}$, $l = 1, \ldots, L$,
- $b^l_n \leq B_l$, $C^l_n \leq C_l$.

Note that if there were M_l pre-existing connections at link l, then the new loss probability $L(b^l_n, C^l_n)$ is derived using (1) for the set of pre-existing users to whom we have added the new user u. In general there will either be no pair (b^l_n, C^l_n) which can satisfy all of these constraints, or there will be many.

Allocating buffer and bandwidth at link l

In order to make a choice of the new values of buffer size and bandwidth, BABA will seek out an allocation which is "closest" to the previous allocation – where closeness will be defined using the Euclidean distance:

$$D(C) = \sqrt{(b - B_o)^2 + (C - C_o)^2} \tag{5}$$

The purpose of remaining close to the preceding allocation is two-fold:

- To avoid allocating excessive resources,

- To reduce disruption in network operation due to the new incoming user.

Note from (1) that any pair (b, C) which satisfies the loss probability constraint must satisfy the following relationship, written by representing b as a function of C:

$$b(C) = \frac{1}{z} ln(\frac{P^*_{ln}(C - m_R)\sqrt{2\pi}}{\sigma_R} e^{\frac{(C-m_R)^2}{2\sigma_R^2}}) \tag{6}$$

The new allocation (b^l_n, C^l_n) will then be the pair $(b(C), C)$ which minimizes the cost function:

$$K^l = \xi_b(b(C) - B_o)^2 + \xi_C(C - C_o)^2 \tag{7}$$

where the constants ξ_b and ξ_C:

$$\xi_b = \begin{cases} 0, & B_o = B_l \\ 1, & \text{otherwise} \end{cases}$$

$$\xi_C = \begin{cases} 0, & C_o = C_l \\ 1, & \text{otherwise} \end{cases}$$

are used to guarantee that we are not coming up with an infeasible allocation which exceeds available capacity.

Minimizing of the cost function K^l expresses a tradeoff between the users' QoS requirements and the network's general efficiency. Note that although we will be minimizing with respect to a single variable C, we will be in fact searching for a minimum in the (b, C) space, since b and C are functionally related.

The minimization procedure is conducted by using the gradient descent rule which guarantees that each new value of the parameter C improves on the previous values with respect to the cost function K^l:

For every link l along the selected route
Update $M_l = M_l + 1$
while $|K^l_{*new} - K^l_{*old}| > \epsilon$
Do
$K^l_{old} \Leftarrow K^l_{new}$
$C \Leftarrow C - \eta^l_c \frac{\partial K^l}{\partial C}|_{old}$ (8)
Calculate $b(C)$
Check constraints
Calculate K^l_{new}
End
Return (b^l_n, C^l_n)

Here, $\epsilon > 0$ stands for an acceptable error level concerning the cost. Also $\eta_c^l > 0$ is the gradient descent rate for C which determines the speed of convergence. We use of an *adaptive gradient descent rate*, where η_c^l decreases gradually as long as the condition $|K_{old} - K_{new}| > \epsilon$ is met during the computation. In this way both speed of convergence and algorithm stability will be enhanced.

To perform the update in (8), we calculate the sensitivity (or partial derivative) of the cost function with respect to the parameter C. From the cost function in Equation (7), we obtain:

$$\frac{\partial K^l}{\partial C} = 2\xi_b(b(C) - B_o)\frac{\partial b(C)}{\partial C} + 2\xi_c(C - C_o) \tag{9}$$

Using (6) we have:

$$\frac{\partial b(C)}{\partial C} = -\frac{b(C)}{C - m_R} - (\sum_{u=1}^{M_l} \frac{R_u^2\sigma_u^2}{\alpha_u + \beta_u})(\frac{1}{\sigma_R^2} + \frac{1}{(C - m_R)^2}).e^{\frac{(C-m_R)^2}{2\sigma_R^2}} \tag{10}$$

Thus the cost sensitivity from above derivation is a simple algebraic expression. This fact makes our BABA algorithm very attractive, since simplicity is a highly favorable aspect of a real-time algorithm such as BABA.

Link and path level BABA

What we have described above is the manner in which information about the user, and about the path from s to d will be used by the BABA algorithm, individually on each link.

The relationship between each link level computation and the path as a whole is described in Figure 4.

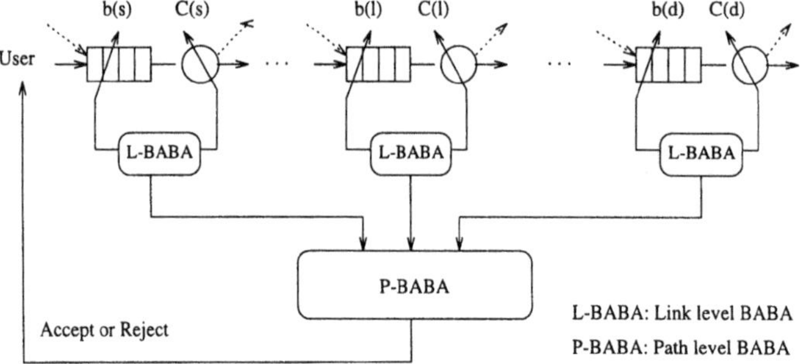

Figure 4 BABA algorithm's two level hierarchical structure.

The BABA algorithm is a source node control scheme, in the sense that the source node will compute the appropriate maximum allowable loss probabilities on each link, and then it can request each link along the chosen path to carry it out. In this sense it may be viewed as a distributed algorithms since much of the computation can proceed on each individual link separately without needing cross information from other links along the chosen path.

Once BABA computes the Bandwidth C_n^l and buffer space b_n^l for each link in view of the new incoming user, the access of user to that particular path will be rejected if any one of these computations is unsuccessful in coming up with an allocation which does not exceed the resource constraints. Otherwise the user u will be accepted.

We have not considered in this paper the case when several routes or paths may be available. This will be considered in future work. We may consider however, that in this case the algorithm could be run independently for each path with a decision being taken to admit the user to the path which seems to provide the best performance at lowest cost.

User disconnection

When a user terminates its connection, the BABA algorithm may once more be invoked in order to reduce the resource allocation on each link of the path that the user was utilizing. This will be carried out as when the connection was being established. First, new maximum allowable cell loss rates will be computed for each link. Then the closest pair (b, C) will be computed to the preceding allocation at each link, which respects the cell loss constraint of the link.

3 NUMERICAL STUDY AND SIMULATION RESULTS

Since the advantage of statistical multiplexing is to increase the number of connections which the network can handle with limited resources and without significant degradation of QoS, in this section we compare BABA with the following two existing policies:

1. **The Peak Rate Policy (see for instance (Baiocchi *et al.*, 1994))**
   Here, bandwidth is assigned to each connection according to its declared peak rate R_u. The total bandwidth allocate to M_l users is then:
   $C_p = \sum_u^{M_l} R_u$;
2. **The Equivalent Bandwidth Policy (see (Guerin *et al.*, 1992))**
   In Guerin *et al.*, 1992) the following equivalent bandwidth formula is proposed to perform bandwidth allocation for admission control:
   $C_e = \min\left\{ m_R + \alpha' \sigma_R, \ \sum_{u=1}^{M_l} \hat{c}_u^l \right\}$
   where
   $\alpha' \simeq \sqrt{-2 ln P_{ln}^* - ln 2\pi}$, $\quad c_u^l = R_u \frac{y_u^l - b + \sqrt{[y_u^l - b]^2 + 4 b a_u y_u^l}}{2 y_u^l}$, $\quad y_u^l = (-ln P_{ln}^*)(\frac{1}{\beta_u})(1 - a_u) R_u$

The comparisons have been carried out with the following examples of artificial traffic patterns:

- A) Homogeneous traffic with high activity ($a_u = 0.6$) of each individual On-Off source with $1/\alpha_u = 0.0105(sec)$, $1/\beta_u = 0.0045(sec)$, $R_u = 15(Mbits/sec)$.
- B) Homogeneous traffic with low activity ($a_u = 0.3$) of each individual On-Off source with $1/\alpha_u = 0.0101(sec)$, $1/\beta_u = 0.009(sec)$, $R = 15(Mbits/sec)$;
- C) Heterogeneous traffic which randomly combines the above two types of On-Off sources.

In order to carry out a reasonable comparison of other algorithms (such as Effective Bandwidth and Peak Rate Allocation) with the algorithm which we propose, we need to keep in mind that BABA is a resource allocation scheme which combines *both* bandwidth and buffer space allocation, while existing policies consider them separately – and in general only consider bandwidth allocation for fixed buffer size. Thus to make meaningful comparisons, we first calculate a set of (C, b) for a given M_l using BABA. Then we calculate the bandwidth C_e required by the Equivalent Bandwidth policy, and C_p the bandwidth required by the Peak Allocation policy for the same values of M_l and b. We then compare the bandwidths C, C_e and C_p as well as the observed performance in each case.

Figure 5 (a) shows the buffer space required for above given traffic patterns on the link being examined. The comparison between BABA with equivalent and peak rate policies is provided in Figure 5 (b), (c) and (d). We see that both BABA and the Equivalent Bandwidth policy save significant bandwidth compared to the Peak Rate policy. However BABA is the most bandwidth efficien policy, particularly when the number of connections increases.

3.1 Simulation Results

We have shown the efficiency of BABA by numerically comparing it to others. We now validate its effectiveness via extensive simulations which measure the cell loss and link utilization. We first conduct simulations on a single link, for a maximum allowable cell $P_{ln}^{*} = 10^{-4}$. Simulations runs were independently replicated 200 times, and each run included the transmission of 10^6 cells. Confidence intervals are calculated using the *Student-t* distribution with 98% confidence.

Table I shows cell loss statistics for varying traffic patterns. We can see that BABA does provide sufficient enough resources to satisfy users' QoS, so that the cell loss is less than objective value 10^{-4}. Table II shows that the bandwidth has been efficiently used since the average link utilization is high.

Table I Cell Loss Measured via Simulations

	Homogeneous $a = 0.3$	Homogeneous $a = 0.6$	Heterogeneous Traffic
No. of Users M_l	Cell Loss	Cell Loss	Cell Loss
20	$(1.65 \pm 0.27) \times 10^{-6}$	$(0.00 \pm 0.00) \times 10^{-7}$	$(0.00 \pm 0.00) \times 10^{-7}$
30	$(2.27 \pm 0.37) \times 10^{-6}$	$(0.00 \pm 0.00) \times 10^{-7}$	$(0.00 \pm 0.00) \times 10^{-7}$
40	$(0.00 \pm 0.00) \times 10^{-6}$	$(0.00 \pm 0.00) \times 10^{-7}$	$(0.00 \pm 0.00) \times 10^{-7}$
50	$(2.58 \pm 0.42) \times 10^{-6}$	$(5.85 \pm 0.96) \times 10^{-7}$	$(9.00 \pm 1.48) \times 10^{-8}$
60	$(3.31 \pm 0.54) \times 10^{-6}$	$(6.97 \pm 1.14) \times 10^{-6}$	$(1.83 \pm 0.30) \times 10^{-6}$
70	$(3.10 \pm 0.51) \times 10^{-7}$	$(1.50 \pm 0.25) \times 10^{-5}$	$(2.15 \pm 0.35) \times 10^{-6}$
80	$(1.12 \pm 0.18) \times 10^{-5}$	$(9.15 \pm 0.15) \times 10^{-5}$	$(5.44 \pm 0.90) \times 10^{-6}$

Confidence interval calculations use the *Student t* distribution with 98% confidence.

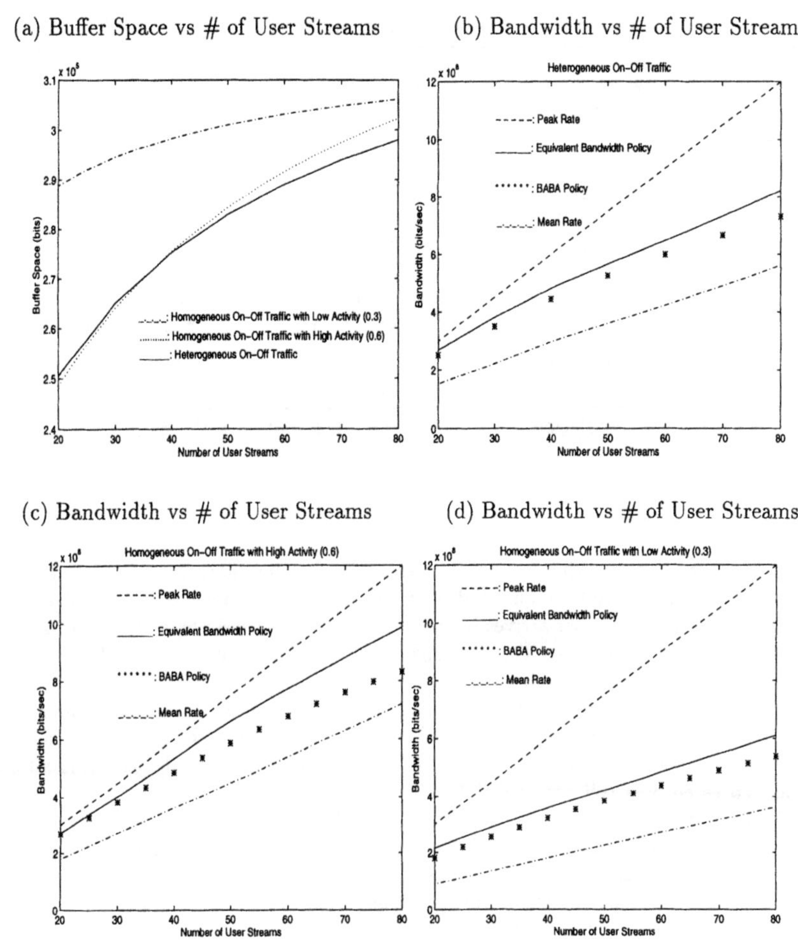

Figure 5 Numerical comparison of BABA and existing policies on a given high speed link: $C_l = 1(Gbits/sec)$, $B_l = 318(Kbits) = 750(Cells)$ and $P^*_{ln} = 10^{-4}$.

Table II Link Utilization Measured via Simulations

	Homogeneous $a = 0.3$	Homogeneous $a = 0.6$	Heterogeneous Traffic
No. of Users M_l	Utilization (%)	Utilization (%)	Utilization (%)
20	50.02±0.066	67.48±0.030	61.18 ±0.152
30	53.34 ±0.064	71.10 ±0.135	63.39 ±0.200
40	57.52 ±0.162	75.45 ±0.023	66.68 ±0.014
50	59.05 ±0.354	76.98 ±0.003	68.73 ±0.062
60	61.49 ±0.002	79.87 ±0.125	71.02 ±0.136
70	64.22 ±0.079	83.26 ±0.217	73.53 ±0.206
80	66.90 ±0.305	82.89 ±0.282	76.80 ±0.194

Confidence interval calculations use the *Student t* distribution with 98% confidence.

4 CONCLUSION

In this paper we propose the new BABA algorithm for the allocation of both bandwidth and buffer space in the links of an ATM source-to-destination connection. This algorithm is invoked each time a new user arrives to the network, and is run independently on each link of the path that the user will take. The algorithm can also be used to decide whether the user can be accepted or rejected.

The algorithm is meant to be run in real-time, and we show that it only uses simple algebraic computations in conjunction with a gradient descent procedure. The idea is to choose the "nearest" resource allocation to the current allocation, while satisfying all users' QoS as expressed by a cell loss probability bound.

BAB is compared to the existing well-known policies of Equivalent Bandwidth Allocation and Peak Rate Allocation both numerically (to obtain the number of connections which may be supported simultaneously for a given cell loss probability) and using simulation results. The comparisons are carried out for different types of homogeneous or heterogeneous On-Off sources. Simulations are carried out with 98% confidence level. These results indicate that BABA will allocate resources in a significantly more economic manner, while respecting the QoS requirements that these other policies will also meet.

Future work will address the use of BABA for multi-path policies, as well as the study of BABA and other policies in the presence of traffic transients.

5 REFERENCES

M. Aicardi, R. Bolla, F. Davoli, and R. Minciardi (1990). Optimization of capacity allocation among users and services in integrated network. In *Proc. ICC'90*, pages 0302–0808.

A. Baiocchi, N. Blefari-Melazzi, F. Cuomo, and M. Listanti (1994). Achieving statistical gain in ATM networks with the same complexity as peak allocation strategy. In *Proc. INFOCOM'94*, pages 374–382.

A. Baiocchi, A. Roveri N. Bléfari-Melazzi, and F. Salvatore (1993). Stochastic fluid analysis of an ATM multiplexer loaded with heterogeneous ON-OFF sources: an effective computational approach. In *Proc. INFOCOM'92*, pages 0405–0414.

R. Bolla and F. Davoli (1993). Dynamic hierarchical control of resource allocation in an integrated services broadband network. *Computer Networks and ISDN Systems*, 25:1079–1087.

J. H. S. Chan and D. H. K. Tsang (1994). Bandwidth allocation of multiple QOS classes in ATM environment. In *Proc. INFOCOM'94*, pages 360–367.

T. Cheng (1994). Bandwidth allocation in B-ISDN. *Computer Networks and ISDN Systems*, 26:1129–1142.

S. Chowdhury and K. Sohraby (1994). Bandwidth allocation algorithms for packet video in ATM networks. *Computer Networks and ISDN Systems*, 26:1215–1223.

Z. Dziong, K. Liao, and L. Mason (1993). Effective bandwidth allocation and buffer dimensioning in ATM based networks with priorities. *Computer Networks and ISDN Systems*, 25:1065–1078.

A. I. Elwalid and D. Mitra (1991). Analysis and design of rate-based control of high speed networks, I: Stochastic fluid models. *Queueing Systems*, 9:29–64.

A. I. Elwalid and D. Mitra (June 1993). Effective bandwidth of general markovian traffic sources and admission control of high speed networks. *IEEE/ACM Transactions on Networking*, 1:329–343.

V. J. Friesen and J. W. Wong (1993). The effect of multiplexing, switching and other factors on the performance of broadband networks. In *Proc. INFOCOM'93*, pages 1194–1203.

E. Gelenbe (1975). On approximate computer system models. *J. ACM*, 22:261–263.

E. Gelenbe and I. Mitrani (1980). *Analysis and Synthesis of Computer Systems.* Academic Press, New York.

E. Gelenbe and G. Pujolle (1976). An approximation to the behaviour of a single queue in a network. *Acta Informatica*, 7:123–136.

R. Guerin and L. Gun (1992). A unified approach to bandwidth allocation and access control in fast packet-switched networks. In *Proc. INFOCOM'92*, pages 0001–0012.

F. P. Kelly (1991). Effective bandwidths at multi-class queues. *Queueing Systems*, 9:5–16, 1991.

H. Kobayashi and Q. Ren (December 1992). A mathematical theory for transient analysis of communication networks. *IEICE Transaction on Communications*, E75-B:1266–1276.

H. Kobayashi and Q. Ren (1993). A diffusion approximation analysis of an ATM statistical multiplexer with multiple state solutions: Part I: Equilibrium state solutions. In *Proc. ICC'93*, pages 1047–1053.

J. Medhi (1991). *Stochastic Models in Queueing Theory.* Academic Press, New York.

G. Meempat, G. Ramamurthy, and B. Sengupta (1991). A new performance measure for statistical multiplexing: Perspective of the individual source. In *Proc. INFOCOM'93*, pages 531–538.

J. W. Roberts and A. Gravey (1991). Recent results on B-ISDN/ATM traffic modeling and performance analysis - a review of ITC 13 papers. In *Proc. GLOBECOM'91*, pages 1325 –1330.

J. Sole-Pareta and J. Domingo-Pascual (1994). Burstiness characterization of ATM Cell streams. *Computer Network and ISDN Systems*, 26:1351 –1363.

K. Sriram (1993). Methodologies for bandwidth allocation, transmission scheduling, and congestion avoidance in broadband ATM networks. *Computer Networks and ISDN Systems*, 26:43–59.

T. E. Tedijanto and L. Gun (1993). Effectiveness of dynamic bandwidth management mechanisms in ATM networks. In *Proc. INFOCOM'93*, pages 358–367.

M. Wong and P. Varaiya (1994). A deterministic fluid model for cell loss in ATM networks. In *Proc. INFOCOM'93*, pages 395–440, 1993.

N. Xiao, F. F. Wu, and S. Lun. Dynamic bandwidth allocation using infinitesimal perturbation analysis. In *Proc. INFOCOM'94*, pages 383–389.

F. Yegenoglu and B. Jabbari (1994). Characterization and modeling of aggregate traffic for finite buffer statistical multiplexers. *Computer Networks and ISDN Systems*, 26:1169–1185.

21

ATM traffic analysis and control for ABR service provisioning

N.M. Mitrou, K.P. Kontovasilis and E.N. Protonotarios

National Technical University of Athens
Computer Science Division,
Heroon Polytechneiou 9, Zografou
GR 157-73, Athens, GREECE
Phone: +30-1-7721639, Fax: +30-1-7757501
e-mail: mitrou@phgasos.ntua.gr

Abstract

In this paper the main traffic analysis and control problems related with the ABR service are addressed, modelled and answered on the basis of effective rates defined for the multiplexed connections. Emphasis is given to an adaptive shaping function as a response to congestion feedback. A simple example demonstrates the proposed shaping scheme.

Keywords

ATM, ABR service, fluid-flow modelling, effective rate, adaptive shaping

1 INTRODUCTION

The Available-Bit-Rate **(ABR)** service is being designed as a low-cost transport service over ATM, which will be using those network resources (link bandwidth, in particular) that are left available by other high-priority connections (belonging to the Guaranteed-QoS or **G-QoS** class). The penalty paid is the lack of any QoS guarantee in terms of transport delay figures, but this may be perfectly affordable by many applications, like file transfer, LAN interconnection, even by some real-time video applications (e.g. MPEG-coded video stream transfer).

Although the idea sounds quite simple, there are several difficulties related not only to the implementation of an ABR service, but also to the analysis and control of the carried traffic. Apart from a priority mechanism for serving the two classes, large buffers are required at each multiplexing/switching stage to store the ABR traffic, in order to maintain the stringent cell-loss requirements that are usually imposed by the ABR-serviced applications. A feedback mechanism is also necessary to inform preceding nodes/terminals about the congestion conditions along upstream nodes. Additional enhancements are further necessary at the terminals to be able to adapt their traffic profile to different network congestion conditions.

The purpose of this paper is to cope with some of the basic traffic engineering issues related to an ABR service implemented on ATM. Bandwidth allocation, Connection Admission Control (CAC) and traffic shaping are among these issues (CCITT, 1992). Our approach is based on the fluid-flow modelling of the multiplexed traffic (Anick 1982, Baiocchi 1992, Kontovasilis 1994, Kontovasilis 1995, Kosten 1984, Mitra 1988, Mitrou Stern 1991). The large buffer size and the very small overflow probabilities required by an ABR service fully justify this choice. Simple queuing arguments and basic fluid-flow-analysis results lead to the assessment of the performance of an ABR multiplexer, which is further utilised to handle the main traffic control problems outlined above.

A central point in the traffic analysis and control methods considered in this paper is the calculation of an **effective** or **equivalent rate** for each of the multiplexed ABR-class streams, in terms of the main traffic and QoS parameters (Elwalid 1993, Guerin 1991, Kelly 1991, Kesidis 1993, Mitrou - St. Peters. 1995). In the case of ON/OFF Markovian streams, a closed-form expression greatly simplifies most of the traffic control problems (Mitrou, St. Peters. 1995). Traffic shaping, in particular, is tackled in the above context through changing the source profile (e.g. the peak or the mean rate) according to the loading/congestion conditions prevailing along the connection's path. The calculation of the appropriate profiles is based on the inversion of the effective-rate formula.

Handling the statistical multiplexing problem on the basis of effective rates, implicitly solves one of the major issues arising in an ABR service environment: that of the **fair sharing** of the available bandwidth among the multiplexed connections and, even further, **charging** each user accordingly.

In section 2, following this introduction, the implementation and modelling assumptions concerned with the ABR service are presented and some basic results from the related traffic theory are reviewed. The latter are drawn either from the general queuing theory or as more specific derivations of the fluid-flow method. Section 3 addresses the main traffic control problems related with an ABR service, and proposes solutions to them, under the light of the results of section 2. In section 4 the solutions to the main control problems (bandwidth allocation, connection admission control and shaping) proposed in section 3 are implemented in terms of approximate analytical formulas, relating the effective rate with the other traffic, system and QoS parameters. In the same section a simple example is presented which illustrates a proposed adaptive shaping scheme. Finally, section 5 draws some concluding remarks.

2 MODELLING ASSUMPTIONS AND SYNOPSIS OF RELEVANT QUEUING RESULTS

Figure 1 shows a single multiplexing stage with two classes sharing the same link capacity, C. One class is of the guaranteed-QoS type having an absolute time priority on the link, i.e. cells of this class are always served first, if present in the queue. The second class is of the ABR type, i.e. it is only served if no cells from the other class are waiting for transmission. Either a common buffer or separate buffers may be used for the two classes. In the sequel a common buffer configuration is assumed. In the case, however, of implementing the ABR service by an *add-on* equipment to existing single-class switches, a separate buffer is obviously necessary.

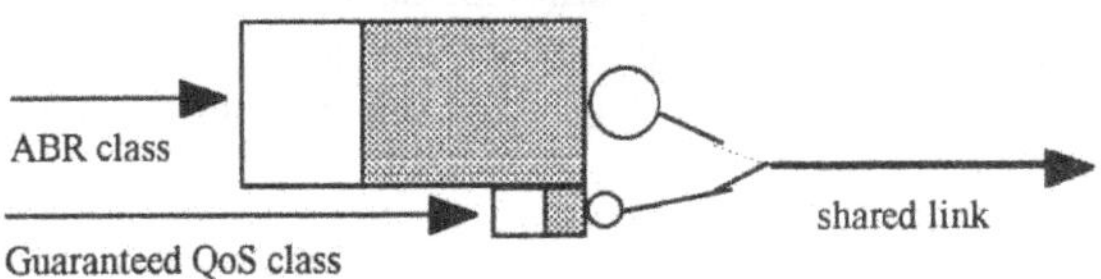

Figure 1 ABR service supporting multiplexer

We now list some assertions that will be of help later on:

Assertion 1

The total buffer occupancy distribution does not depend on the time priority scheme used, as far as the server keeps on working in the presence of cells waiting for transmission (no matter from which class).

Although rather obvious, the above statement is very useful in many situations, since changing the priority scheme may give rise to easier models to solve. Specific applications of it will be exposed later.

Assertion 2 **(Finite- versus infinite-buffer modelling)**

The buffer occupancy complementary probability distribution (CPDF) of an infinite-buffer configuration is always an upper bound to the respective finite-buffer one, thus providing a conservative estimate of the overflow probability or delay percentiles in the latter. The larger the buffer and/or the smaller the overflow probability enforced on the finite boundary, the tighter the bound becomes.

Assertion 2 stems from the fact that in an infinite-buffer system information is always stored and never rejected, thus the probability of exceeding a certain buffer threshold is higher than in the finite variant. The infinite-buffer boundary conditions are always much easier to solve and give a solution independent of the buffer dimension. Furthermore, the ATM systems are usually engineered for small overflow probabilities, in which case the infinite-buffer solution is quite satisfactory in practice.

Assertion 3 **(Buffer occupancy seen by arriving cells)**

The buffer occupancy CPDF seen at instances of cell arrivals $H(x)$ and the global CPDF at a random time instance $G(x)$, are related by the simple relationship

$$H(x)=\frac{G(x)}{\rho}, \tag{1}$$

where ρ is the average load, normalised over the output link capacity.

This relationship, which is a kind of **Little's formula** for distributions, has been proven for Markovian fluid models in (Kontovasilis, *Annals* 1994). However its validity is conjectured to be more general (because it involves only the occupancy distributions and the mean load).

Assertion 4 **(Buffer occupancy seen by different classes in heterogeneous mixes)**
In a heterogeneous traffic multiplexing environment, arriving cells of different classes observe, in general, a different CPDF at the buffer. If (1) holds true, then the following relationship does also hold

$$G(x) = \sum_i \rho_i H_i(x) \tag{2}$$

where ρ_i is the normalised load of and $H_i(x)$ the occupancy CPDF seen by class i.

The proof for Markovian fluid models is given in (Kontovasilis, *Annals* 1994), as for the previous assertion.

Assertion 5 **(Mixing of Constant Bit Rate (CBR) and non-CBR traffic)**
A CBR component sees always the global CPDF, *G(x)*, while the remaining traffic (provided that (1) holds true) sees a CPDF equal to

$$H_{non-CBR}(x) = \frac{1-\rho_{CBR}}{\rho_{non-CBR}} G(x). \tag{3}$$

Proof

Due to Assertion 1 above, the buffer occupancy CPDF does not depend on the time priorities assigned to the different classes of the multiplexed traffic. If we assign a high priority to the CBR traffic, then this is equivalent to a system loaded by only non-CBR traffic and featuring an output link of $C(1-\rho_{CBR})$. Thus at an instance of (now non-CBR only) cell arrival we have, using (1),

$$H(x) = \frac{G(x)}{\dfrac{C\rho_{non-CBR}}{C(1-\rho_{CBR})}} = \frac{1-\rho_{CBR}}{\rho_{non-CBR}} G(x).$$

Now, returning to the original setting, *H(x)* as just computed is $H_{non-CBR}(x)$ and from (2):

$H_{CBR}(x) = \dfrac{G(x) - H_{non-CBR}\rho_{non-CBR}}{\rho_{CBR}} = G(x)$, which completes the proof.

We now state two assumptions upon which further developments are based:

Assumption 1:
The delay requirements of the G-QoS class impose a small buffer occupancy for that class.

This will be assured by the Connection Admission Control functions, through keeping the load of the G-QoS class low enough to maintain the affordable delay percentiles. The simplest treatment of this issue consists in allocating peak rates to G-QoS class connections. In that case only a small buffer is required, just to absorb the cell-level congestion stemming from the simultaneous cell arrivals.

Assumption 2

A large buffer is provided for the ABR class to avoid cell losses or to keep the probability of such losses acceptably low.

From Assumption 1 above, the buffer occupancy distribution for the ABR class is approximately equal to the total occupancy distribution and, by virtue of Assertion 1, it can be calculated by applying any analysis method for the aggregate traffic (from both classes), with or without priorities. The fluid-flow method with an infinite buffer, in particular, is ideally suited for this case, due to the large dimension of the buffer and the very low target overflow probability.

3 TRAFFIC CONTROL FOR AN ABR SERVICE

This section addresses the main traffic control problems related with an ABR service and proposes some solutions to them, under the light of the discussion in section 2. Connection Admission Control (CAC) is first considered for the two classes (G-QoS and ABR), in the absence of any traffic-profile-adaptation mechanism at the terminals. Then, we assume that the latter have the ability to change (shape) their traffic profile according to the congestion conditions prevalent in the network nodes. The related function is called **Traffic Shaping** and is examined in sub-section 3.2. In this case CAC is greatly simplified for both classes.

3.1 Bandwidth Allocation and Connection Admission Control

The role of the CAC is to ensure that the acceptance of a new connection through a series of multiplexing/switching stages (nodes), as the one depicted by Figure 1, will not violate the QoS contract for the already-established connections through the same nodes. This is equivalent to answering a bandwidth allocation problem at each multiplexing stage, formulated as follows: *Given a number of connections and a QoS figure, calculate the bandwidth required to serve this group of connections.* We assume the following **guaranties** for each connection of the two classes:

i) *for the G-QoS class*

(a) a certain delay percentile: $\Pr\{\text{delay}>D\}<\varepsilon_D$

(b) a certain maximum cell-loss probability*

* When a common buffer with the ABR class is used, a common cell-loss criterion is set for both classes and guarantying (iia) usually suffices. If a separate buffer is devoted to the G-QoS class, usually small, then a separate cell-loss criterion becomes necessary also for this class.

ii) *for the ABR class*

(a) a certain maximum cell-loss probability

(b) a minimum throughput (mean rate).

Based on the discussion of section 2, the following **acceptance conditions** must be checked:

I) *for a new G-QoS connection*

1. the already established G-QoS connections and the new one should form a set that respects (ia) when served by the total output capacity C.
2. the new connection and the already established ones from **both classes** should form a set that respects (iia) and (iib) when served by the total output capacity C.

II) *for a new ABR connection*

1. the new connection and the already established ones from *both classes* should form a set that respects (iia) and (iib) when served by the total output capacity C.

A very simple peak-rate-based acceptance condition can be adopted for the G-QoS class, in which case the delay for that class comes only from the cell-level congestion (usually quite small, satisfying (ia)). Then, the challenging issue is the satisfaction of the second condition above, aiming at the ABR class guarantee. In the remaining part of this section we confine ourselves to that condition, with the assumption of fixed, non-adaptable traffic profiles for the ABR connections (i.e. no capabilities for adaptive shaping).

Checking the acceptance condition (I.1 or II.1) can be based on the *multiplexing of two different traffic classes served without priorities and sharing the same buffer*. The following approaches to that are applicable (Mitrou, *ETT* 1994):

A) *Logical Partitioning of the output capacity*

It consists of calculating the required capacity for each of the two classes with the requirement of (iia), adding the two capacities and checking whether the sum is less than or equal to C. This can be done by using either any analysis method available or approximate formulas.

Obviously this is a conservative approach since no multiplexing between the two classes is considered. In the extreme case where the G-QoS connections are of the CBR type then, indeed, this class does not contribute to the multiplexing gain and the above procedure becomes exact. In this case, equations (1), (3) are directly applicable.

B) *Using an heterogeneous-traffic-analysis method (Baiocchi 1992, Kontovasilis - Annals 1994, Stern 1991) to assure (iia)*

This approach suffers from a large computational overhead that may stress the time-constraints of the Connection Admission procedure.

C) Allocating peak rates to the G-QoS connections together with shaping of the ABR streams to ensure adaptation to the changing availability of bandwidth

This is a promising approach that matches well to the effective rate schemes; it is discussed further in section 3.2.3.

The CAC (as well as the other traffic control functions considered later) is facilitated by using *effective rates* to characterise the bandwidth requirements of each connection in the given multiplexing environment, with the specified QoS requirements (see also the next sub-section). In section 4 an approximate closed-form expression relating the *Effective Rate* of a traffic stream with the basic traffic and QoS parameters is given, which greatly simplifies the CAC procedures outlined above.

3.2 Traffic shaping for the ABR service

Traffic shaping aims at producing profiles which exhibit certain parameter values. The simplest traffic shaping operation is the *peak-rate enforcement*, keeping the peak rate of a stream less than or equal to a pre specified value (Guillemin 1992). Other shaping operations, like mean or maximum burst-size enforcement or burstiness enforcement, are applicable on bursty traffic streams.

Here we examine traffic shaping functions that are necessary to enforce specific effective rate values, i.e. to produce streams with a specific multiplexing attitude. By definition, the ABR service utilises a bandwidth that is not constant, in general, but varies according to the aggregate rate fluctuations of the G-QoS connections. Moreover, the ABR connections themselves are bursty in nature and have variable rate requirements. It becomes therefore obvious that the notion of an effective (or equivalent) bandwidth is ideally suited for this case.

3.2.1 Effective Rate enforcement

There exist various terms (effective rate, effective bandwidth, equivalent bandwidth etc.) in the literature, to address the notion of a descriptor for the bandwidth requirements of a bursty traffic stream for a certain buffer overflow probability or a delay percentile (Elwalid 1993, Guerin 1991, Kesidis 1993, Mitrou - *ETT* 1994) . Some of the definitions are concerned with each stream in isolation (Elwalid 1993, Guerin 1991, Kesidis 1993), in which case any gain comes only from buffering, while other consider them in a multiplexing environment (Mitrou - *ETT* 1994), where an additional gain comes from multiplexing. Depending on the buffer size (compared to the burst size) and the burstiness of the multiplexed streams one or the other of the two parts may prevail.

Here we consider the more general definition of the effective rate (Mitrou - *ETT* 1994):

> Given that N traffic streams of a certain fluid class are multiplexed together, there can be found a unique output rate C_N, such that $\Pr\{\text{ queue} > V_b \} = p$ (specified). Then $R_e = C_N / N$ is defined as the **effective rate** (or **effective bandwidth**), required by each stream within this multiplexing environment.

Calculating the R_e as a function of N and making a suitable interpolation between the discrete points (C_N, C_N/N), N=2,3,... (a linear interpolation is sufficient in the context of this work), a continuous curve is derived allowing the definition of R_e for a fixed output rate (instead of fixing the number of multiplexed streams). The continuous alternative permits, in principle, an extension to cases where heterogeneous streams share a common output link.

To simplify things, we further assume streams of the ON/OFF type. This assumption is not unrealistic in an ABR service environment, where the traffic sources either transmit at a constant rate or remain silent, as a result of the shaping operation. In that case the effective rate is a function of the form

$$R_e = f([V,r,c],[V_b,C],p), \tag{4}$$

where $[V,r,c]$ is the **traffic descriptor** of the stream (in terms of its mean rate, r, the peak rate, c, and the mean burst size, V), $[V_b, C]$ is the **multiplexer's dimensions** (in terms of the buffer size, V_b, and the output link rate, C) and p is the **QoS requirement**. A traffic descriptor equivalent to $[V,r,c]$ above is the $[V,B,c]$, where $B \equiv c/r$ is the **burstiness** of the stream.

As far as the shaping function is concerned with, each of the three traffic parameters given above can be controlled so as to keep the effective rate of the stream constant. This way, different shaping schemes can be devised.

Peak-rate shaping for effective rate enforcement

Equation (4) is solved for c to yield

$$c = f_1(R_e,V,r,V_b,C,p). \tag{5}$$

By fixing the independent parameters in (5) we can determine and enforce a specific value for the peak rate.

Mean-rate shaping for effective rate enforcement

We can instead solve (4) with respect to the mean rate, r

$$r = f_2(R_e,V,c,V_b,C,p). \tag{6}$$

The mean rate, calculated by (6), can be regulated by controlling the silence intervals between successive bursts, while the peak rate is kept constant. It must be noticed that, while c is a deterministic parameter, r is a statistical average. If, for the purpose of shaping, we enforce a silence between bursts that is deterministically calculated by (6), the assumptions under which (4) has been derived may be violated. A usual assumption, for example, is for exponentially distributed bursts and silences. It has been found, however, that deterministic values for the V and/or the r (or, more generally, hypo-exponentially distributed values) lead to a better performance than their exponential or hyper-exponential counterparts (Mitrou - Bradford

1994). Thus, as far as the multiplexing behaviour of the shaped streams is concerned, we are on the safe side by enforcing constant silence values.

For both shaping schemes described above the principal problem focuses on deriving a formula like (4) and inverting it to get (5) or (6). In the next section approximate formulas of the type of (4) will be presented, which provide a solution to that problem.

3.2.2 Adaptive shaping

When the traffic, system and QoS parameters involved in equations (4) through (6) remain constant or stationary, the values calculated for the controlled parameters (c or r) are kept constant for the life-time of a connection. This is referred to as **fixed shaping**.

The above case, however, of constant or stationary parameters, is only the exception. Hardly one finds a source that can guarantee the stationarity of the volume of the produced bursts, for example. Also the number of the multiplexed connections changes dynamically. It is therefore necessary to apply an **adaptive shaping**, which will take into account the short-term parameter values of the produced traffic as well as the loading conditions at the multiplexers and change the traffic profile accordingly.

The application of an adaptive shaping in ABR service provisioning is shown in fig. 2. According to this figure, each source of the ABR type gets a feedback from the network about the loading and congestion conditions prevailing at ABR switches. This information may be generated periodically or after a significant change has occurred (e.g. the number of established connections has changed significantly or the content of the ABR buffer has exceeded some threshold). For a logical partitioning of the bandwidth between G-QoS and ABR class, as discussed in 3.1, the information that has to be fedback to the shaper is the available for the ABR class capacity C_{av}, the available at that time buffer capacity V_{av}, the target overflow probability p, and the effective rate allocated to the source. The shaper utilises this information, along with the one extracted by monitoring its own output traffic, to change the regulated parameter levels accordingly.

3.2.3 CAC combined with adaptive shaping

If the terminals are equipped with an adaptive shaper, the CAC algorithm may be significantly simplified, providing for e.g. **peak-rate allocation to the G-QoS connections and effective-rate allocation to the ABR class.** It is up to the shaping mechanism to adapt to the loading and congestion conditions, in order to enforce the contracted effective rates.

Since peak rate is allocated to the G-QoS class, admission of new and release of terminated G-QoS connections can be modelled by introducing a discrete finite set of states $J=1,..,K$, each associated with a rate-value c_J, expressing the collective bandwidth requirements of the G-QoS class in that particular state. Without essential restriction on generality, one may assume Markovian transitions between these states, since arbitrary sojourn times may be approximated by Markovian modelling—at the expense of increasing the state-space.

The adaptive shaping consists of tuning one or more parameters of each ABR stream (with possibly different controlling parameters among ABR streams of differing natures) so as to adjust to the fluctuations of the bandwidth consumed by the G-QoS class. It is assumed that, while the parameter values may be freely tuned, the number of states describing each ABR stream remains constant; this is not restricting for all the controls discussed in this paper. It is further assumed that the dynamics governing G-QoS state-changes are much slower, as compared to the burst-dynamics of the ABR connections; this is a fairly natural assumption, given that the G-QoS state-changes correspond to call-level phenomena. Under this assumption, the global model for the compound traffic belongs to the class of **Nearly Completely Decomposable Markovian Fluid Models**.

By employing results from the relevant theory (Kontovasilis 1995) it can be shown (Mitrou - *Telecom. Systems* 1995) that any adaptive control conforming to the description above is stable, in the following sense: given that, for all J, the tuning of the controlling parameter(s) results in satisfying the QoS criterion within a particular G-QoS state J (i.e., by assuming that the G-QoS bandwidth requirements remain time-invariant and equal to c_J) then the quality criterion is respected globally, viz. with G-QoS rate fluctuations present.

This result is independent from the from of control applied, given that a simple loading condition applies, namely that for each state J, the sum of the compound G-QoS rate c_J plus the aggregate mean rate of all ABR streams, as controlled within state J, does not exceed the link capacity. This loading condition is always satisfied when enforcing the connection acceptance conditions in the beginning of section 3.1. In particular, the loading condition is satisfied when the control is based on an effective rate notion. For more details and a proof on this stability issue, see (Mitrou - *Telecom. Systems* 1995).

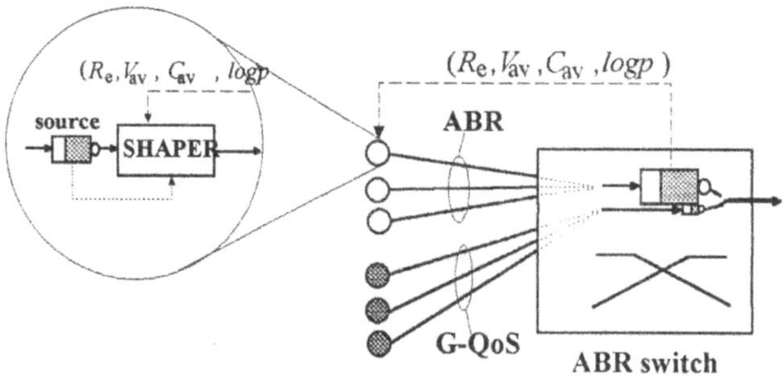

Figure 2 Adaptive shaping of the ABR-serviced traffic

4 APPROXIMATE CLOSED-FORM CALCULATION OF THE TRAFFIC CONTROL

In (Mitrou - St. Peters. 1995) an approximate calculation of the effective rate of ON/OFF Markovian streams is presented in closed form. The formula is derived by fitting experimental data for large buffer sizes and burstiness values (such conditions fit quite well into an ABR service environment).

In particular, by optimising the fitting in the least-squares sense over the range

$$20 < \frac{V_b}{V} < 200, \quad 5 < B < 100, \quad -3 < \log(p) < -9 \tag{7}$$

we get:

$$\frac{c - R_e}{R_e - r} \cong a\sqrt{\frac{C}{c}} + b \tag{8a}$$

$$\begin{aligned} a &\cong [0.075 + e^{0.4\log p - 0.7186} + e^{-0.03\frac{V_b}{V} - 1.3}]B + 0.3, \\ b &\cong [(0.035 + e^{0.3321\log p - 1.28})\frac{V_b}{V} - 1.1]B + (0.043 + e^{0.281\log p - 1.247})\frac{V_b}{V} - 1.7 \end{aligned} \tag{8b}$$

where the log is in base 10.

A cruder approximation is derived by using the asymptotic (dominant-eigenvalue) behaviour of the buffer-content CPDF (Anick 1982). This results in a conservative approximation

$$\ln p \le -\frac{c}{c-r}\frac{1-rN/C}{1-C/cN}\frac{V_b}{V} = -(\frac{c}{c-r})\frac{1-r/R_e}{1-R_e/c}(\frac{V_b}{V}). \tag{9}$$

In (9) the effective rate R_e does not depend on the output capacity, but only on the traffic stream characteristics and the QoS point (buffer size, overflow probability). This is the effective rate as defined in (Elwalid 1993, Guerin 1991), taking into account only the buffering.

For very large peak rate values (i.e. the bursts enter the queue "instantly"), the right-hand side of (9) tends to $-(1-r/R_e)V_b/V$, providing a very rough but useful upper bound of the overflow probability.

By using (8) or (9) the traffic control procedures described in the previous section are directly applicable, although not always in closed form.

Currently, we study the possibility of extending the dominant pole approximation to include the dependence of the overflow probability on the output capacity. In this approach the overflow probability is $G(x) \approx \alpha e^{-z_\infty x}$, where z_∞ is the dominant slope as before (in semilog scale), while α is the Chernoff upper bound to the probability that the cumulative input rate exceeds the link capacity. Study of this approximation for general Markovian fluid models shows that indeed, the dependence of the ER on the link capacity asymptotically occurs in a square-root fashion, as in (8a).

4.1 Approximate, analytical Bandwidth Allocation and Connection Admission Control

Recalling the principle of **Logical Partitioning** from 3.1, we need to calculate the capacity X that must serve a number K of similar* connections in order to maintain a maximum cell loss probability in a buffer of size V_b.

By applying (8) for each of the K connections in the group, we have

$$\frac{c_i - x_i}{x_i - r_i} = a_i \sqrt{\frac{X}{c_i}} + b_i \tag{10a}$$

with x_i denoting the effective rate allocated to the i_{th} connection and

$$\sum_{i=1}^{K} x_i = X. \tag{10b}$$

The above equation can be solved with respect to X, since we know that the x_i's are continuous (decreasing) functions of X. A similar (and simpler) approach can be followed based on (9). For each stream an effective rate is calculated (through solving (9)) and the sum of all these rates gives the required output capacity. If one uses (9) (or its approximation with an infinite rate) instead of (8) then the requirement of similar connections is waived and a heterogeneous environment can be fully accommodated (Elwalid 1993, Guerin 1991).

4.2 Approximate, analytical calculation of the shaping values

By substituting in (8b) B with its equal c/r we can solve (8a) with respect to c (a third-order equation) or r (a second-order equation).

Equation (9) is again more manageable. It gives a second-order equation in c and a first-order one in r.

4.3 An example of rate-based adaptive shaping

The following example is a simple case of rate-based adaptive shaping. A number ($N = 70$) of identical bursty sources are multiplexed on the same link of a capacity C, supported by a buffer of a size $V_b = 3000$ cells. No connections of the G-QoS class are present (equivalently, we could assume that the considered link capacity C is just what is left for the partition of the ABR class). All the rates will be subsequently normalised with respect to C, i.e. $C = 1$ will be assumed.

---

* As mentioned before, only similar connections may be grouped together for the above calculations, since the effective rate has been defined in a homogeneous multiplexing environment. *Similarity* here means *not large differences in the mean burst size and the burstiness.*

The sources generate data packets of an exponentially distributed size with mean V=75 cells, according to a Poisson process such that r = 0.01. With reference to Figure 2, no monitoring of the produced traffic is necessary by the shapers, since the sources are assumed to be stationary with known characteristics. The sources are informed by the network about the loading & congestion conditions at the ABR multiplexer. In this simple example the number of connections is assumed fixed (=70) and a feedback is only issued whenever the buffer content crosses a threshold. The rate adaptation mechanism is a two-state machine, as described by the pseudo code of Figure 3. A hysterisis in switching from c_2 *to* c_1 (i.e. *thr2* < *thr1*) is used to avoid unwanted oscillations around the threshold.

Using (8) we get c_1 = 0.666 (it is rounded up to unity, i.e. c=1, is used in the uncongested state) and c_2= 0.048. Equation (9) would give more conservative rates, i.e. c_1=0.1817 and c_2=0.0213. For the uncongested state, a relatively high overflow probability ($p_1 = 10^{-2}$) has been used for the calculation of the respective rate c_1, in order to increase the

```
initialise
{  Re=1/70; V=75; r=0.01; C=1; Vb=3000;
   thr1=1000; thr2=900;  p1=10^-2; p2=10^-10;
   calculate c1: Re=f(V, r, c1,thr1 , C, p1);          /* using (5) or (6) */
   calculate c2: Re=f(V, r, c2 ,Vb-thr1, C, p2);       /* using (5) or (6) */
   state=uncongested;
   c=min[1,c1];  }
begin
    do
    {      if buffer_content > thr1 & state==uncongested
           {       c=c2;
                   state=congested;  }
           if buffer_content < thr2 & state==congested
           {       c=c1;
                   state=uncongested;  }
    }
    forever
end
```

Figure 3 Pseudo code for the example of rate-based adaptive shaping

throughput. This probability does not affect the ultimately offered QoS to the ABR class; it affects only the rate at which feedback messages are issued and, hence, the signalling overhead induced by these messages.

Figure 4 presents the results from a discrete-time simulation of the above example. We can clearly see the two different slopes of the buffer-content probability distribution, corresponding to the uncongested and congested states. A zero delay for the feedback has been used. In a real system, however, there will be a finite response time, which must be taken into account.

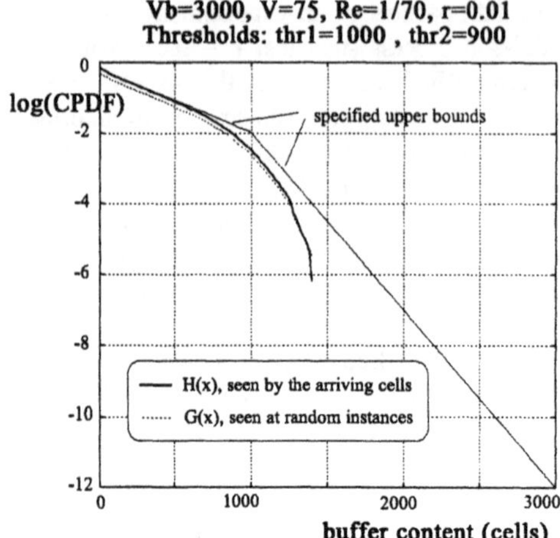

Figure 4 Simulation results of the adaptive shaping example

5 CONCLUSION

The main traffic analysis and control problems related with an Available Bit Rate (ABR) service were addressed, modelled and provided with a solution method. The proposed method is based on the notion of Effective Rates defined for the multiplexed connections and calculated approximately, by using a fluid-flow model for the ATM multiplexer. Simple queuing arguments and results from the application of the fluid-flow method on Markovian traffic streams lead to important and analytically simplifying relations between the QoS figures experienced by the two classes, the ABR and the Guaranteed-QoS.

Inverting the Effective Rate equation yields a solution to the specific problems of Bandwidth Allocation, Connection Admission Control and Traffic Shaping in the considered multiplexing environment. An adaptive, rate-based traffic shaping mechanism was proposed and tested through simulation.

6 REFERENCES

Anick, D., Mitra, D. and Sondhi, M.M. (1982) "Stochastic theory of a data-handling system with multiple sources", *Bell Syst. Tech. J.* 61 1871-1894.

Baiocchi, A., Blefari-Melazzi, N., Roverti, A. and Salvatore, F., (1992) "Stochastic Fluid Analysis of an ATM Multiplexer Loaded with Heterogeneous ON-OFF Sources: an Effective Computational Approach," In *Proc. INFOCOM '92*, pp. 3C.3.1-3C.3.10.

CCITT STUDY GROUP XVIII (1992) Traffic control and resource management in B-ISDN. CCITT Recomendation I.371, Geneva.

Elwalid, A. and Mitra, D. (1993) "Effective bandwidth of general Markovian sources and admission control of high speed networks," *IEEE/ACM Trans. on Netw.* , pp. 329-343.

Guerin, R., Ahmadi, H. and Naghshineh, M. (1991) "Equivalent capacity and its application to bandwidth allocation in high-speed networks," *IEEE JSAC*, vol. 9, pp. 968-981.

Guillemin, F., Boeyer, P., Dupuis, A. and Romoeuf, L. (1992) "Peak Rate Enforcement in ATM Networks," *IEEE INFOCOM '92*, paper 6A.1.

Kelly, F.P. (1991) "Effective bandwidths at multi-type queues," *Queue. Syst.*, vol. 9, pp. 5-15.

Kesidis, G., Warland, J. and Chang, C.-S. (1993) "Effective Bandwidths for Multiclass Markov Fluids and Other ATM Sources", *IEEE/ACM Trans. on Netw.*, Vol.1,No.4, pp. 424-428.

Kontovasilis, K.P., and Mitrou, N.M. (1994) "Bursty Traffic Modeling and Efficient Analysis Algorithms via Fluid-Flow Models for ATM IBCN," *Annals of Operations Research*, Vol. 49, special issue on Methodologies for High Speed Networks pp. 279-323.

Kontovasilis, K.P., and Mitrou, N.M. (1995) "Markov Modulated Traffic with Near Complete Decomposability Characteristics and Associated Fluid Queueing Models," to appear in the *Adv. Appl. Prob.* Dec. 1995.

Kontovasilis, K.P., and Mitrou, N.M. (1994) "Stochastic fluid models for a buffer loaded by Markov modulated traffic with near complete decomposability characteristics", in *IFIP Transactions C-21, "High Speed Networks and their Performance*, H.G. Perros and Y. Viniotis eds., North-Holland, pp. 363-397.

Kosten, L., (1984) "Stochastic theory of data-handling systems with groups of multiple sources," in *Performance of Computer Communication Systems*, H. Rudin and W. Bux, Eds. Amsterdam, The Netherlands: Elsevier, pp. 321-331.

Mitra, D., (1988) "Stochastic theory of a fluid model of producers and consumers coupled by a buffer," *Adv. Appl. Prob.* 20 pp. 646-676.

Mitrou, N.M., and Koukos, A. (1993) "An effective-rate enforcement algorithm for ATM traffic and its hardware implementation", *Proceedings of the IBCN&S*, Copenhagen, April 20-23 1993.

Mitrou, N.M., Kontovasilis, K. and V. Nellas, "Bursty Traffic Modelling and Multiplexing Performance Analysis in ATM Networks: A Three-moment Approach," *2nd IFIP Intern. Workshop on on Performance Modelling and Evaluation of ATM Networks*, Bradford, 4-6 June 1994.

Mitrou, N.M., Kontovasilis, K.P., Kroener, H., amd Iversen, V.B. (1994) "Statistical Multiplexing, Bandwidth Allocation Strategies and Connection Admission Control in ATM Networks,"*European Transactions on Telecommunications*, Vol. 5, No. 2, pp. 161-175.

Mitrou, N.M., Kontovasilis, K.P., and Protonotarios, E.N. (1995) "A closed-form expression for the Effective Rate of ON/OFF traffic streams and its usage in basic traffic control problems", *Proceedings of the Intern. Teletraffic Seminar*, St. Petersburg, June 1995.

Mitrou, N.M., Kontovasilis, K.P., and Protonotarios, E.N. (1995) "ATM traffic engineering for ABR traffic provisioning", to appear in the *Telecommunications Systems Journal.*

Stern, T.E., and Elwalid, A.I. (1991) "Analysis of separable Markov-modulated rate models for information-handling systems", *Adv. Appl. Prob.*, vol. 23, pp. 105-139.

PART EIGHT

Performance and Optimization of ATM Networks

22

Performance Evaluation of Frame Relay, SMDS, and ABR Services in ATM Networks

L. Fratta, L. Musumeci
Dipartimento di Elettronica e Informazione, Politecnico di Milano
Piazza Leonardo da Vinci 32, I-20133 Milano, Italy
tel. +39 2 23993578 - fax +39 2 23993413 -
e-mail: fratta@elet.polimi.it

Abstract

The broadband integrated services digital networks are based on the ATM technology. As these networks will be deployed gradually, the existing high-speed data services, as Frame Relay and SMDS, will continue to exist and, for interoperability purposes, they will also be supported by the ATM networks. Moreover, ATM networks allows to efficiently support a new service, called available bit rate (ABR) service in the ATM Forum, which requires no bandwidth reservation. When this service is requested, data are transmitted through the network with low priority, being the available network resources shaped dynamically among all the active sources. In this paper, we first discuss how to manage FR and SMDS services in the ATM networks, and report the performance evaluation of the proposed interconnection scenarios; then, we present some preliminary simulation results on the ABR service, whose implementation needs a sophisticated reactive congestion control to obtain an acceptable quality of service.

Keywords

Frame Delay, SMDS, ABR, ATM network, rate-based control.

1 INTRODUCTION

The increasing demand for high-speed data switched services has stimulated the introduction of new "fast packet" technologies. In particular, Frame Relay (FR) has appeared appropriate to support new data applications, such as LAN-to-LAN interconnection at 2 Mbit/s. Frame relay [1] is a connection-oriented technology that supports variable-length packets and is designed to improve the efficiency of X.25 packet networks.

In addition, SMDS (Switched Multimegabit Data Service) represents another strategy to meet the demand for high-speed data services up to 45 Mbit/s before the introduction of the broadband integrated services digital networks, B-ISDN. The SMDS interface and associated protocols are based on the connectionless part of the IEEE 802.6 distributed queue dual bus (DQDB) MAN standard [1].

While the scope of FR and SMDS networks is to support only high-speed data applications, the Asynchronous Transfer Mode (ATM) technology has been proposed as a standard for supporting B-ISDN. ATM is based on the concept of a homogeneous network, where all traffic is transformed into a uniform 53-byte packet or cell stream. This allows the network to carry a wide variety of different traffic types [1].

As B-ISDN deployment will be gradual, ATM will not replace Frame Relay and SMDS services in the near future. It is expected that these services will be supported by the ATM networks since the initial phase of their deployment.

The introduction of ATM technology offers new problems and challenges, which have been subject of intense research over the past few years. The ATM Forum has promoted an important work, devoted to accelerate the development and standardization of ATM technology.

Four traffic classes, constant bit rate (CBR), variable bit rate (CVR), available bit rate (ABR), and unspecified bit rate (UBR), have been proposed on the ATM Forum [2]. For CBR and VBR services, the ATM networks guarantee the negotiated quality of service. For example, a CBR service is described in terms of peak cell rate (PCR) and cell delay variation (CDV). For this service, congestion control is managed through admission control and bandwidth allocation procedures. Therefore, if the resources requested are not available, the connection will be rejected at the call setup phase.

For many conventional computer applications, it is not easy to predict the bandwidth requirements. In these cases, it is not required to reserve bandwidth in the network and it is more appropriate to share the available bandwidth among all active users. Such a service, the available bit rate (ABR) service, requires the application of an explicit control scheme to reduce cell losses.

The congestion control for ABR service has been standardized in the ATM Forum's September 1994 meeting. The rate based control for ABR services was preferred to the credit control scheme, which requires extremely large buffers in each switch in the wide-area networks with many VCs and large propagation delays.

The goal of this paper is twofold. First, we discuss some proposals to achieve an efficient management of FR and SMDS services, when supported in an ATM network. In particular, we investigate the impact of FR traffic on the performance characteristics of ATM networks and discuss how to improve the cell loss probability and the fairness due to the presence of the FR traffic in the ATM multiplexer. In addition, we describe the access of the SMDS traffic to the ATM network through a Two Rate Multiplexer (TRM), which allows to dynamically manage the bandwidth of a Virtual Path (VP) towards a Connectionless Server (CLS) through an ATM switch. A model of this TRM is presented together with its performance evaluation. Second, we present our preliminary simulation results on the ATM Forum proposal for the ABR service illustrating its performance characteristics. Finally, we conclude with some considerations on the future extension of this work.

2 FRAME RELAY SERVICE IN ATM NETWORKS

The LAN interconnection based on FR interfaces, can be implemented in an ATM network by the use of dedicated Virtual Paths (VP), dimensioned at the peak bit rate. This approach is simple, but not efficient. More efficient solutions can be obtained by statistically multiplexing the FR traffic with ATM sources. To investigate the impact of FR traffic on ATM multiplexer performance, we consider the network model of figure 1, where the FR multiplexer is colocated with the ATM multiplexer. The FR sources are connected to the FR multiplexer via 2 Mbps links. Each FR source can adjust its current window size between a minimum (W = 1) and a maximum (W = 7) value, according to network congestion status. The generated frames, which can not be forwarded, are stored in the source buffer. The FECN/BECN mechanism, implemented in the FR multiplexer, allows to notify the network congestion to the FR sources. In the FR multiplexer, a transmitted frame has the FECN bit set to one, if the multiplexer is congested. Each transmitted frame is acknowledged by the destination through an ACK frame. This frame is addressed to the source with the BECN bit set to one, if the corresponding frame was received with the FECN bit set to one. An ACK frame is received by the source after an overall round-trip delay τ. In presence of network congestion, each source reduces his window to a value rW, where r is a preassigned reduction factor. Otherwise, the window is increased by one up to the maximum value. The algorithm to measure the network congestion in the FR multiplexer and the algorithm to modify the window in each source are described in more details in [3]. Each FR frame is segmented into 32 ATM cells before accessing the ATM multiplexer.

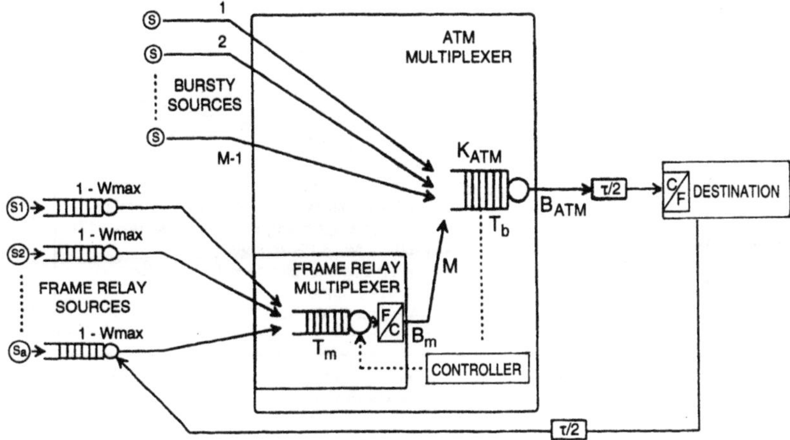

Figure 1 Architecture for the FR access in an ATM multiplexer.

M=10 bursty sources are connected to an ATM multiplexer, which is characterized by a single server with a finite buffer K_{ATM}. The ATM multiplexer has a buffer capacity K_{ATM}=50 cells and a transmission bit rate B_{ATM}=150 Mbps. Each source generates traffic with peak bit rate B_p=34 Mbps, mean burst length L_m=100 cells and burstiness b_b, which is selected in order to obtain a proper average load on each access link, ρ_b. The simulation allows to evaluate the cell loss probability P_e as function of ρ_b.

To investigate the effect of the FR traffic on the ATM multiplexer cell loss probability, one ATM bursty source has been substituted by a 34 Mbps FR multiplexer. We suppose that FECN and BECN bits are also supported by the ATM cell header. Therefore, in our model, both FR and ATM multiplexers can provide explicit congestion information, at the frame and at cell level, respectively, when congestion is detected in the corresponding buffer.

Simulation results confirm that the overall cell loss probability remains the same as in the case of only bursty sources connected to the ATM multiplexer, if the average load of the FR traffic, ρ_{FR}, is equal to ρ_b. As the average load, ρ_{FR}, of the FR multiplexer increases, the cell loss probability also increases, as shown in figure 2.

In order to discuss the cell loss probability fairness, in Table 1 we have reported the simulation results obtained with ρ_{FR}=0.4, which corresponds to a FR traffic load equal to 43% of the total traffic offered to the ATM multiplexer. In these conditions, the overall cell loss probability is less than 10^{-4}.

ATM bursty sources suffer a worse cell loss probability, compared to the FR traffic. This behavior is a consequence of the fact that the FR sources transmit at 2 Mbps and are multiplexed at 34 Mbps. The better performance of FR traffic is obtained at the cost of a cell loss probability, suffered by bursty sources, which is higher than the ATM multiplexer overall cell loss probability. This unfairness penalizes the bursty sources in favor of the FR traffic. Alternative solutions have been investigated [3] in order to reduce the drawback caused by the presence of the FR traffic.

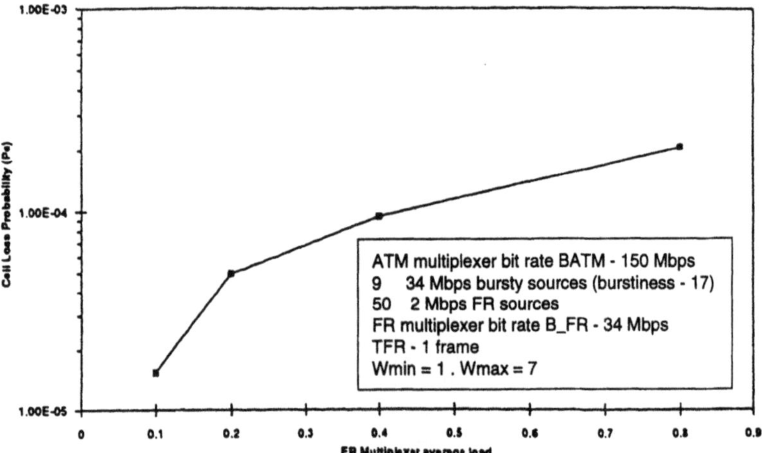

Figure 2 ATM cell loss probability as a function of the FR average load.

Table 1 Simulation results on FR fairness and delay in the basic scheme.

Overall ATM Cell Loss Probability	9.00E-05
Frame Relay Cell Loss Probability	1.50E-05
Bursty sources Cell Loss Probability	1.46E-04
FR Mean File Delay [s]	0.837
SR File Delay Standard Deviation [s]	0.962

A first alternative is to introduce a Burst Length Shaping function (BLS) [4] in the FR multiplexed traffic. The shaping function permits to limit the FR cell burst length to a maximum value L_{max}. Whenever a maximum-length burst occurs, an appropriate silent time is inserted [4]. By reducing the maximum length, L_{max}, of the cell bursts, the mean length of the FR bursts increases, as long as L_{max} is greater than the mean length of the bursts resulting from multiplexing FR sources. Therefore, FR and bursty traffic characteristics become more similar and cell loss unfairness tends to disappear, as shown in Table 2. The main drawback of this mechanism is the increase in the FR end-to-end file transmission delay, as resulting from the comparison between the values of Table 1 and Table 2.

Table 2 Comparison of simulation results on FR fairness and delay.

	BLS (L_{max} = 1200)	BLS (L_{max} = 300)	IFRC	BFRC
Overall ATM Cell Loss Probability	7.00E-05	5.00E-05	4.00E-06	1.00E-06
Frame Relay Cell Loss Probability	8.00E-06	6.20E-05	< 1E-07	< 1E-07
Bursty Sources Cell Loss Probability	1.10E-04	4.20E-05	6.00E-06	1.35E-06
FR Mean File Delay [s]	0.973	1.522	0.879	0.845
FR File Delay Standard deviation [s]	1.01	1.498	1	0.97

The second method, integrated FR Control (IFRC), introduces a control mechanism on the FR multiplexer.

When the ATM buffer occupancy reaches a given threshold value T_b, the FR server is disabled and no FR frame can access the ATM buffer. The FR server is enabled when the ATM buffer occupancy drops below T_b. This method, implemented with a threshold value T_b = 1 cell, produces a dramatic decrease of the cell loss probability in the ATM multiplexer for both FR and bursty sources, even if the fairness is not improved. The reduction of the overall cell loss probability is due to the fact that the FR traffic enters the ATM buffer only when the total load is low (queue length smaller than T_b). This way of operation transfers network congestion situations to the FR multiplexer, which has the means, based on the ECN mechanism, to reduce the source FR traffic. The additional delay incurred by the FR traffic is negligible, as shown in Table 2.

The integrated control can be implemented only if the access ATM multiplexer includes the FR multiplexing stage. If the two multiplexers are remote, a similar method, the Buffer-based FR Control (BFRC), can be implemented by storing the FR cells in an additional FR-dedicated ATM buffer, when the access to the buffer (K_{ATM}) is forbidden. This implementation further enhances multiplexing performance in terms of cell loss probability and transmission delay, because FR traffic control is performed on a per-cell basis, and not on a per-frame basis as in the previous case.

The best performance is provided by the BFRC, which reduces the overall cell loss probabilities of two orders of magnitude. This improvement is obtained by preventing the FR traffic to enter the ATM buffer when it is not empty. The additional delay due to this operation has no effect on the FR file delay.

3 SMDS ACCESS IN ATM NETWORKS

As ATM networks offer connection-oriented services, the provision of connectionless data services requires that each Interworking Unit, which interfaces the users, is interconnected to the ATM network by a permanent Virtual Path (VP) or a permanent Virtual Channel (VC). In this strategy, the ATM network is completely transparent to the connectionless service. This solution is very simple, but its efficiency is very limited if the IWU operates at the peak bit rate, even if the connectionless traffic is bursty. Moreover, the use of permanent connections between all pairs of IWUs is feasible only for small size networks. A more efficient strategy makes use of Connectionless Servers (CLS) in ATM network, as proposed by ITU [5]. In this approach, each IWU may be connected to a CLS through a VP, whose bandwidth can be permanently or dynamically allocated. A Variable Rate Multiplexer (VRM), which selects its output transmission speed, V_o, in relation to the traffic load and the buffer occupancy, presents a high effectiveness.

We proposed a simple implementation of a VRM, obtained with the use of only two values of the output speed V_o [6]. The two rate multiplexer, TRM, is described by the following three parameters:

- the nominal bandwidth, V_N;
- the maximum bandwidth, V_M;
- the utilization factor, F, of V_M.

The bandwidth V_N is permanently allocated in the ATM network, while V_M is used, when needed, for limited periods of time. Its average use is given by F.

The performance of the TRM has been obtained through a PTS (PDU-Transmission Scheduling) algorithm, whose goal is to monitor the traffic and to verify if the lower output speed can be used without violating the declared frame loss probability.

As specified in SMDS, the L3-PDU frames are segmented in cells. The first one, BOM, specifies the length of the frame and the last one, EOM, specifies the end of the frame. At the reception of a BOM in the TRM, the PTS algorithm evaluates which output speed, V_N or V_M, permits to maintain the buffer occupancy, Q, at the minimum value below the permitted threshold (Q_{max}) for all frame being transmitted. If this condition can not be satisfied, the new incoming frame is rejected and the transmission speed remains unchanged. Moreover, at the reception of an EOM, a decision is taken whether to decrease the transmission speed. Details of this algorithm were presented in [6].

The parameter V_N highly influences the performance characteristics, as shown in figures 3, 4, and 5, that summarize the simulation results observed for different values of Q and assuming N sources transmitting at $V_I = 5$ Mbps with an activity A = 0.5 Mbps. The frame length is geometrically distributed with average length L = 5. The frame rejection probability, p_R, (figure 3) presents a maximum, as V_N changes, which also depends on the buffer length. To reduce p_R for a given offered traffic, either a very high value, close to V_M, or a small value, close to the average offered traffic, should be chosen.

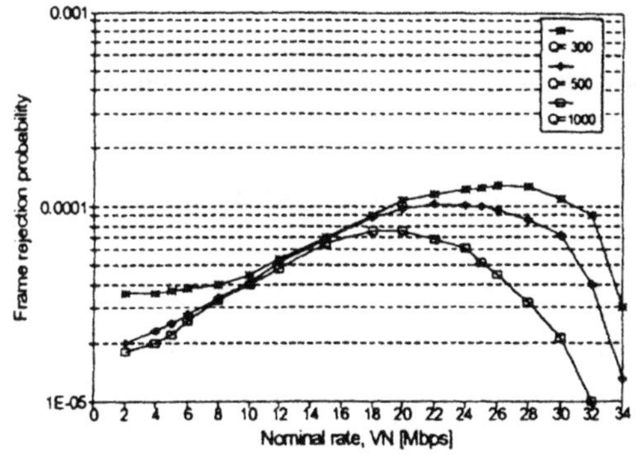

Figure 3 Frame rejection probability versus the nominal rate in the TRM for different values of Parameter Q
[V_I = 5 Mbps, A = 0.5 Mbps, N = 18, L = 5 frames].

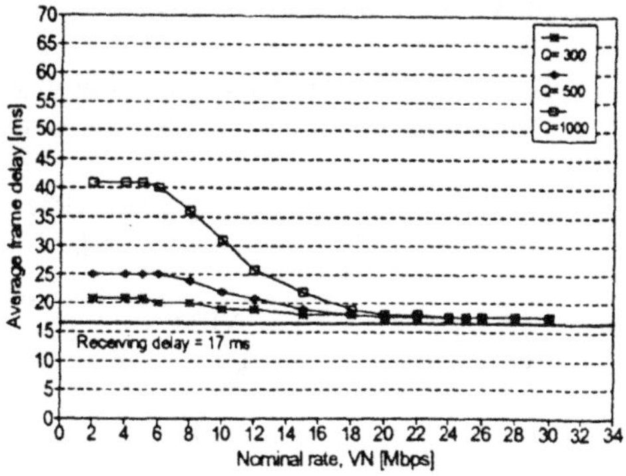

Figure 4 Average frame delay versus the nominal rate in the TRM for different values of parameter Q
[V_I = 5 Mbps, A = 0.5 Mbps, N = 18, L = 5 frames].

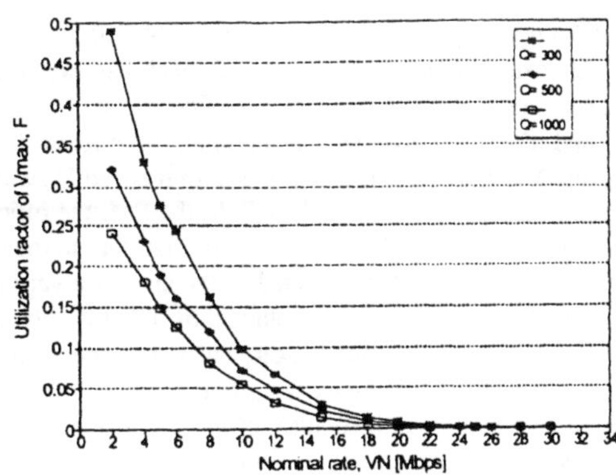

Figure 5 Utilization factor of V_M versus the nominal rate [V_I = 5 Mbps, A = 0.5 Mbps, N = 18, L = 5 frames].

The frame delay, τ, changes with V_N only for values of Q_{max} greater or equal to the average source burst size (figure 4). The utilization factor F is very high for small values of V_N and rapidly decreases as V_N increases (figure 5).

For a required p_R, two sets of values for V_N, $\{V_N<V_{N1}\}$ and $\{V_N>V_{N2}\}$, are possible according to the results shown in figure 3. The corresponding values of the delay and the utilization factor are given in figures 4 and 5, respectively. For a given V_N, several choices of the buffer size are possible to satisfy the constraints given in terms of p_R, τ, and F. The best choice of V_N is obtained by minimizing the "Equivalent Bandwidth", $BE(V_N)$, defined as:

$BE(V_N)=[1-F(V_N)]*V_N+F(V_N)*V_M$

To keep the bandwidth allocation as small as possible, it is more appropriate to find solutions characterized by values of V_N about equal to the average offered traffic.

Table 3 Comparison among VRM, TRM, SM, and PM [V_I = 5 Mbps, A = 0.5 Mbps, L = 5 frames, Q_{max} = 1000 cells, $p_R < 5\ 10^{-5}$].

	VRM	TRM	SM	PM
Vo [Mbps]	5, 10, 15, ..., 34	10,34	34	34
N	12	18	23	7
N A [Mbps]	6	9	11.5	3.5
F	---	0.06	1	1
BE [Mbps]	6	11.4	34	34
E	1	1.26	2.95	10

In Table 3, some numerical results permit to compare the TRM with the Statistical Multiplexer (SM) and the Peak Multiplexer (PM), which assigns the peak bandwidth to each source. To compare the different schemes, we have defined the efficiency, E, as the ratio between the equivalent bandwidth and the total average throughput. A reference multiplexer, called Variable Rate Multiplexer (VRM), that can adjust the output speed V_o within a set of values, is also taken into account. Its efficiency is equal to 1. In the PM, the efficiency E equals the source burstiness. Both the SM and the TRM schemes present an intermediate behavior. The higher multiplexing gain provided by SM is obtained at the cost of a very low efficiency (E = 2.95). On the contrary, the efficiency (E = 1.26) obtained by TRM is only 26% worse than the optimum value. The PM is the simplest multiplexer, but it provides too poor performance. TRM and SM present the same level of complexity, as both require a policing mechanism.

The effective use of the TRM requires the control of its parameters in order to guarantee the ATM network the declared traffic characteristics. While the control of V_N is relatively simple, the control of F presents some problems. In fact, the utilization factor, theoretically defined on an infinite time, must be measured and controlled on a finite time interval. Furthermore, such an interval must be short enough to guarantee that cell bursts, transmitted at V_M by the TRM, are shorter than a given value T_M declared to the ATM network.

In [6], a "Time-Credit" mechanism has been proposed, based on a time-window of fixed length W. According to this mechanism, the TRM is allowed to transmit at V_M within W for a total time up to F*W. Once the Time Credit F*W has been spent, only the transmission speed V_N can be used for the remaining time. The performance of the Time-Credit mechanism is shown in figure 6, where the frame rejection probability, p_R, is given as function of W for different values of F. For a given F, p_R increases as W decreases. As a consequence, to guarantee low values of p_R with small value of the window, it necessary to use high value of the utilization factor, that is, in other words, to work at the peak bit rate.

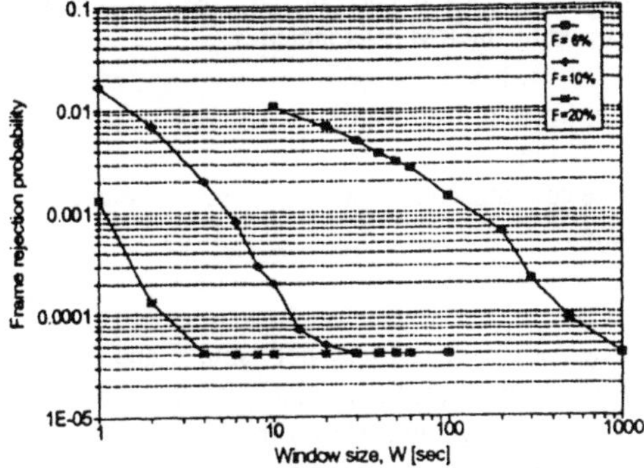

Figure 6 Frame rejection probability vs the window size in the TRM for different value of parameter F [V_I = 5 Mbps, A = 0.5 Mbps, N = 18, L = 5 frames, V_N = 10 Mbps, Q = 1000 cells].

4 ABR SERVICES IN ATM NETWORKS

Initially, CBR services, described in terms of peak bit rate and cell delay variation, will be provided by ATM networks. In addition, ABR services are also supported in an ATM environment for conventional computer communications, such as in the current TCP/IP, where no bandwidth reservation is required. As cells may be lost during the congestion period due to the buffer overflow in the ATM switches, it is necessary, even if large buffers are considered, to use reactive congestion control to prevent such cell losses in ABR services.

According to the ATM Forum indication, we only consider the rate-based control, which is very appropriate to work both in LAN and WAN scenarios. Such kind of control requires only the definition of the end-system behavior and leaves flexibility in both the design of the switch architecture and in the selection of traffic management strategy [7].

4.1 ATM Forum Rate-based Control

The rate-based control, implemented using traditional FECN and BECN schemes [8], presents an unsafe behavior as notification cells can be delayed or lost when a severe congestion is experienced in the network. In fact, if no congestion notification is received, the source continues to increase its transmission rate at regular time intervals up to the allowed peak value.

To overcome this serious drawback, the ATM Forum has proposed a proportional rate control algorithm (PCRA), where a source increases its transmission cell rate only at the reception of an explicit notification, indicating no congestion in the network [7, 8]. Otherwise, the source gradually reduces its cell rate until the minimum allowed value is reached.

In the following, we describe an enhanced version of the PRCA, E-PRCA, reported in [9], but modified in some parts to simplify the functions requested at the ATM switches.

The operation of E-PRCA is based on the information carried by the resource management (RM) cells. Before the transmission begins, the source sets the allowed cell rate (ACR) to the initial cell rate (ICR) value, negotiated during the virtual connection setup. Then, the data transmission phase, consisting of a sequence of one RM cell followed by N data cells, can start. The information carried by the RM cells includes:

- DIR, RM cell direction (forward/backward);
- CCR, current source cell rate;
- MCR, minimum source cell rate, chosen at the connection setup;
- CI, congestion indication;
- ER, explicit cell rate.

CCR and MCR are set by the source, DIR by both source and destination, while CI and ER are set by source and network switches. ER is initially set equal to the peak cell rate (PCR) by the source and possibly reduced by any node on the VC, which has an available bandwidth smaller than the current value of ER. The CI flag is set to "0" by the source and changed to "1" by any congested switch.

The data cell transmission rate is decreased at each RM cell, i. e. every N cells, according to the relation:

$$ACR=ACR*[1-(1/RDF)]$$

where the value of RDF (Rate Decrease Factor) is negotiated at connection set up. The data cell rate is increased at the reception of an RM cell carrying no congestion information, of a value equal to the last decrease value plus an amount, AIR (Additional Increase Rate), negotiated at the connection set up.

4.2 Reference Network Model

In order to investigate the E-PRCA performance, we have simulated a simple network model shown in figure 7.

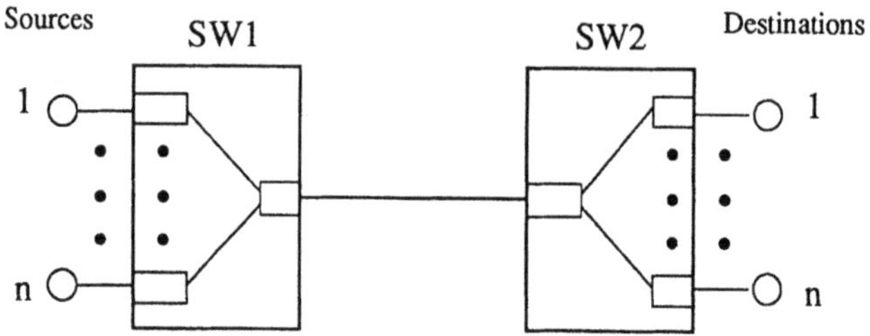

Figure 7 Reference network model for E-PRCA performance evaluation.

Traffic sources are attached to switch 1, while destinations are attached to switch 2. Each source has one virtual channel (VC) for a unique destination. Distance D1 in the access loop, and distance D2 between the two switches can be varied to capture the effect of the propagation delay on the control performance. Distance D1 may be different for different sources.

Each switch is modeled as a buffered switch with buffers at each input and output port [10]. Internally, cells are forwarded from input buffer to output buffer according to a random in random out scheduling to avoid unfairness. A backpressure policy is enforced between the output and input buffers to avoid cell loss at output buffers.
Preliminary simulation results have been obtained assuming MCR = 0.1, PCR = 1, ICR = 0.1, and AIR = 0.01. These values are normalized to the channel rate. Furthermore, we have considered N = 32 and RDF = 128, as widely used in the literature.
In all our simulations, persistent sources, i.e. sources transmitting at ACR, have been assumed in order to stress the control capability.

The parameter ER has been ignored. Only CI is used to notify congestion. Two policies have been considered to recognize a congested switch. In the first one, the queue length policy, QLP, the congestion is detected when the queue length at an input buffer exceeds a preassigned threshold T_2. The congestion terminates when all queue lengths become smaller than a preassigned threshold T_1 ($T_1 < T_2$).

In the second, the length gradient policy, LGP, the congestion is detected taking into account also the gradient of the queue lengths [11]. This mechanism allows to react faster to a suddenly emerging congestion.

Both alternatives have been simulated and the results will be discussed and compared in the next section.

4.3 Numerical results

The main goal of our simulation is to investigate the performance of the different alternatives discussed in the previous paragraphs and to evaluate the impact of various parameters. To compare the effectiveness of the two strategies for congestion detection we consider the network reference model of figure 7 with three homogeneous sources, which are activated at different times.

All sources have the same distance from the switch (10 km) and their transmission rate can reach the maximum value (155 Mb/s). Figure 8 and 10 show the normalized transmission rate permitted to each source versus the observation time. If only source A is active, the transmission rate grows up to the channel speed. As soon as source B is activated, the channel capacity is equally splitted among the two sources. However, strong variations on the transmission rate are observed. This is due to the delay to recognize the congestion situation in the switch and to activate the rate control in the source A. A similar behavior is observed, when source C is activated.

The improvement obtained by using the gradient policy reduces the variation in the transmission rate as shown in figure 10.

The queue length behavior, shown in figure 9 and 11, reflects the same characteristics in terms of variations. Also regarding this performance, the gradient policy is preferable as it reduces the average queue length by a factor of two. This gain allows to significantly reduce the buffer size, even if a large queue length variation has been measured during transient periods, corresponding to the activation of new sources.

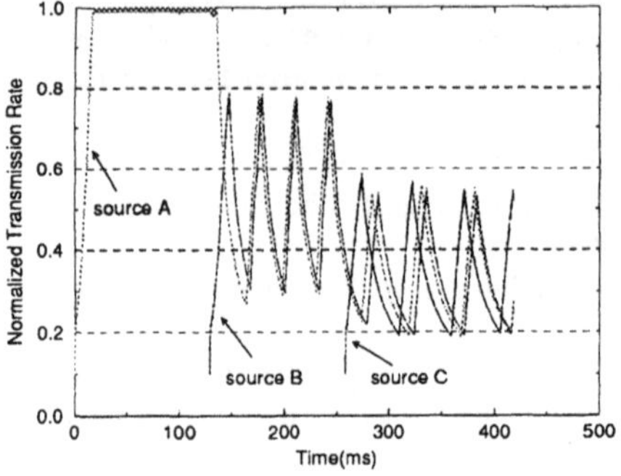

Figure 8 Normalized transmission rate for three homogeneous sources using queue length policy congestion control.

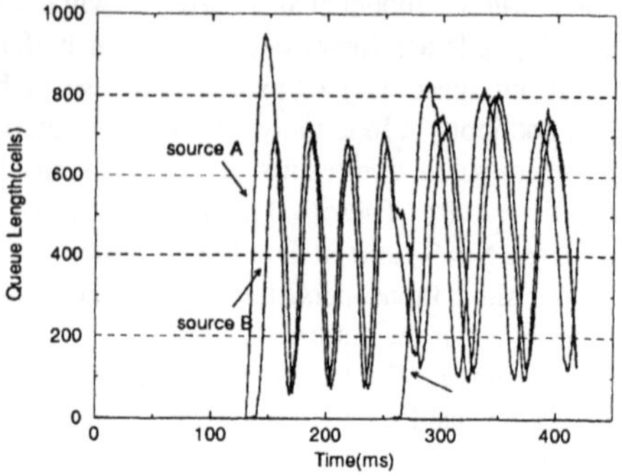

Figure 9 Queue length for three homogeneous sources using queue length policy congestion control.

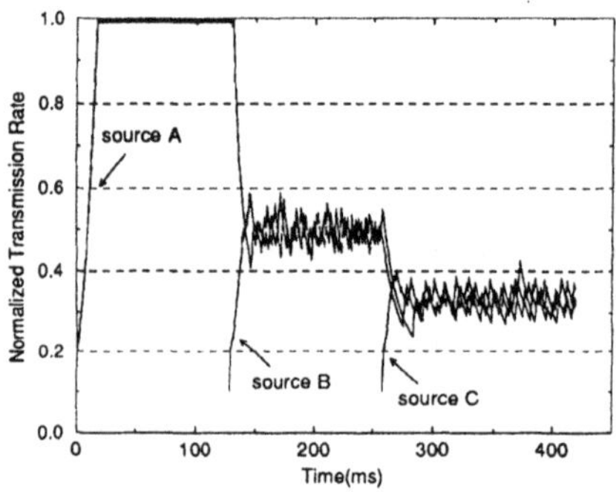

Figure 10 Normalized transmission rate for three homogeneous sources using length gradient policy congestion control.

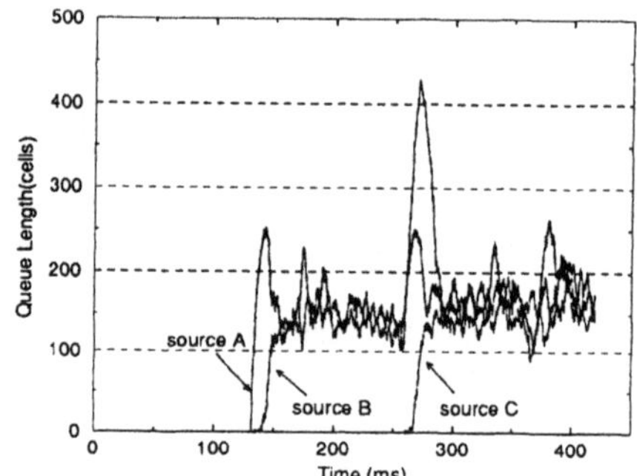

Figure 11 Queue length for three homogeneous sources using length gradient policy congestion control.

A direct comparison between the two techniques, QLP and LGP, is shown in figure 12 and 13, where only the behavior of source A is reported. In all cases, a fair behavior among the different sources is obtained.

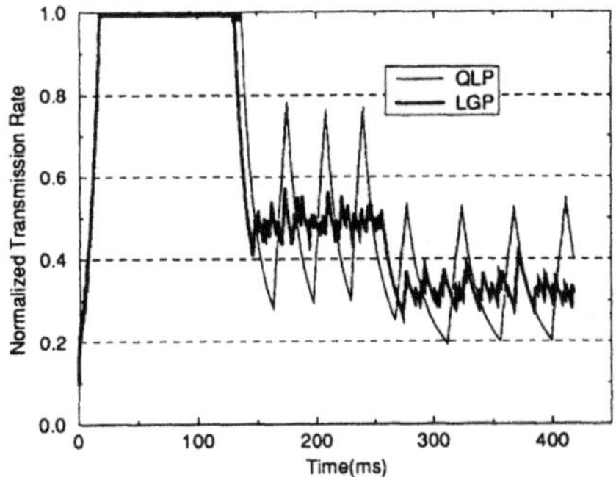

Figure 12 Normalized transmission rate for the source A. A comparison between QLP and LGP.

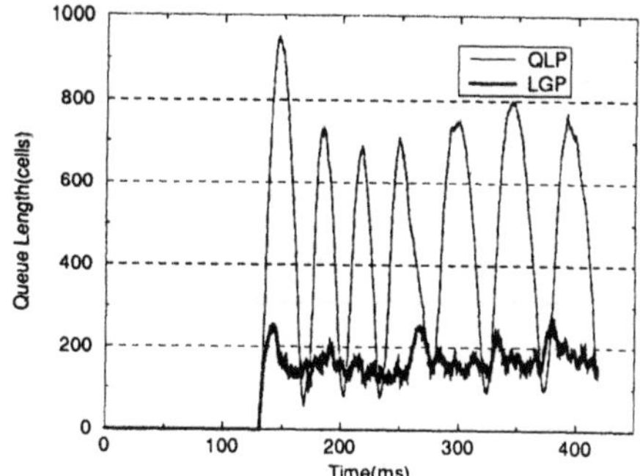

Figure 13 Queue length. A comparison between QLP and LGP.

The choice of ICR has impact on the system performance. High values produce (figure 14) larger variations and therefore larger buffer are necessary.

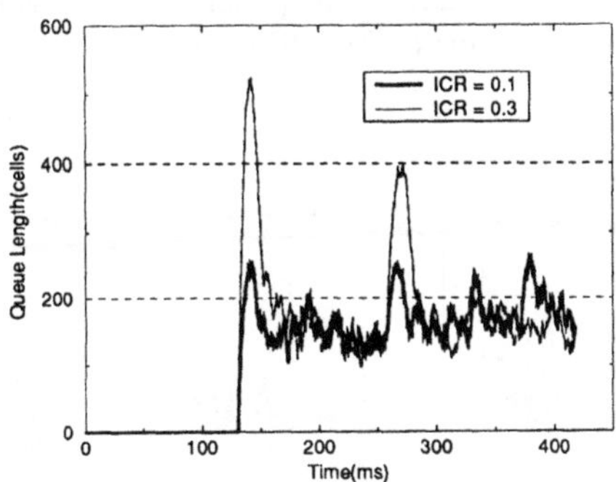

Figure 14 Queue length using GLP. A comparison between ICR = 0.3 and ICR = 0.1.

From these preliminary results, it is evident that, no matter which policy is used, the buffers required at the ATM switch are much larger than those considered (< 100 cells) in current implementations. The use of small values of ICR is also advisable.

The case of non homogeneous sources has been considered assuming two sources only, one at full channel rate (1) and the other at lower rate (0.3). The ABR service obtained by the faster source is degraded with respect to that of the slower one, which is practically able to work at its nominal bit rate without suffering significant variation in transmission rate and queue length (figures 15, 16).

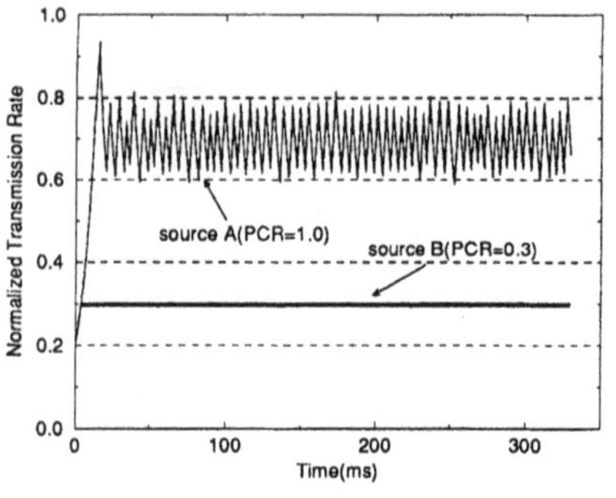

Figure 15 Normalized transmission rate for two non homogeneous sources.

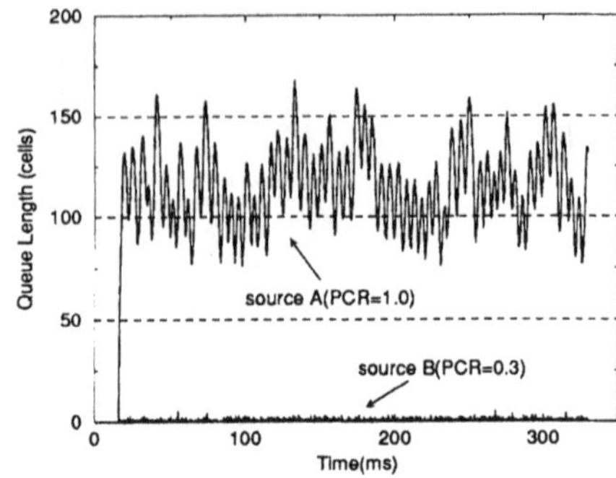

Figure 16 Queue length for two non homogeneous sources.

The effect of the propagation delay is to slow down the source control. This phenomenon has been investigated by assuming two short (10 km) and a long (100 km) access links. Figures 17 and 18 show the behavior of normalized transmission rate and queue length, respectively, for a near and the far source. As expected, the far source suffers from more degradation, having higher variations in both transmission rate and queue length. In a multinode network, not considered in this paper, the effect of different propagation delays may be even more critical. To attenuate such a degradation, mechanisms that provide a faster feedback are under investigation.

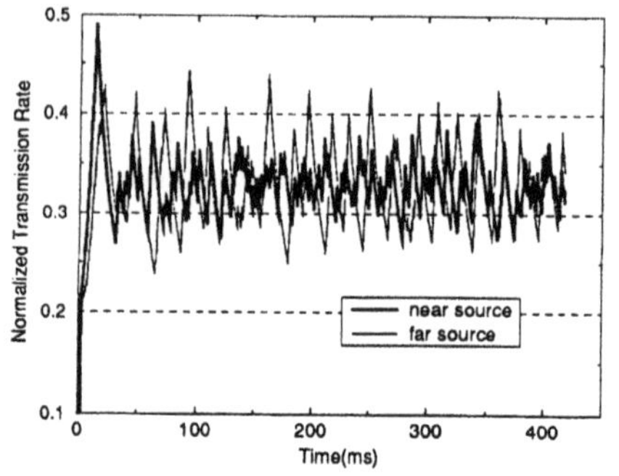

Figure 17 Normalized transmission rate. A comparison between near and far sources.

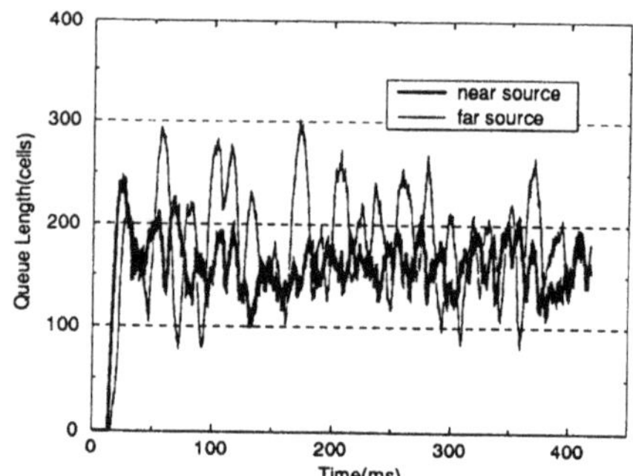

Figure 18 Buffer length. A comparison between near and far sources.

5 CONCLUSIONS

Frame Relay and SMDS are fast packet technologies, designed to support high-speed data applications, and, in particular, LAN-to-LAN interconnection.

As Frame Relay provides a connection-oriented service, it can be implemented on ATM networks based on dedicated virtual path (VP) or, more efficiently, on virtual paths shared with traffic generated by other ATM sources. We have studied this second approach to evaluate the impact of FR traffic on the ATM multiplexer. Simulation results have shown that FR traffic produces an unfair distribution of cell loss probability among different kinds of traffic. To enhance multiplexing performance, different multiplexer structures have been proposed. In particular, the use of additional buffers, dedicated to FR traffic inside the ATM multiplexer, permits to improve the cell loss probability without increasing the end-to-end file transmission delay.

The use of Connectionless Server (CLS), colocated with the ATM switches, is a possible architecture to implement SMDS in an ATM network. In this scenario, each Interworking Unit (IWU) may be connected to a CLS through a permanent VP. Because of the intrinsic bursty characteristic of the offered traffic, the statistical multiplexing operating at the peak bit rate may result inefficient. We have proposed a more efficient Two Rate Multiplexer (TRM), which can select the output speed between two values, the nominal speed and the maximum speed, in order to satisfy the preassigned QoS requirements. The performance of TRM presented in this paper show the superiority of this approach with respect to the Peak Multiplexer and the Statistical multiplexer.

The ATM Forum has recently introduced a new class of ABR services, which require no bandwidth reservation. If no congestion control is enforced, the cell loss becomes very high during overload network conditions. The rate-based control, namely the proportional rate control algorithm (PCRA), has been proposed by ATM Forum. The PCRA is based on the concept that, in absence of an explicit notification indicating no congestion in the network, each source gradually decreases its cell rate until the permitted minimum value is reached. The simulation model considered consists of two ATM switches connected through a VP at 155 Mbit/s. Each switch is equipped with buffers at each input and output port. The main goal of our simulation has been to investigate the system performance, taking into account the influence of the various parameters. Two policies have been assumed to detect the congestion status in the switch. In the first, only the queue length is considered, while in the second, the gradient of the queue length is also taken into account, in order to react faster to an emerging congestion.

With homogeneous sources, simulation results show that a fair behavior among different sources is obtained. Variations on the transmission rate and on the queue length are present during the system operation. These variations are drastically reduced when the gradient policy is used. Moreover, the ICR value also influences the system performance, producing higher variations on the queue length, if it is set at higher values. These preliminary results confirm the need to increase the buffer length in ATM switches (at present, ATM switches are equipped with buffer of less 100 cells) if ABR services must be provided. In this context, it is also advisable to choose small values of ICR.

With heterogeneous sources simulation results show that the performance characteristics are more complicated to manage. The faster source is subject to strong variations of transmission rate and buffer length. The same behavior is observed for a source connected to the switch through a long link (100 km as opposed to 10 km, in our example).

In a multiswitch network, not considered in this paper, the effect of different propagation delays is expected to be even more critical. Further studies are in progress to investigate the possibility to adopt a policy in the ATM switch for handling the parameter ER. Improvements are expected but at the cost of more complex functions to be performed by ATM switches. The trade off will be carefully studied.

6 REFERENCES

[1] Lee, B.G., Kang, M. and Lee, J. (1993) *Broadband telecommunications technology.* Artech House. Boston-London.

[2] Iwata, A., Mori, N., Ikeda, C., Suzuki, H. and Ott, M. (1995) ATM connection and traffic management schemes for multimedia internetworking. *Communications of ACM.* February 1995, **38**, Number 2.

[3] Fratta, L., Musumeci, L., Passalia, A. and Rigolio, G. (1993) Performance of frame relay services on ATM networks. Proceedings of the Fifth International Conference on Data Communication Systems and their Performance. Raleigh, North Carolina, USA. October, 1993.

[4] Fratta, L., Gallassi, G., Musumeci, L. and Verri, L. (1992) Congestion control strategies in ATM networks. *ETT*, **3**, 2, Mar.-Apr. 1992, pp. 183-193.

[5] CCITT, Recommendation I. 364: Support of broadband connectionless data services on B-ISDN. Study Group XVIII, Geneve, June 1992.

[6] Cappellini, G., Decina, M., Fratta, L. and Musumeci, L. (1993) SMDS variable rate access in ATM networks. Globecom '93. Houston, Texas. Dec. 1993.

[7] Bonomi, F. and Findick, K.W. (1995) The rate-based flow control framework for the available bit rate ATM service. *IEEE Network.* March/April 1995.

[8] Siu, K.and Tzeng, H. (1995) Intelligent congestion control for ABR service in ATM networks. *Computer Communication Review.* April 1995.

[9] ATM Forum/94-0974. New pseudocode for explicit rate plus EFCI support. October 1994.

[10] Kolarof, A. and Ramamurthy, G. (1994) Comparison of congestion control schemes for ABR services in ATM local area network. *Globecom '94.* San Francisco, California. Dec. 1994.

[11] Pozzi, D. (1995) "Preliminary results on ABR Service". DEI Internal report, June 1995.

Luigi Fratta received the Doctorate in electrical engineering from the Politecnico di Milano, Milano, Italy, in 1966. From 1967 to 1970 he worked at the Laboratory of Electrical Communications, Politecnico di Milano. As a Research Assistant at the Department of Computer Science, University of California, Los Angeles, he participated in data network design under the ARPA project from 1970 to 1971. From November 1975 to September 1976 he was at the Computer Science Department of the IBM Thomas J. Watson Research Center, Yorktown Heights, NY, working on modeling analysis and optimization techniques for teleprocessing systems. In 1979 he was a Visiting Associate Professor in the Department of Computer Science at the University of Hawaii. In the summer of 1981 he was at the Computer Science Department, IBM Research Center, San José, CA, working on local area networks. During the summers of 1983 and 1989 he was with the Research in Distributed Processing Group, Department of Computer Science, U.C.L.A., working on fiber optic local area networks. During the summer of 1986 he was with Bell Communication Research working on metropolitan area networks. At present he is a Full Professor at the Dipartimento di Elettronica e Informazione of the Politecnico di Milano. His current research interests include computer communication networks, packet witching networks, multiple access systems, modeling and performance evaluation of communication systems, and local area networks.
Dr. Fratta is a member of the Association for Computing Machinery and the Italian Electrotechnical and Electronic Association.

Luigi Musumeci is an Associate Professor at the Electronic and Information Department of the Politecnico di Milano.
He received his degree in Electrical Engineering from the Politecnico di Milano in 1961.
From 1968 to 1986, he was at Italtel, where he was responsible for the design and implementation of ITAPAC the Italian packet network.
His research interests cover packet data networks, wireless networks, and B-ISDN. He is the author of over 40 articles on these topics.

23

The Statistically Correct Approach to Distributed Simulation of ATM Networks

C.D. Pham and S. Fdida
Laboratoire MASI
Université Pierre et Marie Curie, 4 place Jussieu 75252 Paris Cedex 05, France. Telephone: (33-1) 44-27-75-12. Fax: (33-1) 44-27-62-86.
e-mail : {pham,fdida}@masi.ibp.fr

Abstract
We introduce a statistically correct approach in which events can be processed in partial time order. Upon reception of an outdated message, a process does not always attempt to cancel the bad computations it has performed since the arrival of this message but uses local estimations instead. These local estimations are based on saved historical information. A control should decide whether to use local estimations or the ultimate solution of rolling back that should only be done when the use of estimations may alter too much the results. To validate our approach, a distributed simulation of an ATM network is performed applying this new idea. We focus on the mean buffer length of each switch outputs and find the results very close to those provided by a correct simulation.

Keywords
Distributed simulation, Optimistic scheduling, Local estimations, ATM networks

1 INTRODUCTION

Distributed simulation often requires a physical process of a given system to be simulated by a Logical Process (LP). The interactions between the different physical processes are represented by timestamped messages exchanged between LPs. The problem of synchronization arises since the different LPs may advance at different rate. A time error occurs when a message arrives at a receiving LP and is outdated, or old, according to the local virtual time. Existing mechanisms for parallel simulation must ensure correct synchronization between LPs in order to obtain sequential-like simulations. These mechanisms fall in two categories: conservative and optimistic. Conservative algorithms avoid all cases of time errors (Chandy,1979). They require that execution of an LP is halted until it is certain that no time errors could happen any more. Unfortunately, this simple approach could lead to a deadlock, so null-messages that carry no physical signification are used to advance artificially the simulation time in order to avoid the deadlock of processes. On the other hand, optimistic algorithms allow LPs to always process available input messages, but implement a rollback mechanism in order to recover from time errors that can now occur (Jefferson,1985). Such mechanisms are required because these traditional approaches aim at preserving causality and so messages have to be processed in increasing timestamp order.

Ensuring causality compels conservative approaches to work with static configurations. Therefore the simulation of large systems is difficult. In addition, a lot of null-messages is necessary for deadlock avoidance when the number of real messages in the system is small. This high null-message/real-message ratio dramatically affects the performance of the simulator. For an optimistic approach the rollback overhead greatly limits the performance of the protocol. Moreover, the well-known instability of the Time Warp cancellation strategy introduces a high risk of cascaded rollbacks. A good survey of conservative and optimistic approaches, and their performance, can be found in (Fujimoto,1990). Briefly, one can improve the performance of these protocols by adding some optimism to the former while limiting the too much of optimism and minimizing the overhead of the rollback mechanism in the latter. However, all these improvements do not call the causality constraints into question. In this work, we investigate a different approach that relaxes the causality constraint and allows inaccuracies to occur.

In the terminology defined in (Reynolds,1988), accuracy requires that events be ultimately processed in increasing timestamp order. If this is not the case then the protocol is said to be inaccurate. Time Warp is accurate because its rollback mechanism corrects the bad computations. In (Theofanos,1984) the author showed for simulation of queuing networks that bad computations do not affect dramatically mean value statistics. Going a step farther leads us to think that in many cases some useful results can be obtained without the total time ordering established by conventional approaches. This paper introduces the inaccurate *statistically correct approach* in which some events can be processed in partial time order without rollbacks. In this approach, upon reception of an outdated message a process does not always attempt to undo the bad computations it has performed since the arrival of this message but uses local estimations instead. A control should decide whether to use local estimations or the ultimate solution of rolling back. The heavy rollback mechanism should only be used when utilization of local estimations may alter too much the results. Thus, the results obtained by a simulation are *statistically correct* when compared to those provided by a conventional approach.

In order to demonstrate our approach, a distributed simulation of two ATM networks configuration, with exponential and ON/OFF traffic sources, is performed applying this new idea. We focus on the mean buffer length of each switch outputs and find the results very close to those provided by a correct simulation.

This paper is organized as follows. The next section details the new approach that introduces the idea of a statistically correct simulation. Section 3 presents the test cases and the computation of local estimations. Section 4 presents the preliminary results. We conclude in section 5 with indications of future directions for this work.

2 THE STATISTICALLY CORRECT APPROACH

Now we need to give some basic definitions in order to introduce the statistically correct approach. The term simulation refers to distributed discrete event simulation and we assume that the reader is familiar with Logical Process (LP) and process local virtual time (LVT).

2.1 Definitions

Definition 1 (Very-correct) *A very-correct simulation is a simulation in which events are processed by an LP in increased timestamp order with no time errors.*

Definition 2 (Correct) *A correct simulation is a simulation in which some time errors have occurred but have all been corrected by a rollback mechanism.*

Definition 3 (Statistically correct) *A statistically correct simulation denotes a simulation in which some time errors have occurred but not all of them have been corrected.*

Definition 4 (Rollback-free sequence) *Let φ be a finite sequence of n events $< e_i, t_i >$ where t_i is the timestamp of event e_i. A sequence is said to be* rollback-free *if $\forall i, t_{i+1} \geq t_i$. On the other hand, a k-*rollback *sequence is a sequence where there are k t_i such as $t_i < t_{i-1}$. In extension, a* rollback sequence *is a k-rollback sequence where $k \geq 1$.*

A conservative algorithm typically produces very-correct simulations, whereas an optimistic one like Time Warp is more likely to produce correct simulations. Our statistically correct approach is undoubtedly on the optimistic side and can be classified as "very optimistic". Consequently we will compare its behavior to the Time Warp algorithm.

When developing our new statistically correct approach, we find very interesting to give it a tunable behavior. Therefore we wanted to meet the following requirement:

- A statistically correct approach must be able to produce correct and statistically correct simulations.

To meet this requirement, a statistically correct approach must establish a control in order to span the continuum of correctness from correct to statistically correct. Let Φ_c and Φ_s be the controls for obtaining correct and statistically correct simulation with a statistically correct approach. Now we can introduce the notations TW_φ and $ST_\varphi(\Phi)$ to denote the simulation of an LP using respectively the Time Warp protocol and our statistically correct approach applying control Φ, this under the sequence φ. Simulation of an LP under a sequence φ means that the LP successively processes event i of φ, $i = 1..n$. For notation simplicity we will write $a = b$ if the result of simulation a is the same that the result of simulation b. Finally the following statements can be written:

1. if φ is a rollback-free sequence then TW_φ and $ST_\varphi(\Phi)$ are both very-correct simulations whatever the control Φ is. We will have $TW_\varphi = ST_\varphi(\Phi)$.
2. if φ is a rollback sequence then TW_φ and $ST_\varphi(\Phi_c)$ are both correct simulations and we have $TW_\varphi = ST_\varphi(\Phi_c)$. On the other hand, $ST_\varphi(\Phi_s)$ is a statistically correct simulation. In most cases, we will have $TW_\varphi \neq ST_\varphi(\Phi_s)$.

For an optimistic approach, rollback is the sole means to obtain a correct simulation. Therefore the purpose of the control Φ is to decide whether to rollback or not, when a time error occurs. It is obvious that Φ_c is the Time Warp's control requiring a rollback at each time error, whereas Φ_s can be based on some assumptions triggering a rollback only on some specific cases.

Since a rollback-free sequence does not generate time errors, simulations using any synchronization protocol, existing or to come, under such a sequence are all very-correct simulations. Statement 1 says that under a rollback-free sequence our statistically correct approach provides results that are identical to those provided by a conventional Time Warp approach, independently of the control. Now if φ is a rollback sequence the Time Warp algorithm produces correct simulations by means of its rollback mechanism. On the other hand, we wanted our statistically correct approach to produce statistically correct simulations depending on the control Φ. This means that upon reception of a bad message the heavy rollback mechanism is not systematically used to cancel the bad computations. Of course in order to obtain coherent results, alternatives to rollback must be found. We propose utilization of local estimations and a different time evolution scheme as described in the next sections.

2.2 Local estimations

The life of a process consists of receiving and processing incoming messages. Processing a message often changes the process state and may generate output messages to other processes. When a bad message m_{past} with timestamp t_{past} arrives at time t_{now}, Time Warp has to roll back. This is done by restoring the process state before t_{past}, by jumping back to time t_{past} and by sending anti-messages to cancel messages produced by optimistic scheduling. Only then all events since t_{past} are re-executed. This is required because the processing of m_{past} may change the evolution of the process making all states and all produced messages since t_{past} erroneous.

There is no way but to roll back if one wants to correct the bad computations propagated by bad produced messages. Now if one accepts the errors introduced by not cancelling produced messages then there is no need to roll back, local estimations can be used instead to avoid the overhead of rolling back. This is the key idea expressed by our statistically correct approach. Estimations should be designed to produce correct output messages resulting from the processing of m_{past} and to re-evaluate the process states from t_{past} to t_{now}.

We propose local estimations based on historical information. Such historical information can be included within the process state and periodically saved in a Time Warp-like fashion. When a past message m_{past} arrives, the process virtual time does not change and saved historical information before t_{past} are used to process m_{past} and to generate output messages. Re-evaluation of saved states is also performed to take into account the changes introduced by m_{past}. The way these estimations are computed is quite application dependent. We give in section 3 the estimations used for the simulation of an ATM network.

2.3 Proposed controls

The main guideline for the statistically correct approach is to reduce the cost of process synchronization while providing useful results. Use of local estimations has the advantage of avoiding rollbacks but can unfortunately guess wrong and alter too much the results. One solution is to establish a control that decides when the process should roll back and when it can use local estimations.

At this point, we must mention that our correctness metric uses the results obtained by a very-correct or correct simulation–using respectively a conservative and an optimistic approach–as a reference. Actually, in deterministic simulations one can obtained identical results from one execution to another if the sequence of events applied at the input is the same for all executions. So, when all simulations are deterministic, we can define a simple correctness metric by comparing the results provided by our statistically correct approach to those provided by a conventional approach. If both results are identical then we say that maximum of correctness is achieved. In the same manner, a cost metric can be defined by taking the cost of the Time Warp control as a reference.

Therefore, at the uppermost bound of correctness and cost, e.g. $correctness = cost = 1$, we have the Time Warp control Φ_c that provides correct simulations. To obtain "best effort" results, e.g. $correctness \in [0, 1]$, but at a minimum cost, e.g. $cost = 0$, we have:

- *The Null control*, noted Φ_s^{null}. With the null-control a process always uses local estimations and does not use rollback at all.

Between these two end-points, a control should guess when the use of local estimations may give bad results. Let us define the distance of an error due to an incoming message $< e_i, t_i >$ as $d = LVT - t_i$, LVT being the local virtual time of the receiving LP. Some observations suggest to use this distance as a criterion for rolling back. This is motivated by: the more this distance is, (1) the more the estimations may guess wrong because too much changes may have occurred, (2) the more wrong messages have been sent and (3) the more the overhead of state re-evaluation may outpace the overhead of a rollback. This leads us to define the following control:

- *The Distance-Based control*, noted $\Phi_s^{d_0}$. With the distance-based control a process uses the rollback mechanism if the distance d of an error is greater than d_0.

Note that $\Phi_c = \Phi_s^{d_0}$ when $d_0 = 0$. Figure 1 shows the continuum spanned by the statistically correct approach. Under a general sequence φ one can achieved maximum of

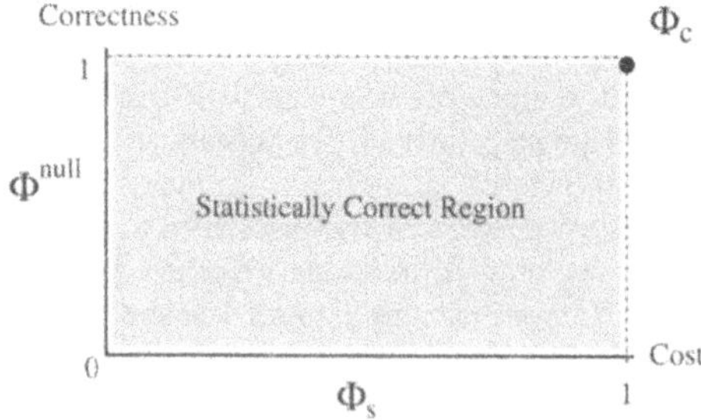

Figure 1 Spanned continuum of correctness from correct to statistically correct for the statistically correct approach.

correctness but also at maximum cost by using $ST_{\varphi}(\Phi_c)$. This is equivalent to a Time Warp simulation. On the other hand, $ST_{\varphi}(\Phi_s)$ provides the possibility to reduce the simulation cost while keeping a reasonable level of accuracy. Simulations on the y-axis always use the null-control. If maximum of correctness is however obtained, then the simulation is a very-correct one–it is equivalent to say that the simulation was run under a rollback-free sequence. Simulations on the x-axis must be rejected since they use a Φ_s control that introduces some cost but produces no correct results! We believe that practically this should not happen. Now, the shaded region represents simulations using any combination of Φ_s control and φ sequence. Of course one would rather like to obtain simulations in the upper-left corner that achieve maximum of correctness at a minimum cost.

2.4 Local virtual time evolution

In the Time Warp algorithm, a time error causes the process virtual time to jump back to the time of the error in order to run forward again in the right sequence. Now since local estimations are used instead, when a time error occurs the process virtual time need not to jump back. The processing of a past message is done in what we call an *estimation environment.* The first time error blocks the process virtual time at time t_{now} until a message with timestamp t greater than t_{now} is received. Since the virtual time is blocked, subsequent messages with timestamps lesser than t_{now} can also produce time errors. We call this kind of error *secondary time error* as opposed to *first time error.*

Figure 2 shows the possible evolution of virtual time for (a) a conventional approach and (b) the statistically correct approach. In the last case a white point indicates the

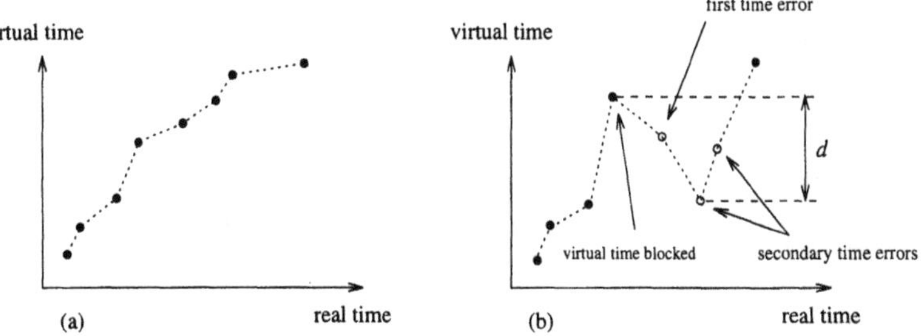

Figure 2 Evolution of local virtual time for (a) a traditional approach and (b) the statistically correct approach.

processing of a past message, relatively to t_{now}, in an estimation environment. With the $\Phi_s^{d_0}$ control the distance d of a time error, first or secondary, can not exceed a threshold d_0 while it is unbounded with the Φ_s^{null} control.

3 TEST CASES AND LOCAL ESTIMATIONS

3.1 Test cases

Network configuration a consists of an ATM switch with three traffic sources (figure 3). The switch has output buffers and the following switching strategy: in one time slot each cell that arrives on an input port is switched to the corresponding output. The choice of the output follows some fixed routing probabilities. Only one cell can arrive per time slot on an input but severals cells can be switched to one output in the same time. In this case they are buffered in the output queue. On the other hand, for all outputs the head cell in the queue is transmitted on the link in each time slot. In this model, the switch has three inputs and two outputs. Traffic sources continually generate cell reception messages $< RECV, t >$ according to a given description. Upon reception of a $< RECV, t >$ message, the switch schedules for itself a $< SEND, t_d >$ event where t_d is the date of departure of the cell. A cell from a given input $i = 1..3$ is switched to an output $j = 1, 2$ with probability p_{ij} and sent out by scheduling at the receiver a $< RECV, t' >$ message. We set for the simulation $p_{ij} = 0.5$ for all i and j. In configuration b a network of five switches

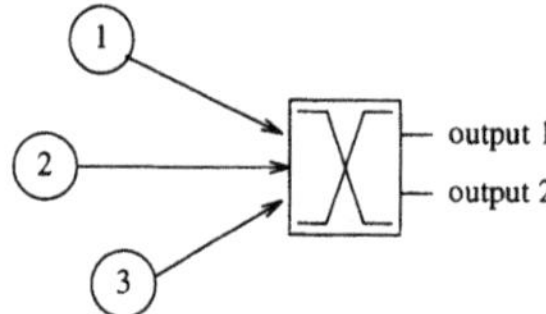

Figure 3 Network configuration a.

is simulated (figure 4). Only one traffic source per switch is considered that generates cell reception message only up to time $t_{endsource}$ lesser than the simulation time. Since there is no sink process, this limits the number of cells in the network. When a switch a receives a cell from another switch, it sends it to the output $j = 1, 2$ with probability p_{aj}. When traffic sources do not generate cells anymore, the number of cells in the network remains constant and all switches keep sending existing cells to each others until the end of the simulation. We set for the simulation $p_{aj} = 0.5$ for all switches a and output j, and $t_{endsource} = 40$. Propagation delay between two switches takes 2 time slots. Three test cases are considered and listed below:

1. Test case 1 considers the network configuration a with exponential sources. The three traffic sources continually generate cell reception messages. In order to generate at most one cell per time slot on a link, the cell inter-arrival time δ is computed as follows: $\delta = ceil(\tau)$ where *ceil()* is the rounding toward positive infinity function and τ an exponentially distributed variable.
2. Test case 2 considers the network configuration b where all sources are exponentials as described above.
3. Test case 3 considers the network configuration a with three ON/OFF sources. Silent and burst periods are both exponentially distributed. While in a burst period, a source generates a cell at each time slot.

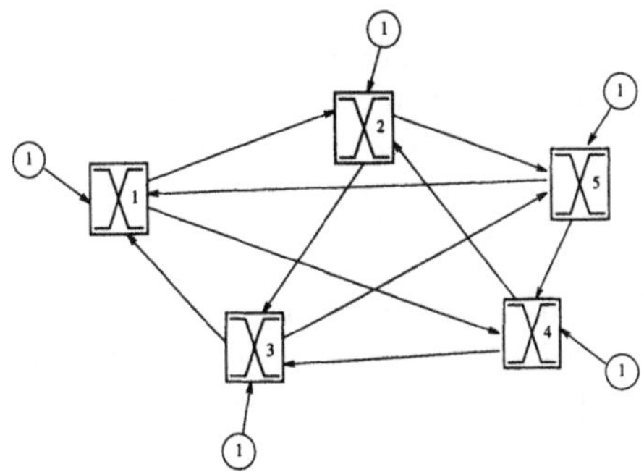

Figure 4 Network configuration *b*.

The state of a switch process is defined by a vector $S = (o_1, o_2, last_1, last_2, n_r, n_s)$. o_j and $last_j$ are respectively the number of cells waiting in the output queue j and the *last time* that a cell has been switched to output j. n_r and n_s are respectively the number of received and sent cells. State saving for local estimations purpose is done after every processing of an $< RECV, t >$ event. The processing of self-generated events such as $< SEND, t >$ does not cause the process to save its state. This choice keeps the number of historical information to a minimum.

3.2 Computation of local estimations

Let L_s be the list of saved states and $S_i = (o_1^i, o_2^i, last_1^i, last_2^i, n_r^i, n_s^i)$ be the ith saved state. The timestamp of S_i is noted T_i. Index i increases as T_i increases. Upon reception of an outdated cell reception event $< RECV, t >$, $t < LVT$, we use the following estimation algorithm at the switch process:

1. Gets the first state before time t from L_s. The retrieved state is S_i with $T_i \leq t < T_{i+1}$.
2. Determines the output j and derives the date of departure t_d of this new cell by computing its waiting time. Then schedules an $< SEND, t_d >$ event for itself.
3. Inserts the new state in L_s before the retrieved state S_i (so at index $i + 1$; all state indexes are incremented).
4. Re-evaluates all states S_k with $k > i + 1$.

Since historical information are saved after every $< RECV, t >$ event the retrieved state is exactly the state the first past message would have find if it was received in sequence. For subsequent bad messages the retrieved state can be a re-evaluated one. We explain steps 2 to 4 in what follows.

Let t_w be the waiting time of the cell before the beginning of its processing at the output j. The date of departure t_d of the cell on output j is computed as follows (time is expressed in time slot):

$$\blacksquare \text{ Step 2} \begin{cases} \text{if } t = last_j^i \text{ then } t_w = max\{o_j^i - 1, 1\} & (1) \\ \text{else } t_w = max\{o_j^i - 1 - (t - T_i), 0\}. & (2) \\ t_d = t + t_w + 1 + 1. & (3) \end{cases}$$

Note that if $t = last^i_j$ we must have $t = T_i$ and $o^i_j \geq 1$. On the other hand, if $t \neq last^i_j$ then we can have $t = T_i$ but this is not mandatory. Since we assume that several cells can be switched from different inputs to an output queue in one time slot, relation $t = T_i$ expresses the fact that there is more than one cell switched to an output in one time slot. Relation $t = last^i_j$ is more precise since it indicates that at least one cell was previously switched to output j in the same time slot. So if $t = last^i_j$ and $o^i_j > 1$ (eq. 1) the cell must wait for all the cells in front of it in output j at time t minus the one that will already be sent when the cell arrives at the output queue (figure 5a). This because transmission of the head cell of an output also requires one time slot. Now if $o^i_j = 1$ then the cell must only wait for the cell in front of it in output j because both have been switched in the same time slot (figure 5b).

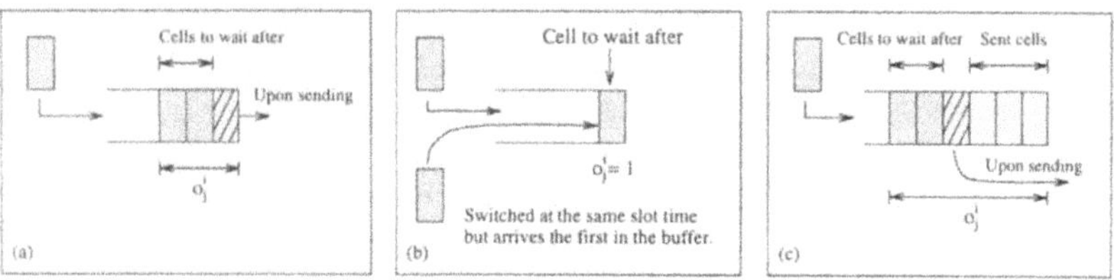

Figure 5 Estimation of the cell waiting time.

If $t \neq last^i_j$ (eq. 2) then the cell must wait for all the cells in output j at time T_i, minus those that were already sent during the interval $t - T_i$, and minus the one that will already be sent when the cell arrives at the output queue (figure 5c). Of course if the output queue is empty then t_w is null. The date of departure t_d is easily obtained by adding to t the switching time (one time slot), the transmission time (one time slot) and the waiting time (eq. 3).

The process state S_{i+1} after the processing of the $< RECV, t >$ message is saved at index $i + 1$ and we have :

■ Step 3
$$\begin{cases} o^{i+1}_j = max\{o^i_j - (t - T_i), 0\} + 1. \\ last^{i+1}_j = t. \\ n^{i+1}_r = n^i_r + 1. \\ n^{i+1}_s = n^i_s + min\{o^i_j, t - T_i\}. \end{cases}$$

If $o^i_j < t - T_i$ then all cells that are waiting in output queue j have been sent during the interval $t - T_i$. In this case $o^{i+1}_j = 1$, e.g. the new cell waits alone in the output queue j. On the other hand, if $o^i_j > t - T_i$ then there are some cells left and the new one waits with the other $o^i_j - (t - T_i)$ remaining cells. In the same way, the number of sent cells is incremented by the number of cells that have been sent between t and T_i.

Table 1 Correct simulation of test 1.

Test case 1	SW1
Buffer 1	0.812
Buffer 2	0.846

Re-evaluation of all states S_k with $k > i + 1$ is done as follows:

■ Step 4
$$\begin{cases} \text{if } T_k - t < o_j^{i+1} \text{ then } o_j^k = o_j^k + 1 \text{ else } o_j^k \text{ is unchanged.} \\ last_j^k \text{ is unchanged.} \\ n_r^k = n_r^k + 1. \\ \text{if } t_d < T_k \text{ then } n_s^k = n_s^k + 1 \text{ else } n_s^k \text{ is unchanged.} \end{cases}$$

Inequality $T_k - t < o_j^{i+1}$ means that the new cell has not been sent yet at time T_k, so o_j^k should be incremented. On the other hand, if $T_k - t \geq o_j^{i+1}$ then the new cell has been sent and o_j^k remains unchanged. In the same manner, if $t_d \leq T_k$ then the new cell has been sent before or at time T_k so n_s^k should be incremented. Note that if $t_d \leq T_k$ then we have $T_k - t \geq o_j^{i+1}$ because both conditions are equivalent.

4 PRELIMINARY RESULTS AND DISCUSSION

Simulations are carried out on a network of SparcStations. Implementation is done in C using the PVM package (Geist and al.,1993) for inter-processes messages passing. In order to easily compare the accuracy of the results provided by our statistically correct approach, deterministic sources are used. This means that from one simulation to another, the sources generate the same sequence of cell reception event $< RECV, t >$. However, the number of time errors from one simulation to another can vary due to the randomness of inter-network message passing delay.

In this paper only the null-control that provides "best-effort" results is studied. The simulation time is fixed to 4000 time slots. The results show the number of first and secondary time errors that are detected, and the switch buffer occupancy. For statistically correct results, the percentage of correctness is also shown.

4.1 Results for test 1 and 2

Tables 1 and 2 show the buffer occupancy for test 1 obtained respectively with a correct simulation and a statistically correct simulation with the null-control. Table 3 and table 4 show the results obtained for test 2. Only one simulation result is shown for test 1 but for all simulations that we have performed we found the results very close to the target values obtained by a correct simulation. We can notice that the number of first time errors is small. This is easily explained by the fact that the system is very simple and that the transfer of cells is only unidirectional from the sources to the switch. This lowers the degree of interactions between the network components and then reduces the number of time errors. Good accuracy is then obtained.

Table 2 Statistically correct simulation of test 1.

Test case 1	SW1
First time error	5
Secondary time error	22
Buffer 1	0.809 → 99.63%
Buffer 2	0.841 → 99.40%

Table 3 Correct simulation of test 2.

Test case 2	SW1	SW2	SW3	SW4	SW5
Buffer 1	2.584	2.588	2.581	2.572	2.578
Buffer 2	2.578	2.574	2.561	2.558	2.558

For test 2 things do greatly differ since interactions between network components are much more complex. In this scenario a switch a can receive from another switch b a cell that it has originally sent. If meanwhile switch a has processed several cell reception events from the other switches then the risk of a time error is quite high. When compared to test 1, we do find that the number of first time errors is far greater. However, and it is encouraging, the results obtained have not been dramatically altered.

For both test cases the values for the buffer occupancy provided by our statistically correct algorithm are always less than the values obtained by a correct simulation. This phenomenon was predictable since the error introduced by not roll backing consists of neglecting the possible amount of extra waiting time of previously sent cells. In the light of the results, this error is kept relatively low.

Table 4 Statistically correct simulation of test 2.

Test case 2	SW1	SW2	SW3	SW4	SW5
First time error	93	109	113	98	112
Secondary time error	135	157	151	137	148
Buffer 1	2.425	2.374	2.421	2.385	2.463
Buffer 2	2.383	2.436	2.401	2.434	2.395
Correctness	93.84%	91.73%	93.80%	92.72%	95.53%
Correctness	92.43%	94.63%	93.75%	94.04%	93.62%

4.2 Results for test 3

Now for test 3 where sources are more sporadic, we have recorded and plotted the evolution of the number of cells in the output queue 1. Figure 6 depicted this evolution for a correct simulation and figure 7 for a statistically correct simulation–always with the null-control. We can notice a missing burst in figure 7. Since record of data is processed on-line, the

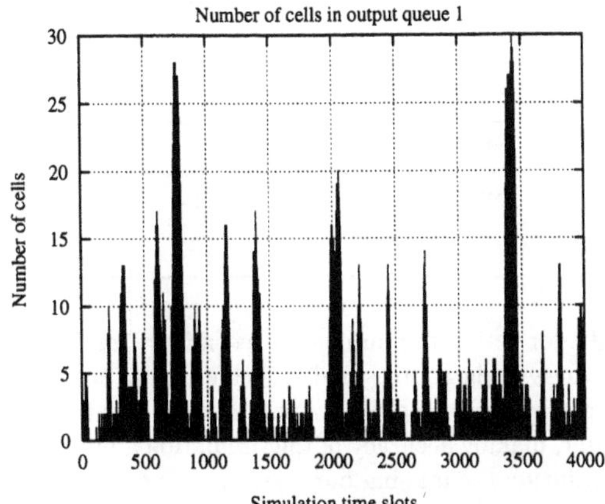

Figure 6 Number of cells in output queue 1—Correct simulation.

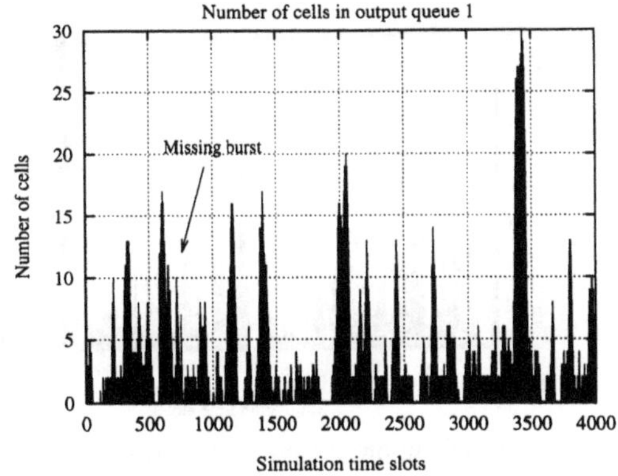

Figure 7 Number of cells in output queue 1–Statistically correct simulation.

only explanation is that the switch process has optimistically gone too far in the future. So when the burst arrives out of sequence the process uses local estimations and state re-evaluation to reconstruct it but does not take the burst into account in the recorded data. Tables 5 and 6 show the buffer occupancy for test 3 obtained respectively with a correct simulation and a statistically correct simulation with the null-control. When compared to

Table 5 Correct simulation of test 3.

Test case 3	SW1
Buffer 1	3.975
Buffer 2	3.938

Table 6 Statistically correct simulation of test 3.

Test case 3	SW1
First time error	2
Secondary time error	123
Buffer 1	3.682 → 92.62%
Buffer 2	3.881 → 98.55%

the previous results for test 1, the number of secondary time errors is much more greater because the burstyness of the sources makes such errors to come in packets. Consequently the errors introduced by neglecting the possible amount of extra waiting time of previously sent cells are greater, making accuracy of the simulation a bit lower. One should note that only the cells sent during the missing burst are concerned. Unfortunately, if the length of the burst is big enough and, if a lot of cells have been scheduled during this burst (figure 8) then accuracy may drop (table 7).

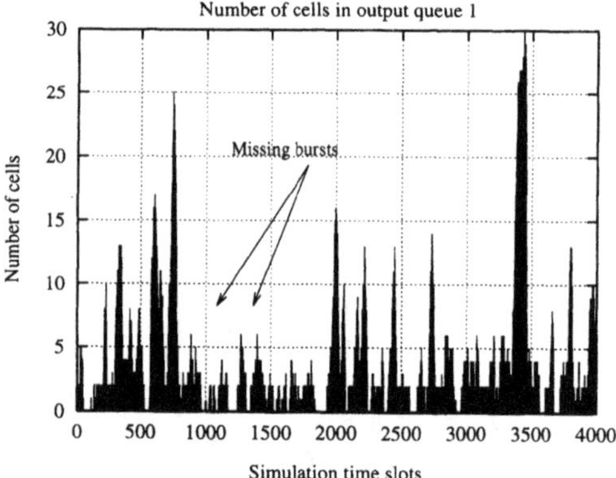

Figure 8 Number of cells in output queue 1–Statistically correct simulation.

In table 7 we can see that the number of secondary time errors is again much more greater. This test shows the limitations of local estimations since the results obtained differ appreciably from those obtained with a correct simulation. Actually, it is not really the number of time errors that affects the simulation accuracy but the number of cells that have been sent during a skipped burst. These cells have been scheduled to leave the switch too early and, in doing so, they affect the collected statistics.

Table 7 Another statistically correct simulation of test 3.

Test case 3	SW1
First time error	9
Secondary time error	317
Buffer 1	3.400 → 85.53%
Buffer 2	3.521 → 89.41%

5 CONCLUSION AND PERSPECTIVES

In this paper, we proposed a new approach called *statistically correct approach* for distributed simulation where total correctness is not always required. This approach provides the possibility to reduce the simulation cost while keeping a reasonable level of accuracy. Early tests with local estimations give favorable results and encourage us to develop this research on. However, we have noticed that the traffic pattern can have some influence on the accuracy of the estimations and further tests should address this point deeper.

The way estimations are obtained is very application dependent. We are conscious that this scheme gives better results when generated messages and process state do not change very much after the processing of the missing events. This is typically the case for ATM networks. Future works would address the use of this approach to larger network of hundreds of nodes. We believe that in order to make such large scale simulations possible in a reasonable amount of time, one has to relax the synchronization constraints and to accept statistically correct results.

We are in the process of implementing the distance-based control added with some heuristics. A lot of works remains to address performance and suitability of this control under sporadic traffic. In addition, we hope that the important issue of determining the simulation accuracy, without comparing the results, could be easier with the introduction of such a control.

Garbage collection, GVT computation and state saving frequency–what to do when there is no consistent historical information available–are also many interesting problems this approach has to deal about and that have not been discussed in this paper.

REFERENCES

Chandy, K.M. and Misra, J. (1979) Distribution Simulation: A Case Study in Design and Verification of Distributed Programs, *Transactions on Software Engineering*, **5**, 440–452.

Fujimoto, R.M. (1990) Parallel Discrete Event Simulation, *Communication of the ACM*, **10**, 31–53.

Jefferson, D.R. (1985) Virtual Time, *ACM Transaction on Programming Languages and Systems*, **3**, 405–425.

Geist, A. and al. (1993) PVM 3 User's Guide and Reference Manual, May 1993.

Reynolds, P.F. (1988) A Spectrum of Options for Parallel Simulation, *Proceedings of the 1988 Winter Simulation Conference*, 325-332.

Theofanos, M. (1984) Distributed Simulation of Queuing Networks, *Master's Thesis*, The University of Virginia, January 1984.

BIOGRAPHY

CongDuc Pham Is a PhD candidate at the MASI Laboratory (Methodologie et Architecture des Systèmes Informatiques), University Pierre et Marie Curie, (Paris, France). He received a Diplôme d'Etude Approfondie (DEA) in computer systems from the University Pierre et Marie Curie in 1993. His research interests focus on parallel discrete event simulation algorithms and their application to the simulation of ATM networks.

Serge Fdida Received the Thèse de 3ème Cycle and Habilitation degrees in computer science from the University Pierre et Marie Curie in 1984 and 1989, respectively. From 1984 to 1989, he was Assistant and then Associate Professor at the University Pierre et Marie Curie. He is currently a Professor at the University René Descartes (Paris, France). His research interests are in the area of modelling and performance evaluation of computer networks and high-speed communication systems architecture and protocols. He is responsible of the "Network and Performance" team of the MASI Laboratory (CNRS). Serge Fdida is the author of many papers in the field and a book on Performance Evaluation. He was the program chairman of the IFIP High Performance Networking '94 Conference. He is a member of ACM, IEEE and AFCET.

24

An Error Monitoring Algorithm for ATM Signalling Links

K. Kant and J. R. Dobbins
Bell Communications Research
331 Newman Springs Road, Red Bank, NJ 07746, USA
Tel: (908)758-5384, Email: kant@perf.bellcore.com

Abstract

This paper proposes an error monitoring algorithm for in-service signalling links at the network node interface (NNI) using the ATM adaptation layer for signalling (SAAL). The algorithm is intended for use on links using the SSCOP (Service Specific Connection Oriented Protocol) of ATM adaptation layer 5. It uses only the information that is either already available from SSCOP or can be obtained easily. It does a better job in monitoring the link quality over a wide range of ATM network parameters (e.g., a wide range of link speeds, message length distributions, and offered loads) than known error monitoring algorithms with only one set of parameter values.

Keywords

ATM Adaptation Layer, Signalling, SSCOP, Error Monitoring, transmit congestion.

1 INTRODUCTION

The provision of switched virtual circuits (SVCs) in an asynchronous transfer mode (ATM) network requires a reliable and efficient signalling protocol. Such a protocol has already been standardized and is known as SSCOP, or service-specific, connection oriented protocol (Quinn 1993). SSCOP is a link level protocol and provides selective retransmission of lost messages (or protocol data units (PDUs)). A "link" in the ATM context is really a virtual circuit designated to carry signalling traffic. The bit rate for this virtual circuit could be anywhere from 64 Kb/sec to 4 Mb/sec depending on the signalling needs. Since the availability of SSCOP links (or virtual circuits) is vital for all services using SVCs, it is important to perform error monitoring on these links. Although the lower layers (e.g., ATM and SONET) do provide a certain degree of error monitoring, an application-level error monitoring is also necessary to provide a tighter control over link quality. The error monitor presented here is a part of the algorithm accepted by ITU as an example of desirable SAAL error monitoring algorithms.

A good design of an error monitor involves several considerations as discussed in (Ramaswami 1993, Kant 1994). This section motivates some of the major considerations.

In telecommunication networks, most network components, including links, are operated in the duplex mode, i.e., normally, the carried load is shared by two identical links. If one of these links is taken out of service, a *changeover* is initiated which transfers all undelivered messages to its "mate" link. When a repaired link is put back into service, a *changeback* process is initiated, which redistributes the traffic over the pair. These mechanisms provide immunity against single failures; however, they do involve a lot of overhead and message delays. Thus, if the link is hit with only a short error burst, it is better to keep the link in service even if the *bit error ratio* (BER) approaches 0.5 (i.e., total garbage) during the error burst. In other words, *an error monitor must be designed to "ride-over" some minimum error burst duration of t_b seconds, irrespective of its severity.* At the same time, the error monitor should not take too long to pull the link out of service, as this not only subjects messages to long delays, but also results in a large transmit congestion (i.e., a large number of messages in the transmit buffer at the time of link taken-down). Large transmit congestion translates into a large buffer size and long changeover time. Thus, *an error monitor should try to keep the transmit congestion as low as possible.*

Let X denote the time to take the link out of service, and $M(X)$ some measure of it (e.g., mean, mean+2σ, 95 percentile, etc.). A typical behavior of $M(X)$ as a function of BER can be seen in the first chart of Figure 5. Clearly, $M(X)$ should increase as BER decreases. More importantly, for a well-designed algorithm, this curve should show a definite *knee*. That is, as the BER goes below the knee, $M(X)$ should increase drastically, so that the algorithm will keep the link in service almost indefinitely. On the other hand, as the BER increases above the knee, $M(X)$ should quickly go down to its minimum value dictated by the error-burst ride-over requirement.

An important concept in locating the knee is one of *sustainable error rate* (SER), i.e., the maximum bit error rate at which the message delay requirements are still satisfied. An error monitor should take the link out of service quickly if the error rate exceeds the SER, and otherwise leave the link in service. That is, the knee should occur at or below SER. It should also be clear that a sharper knee is preferable — ideally, the knee should be "square", i.e., $M(X) = \infty$ for BER<SER, and $M(X)$ =const (determined by the error burst ride-over requirement) if BER>SER. For brevity, a precise characterization of SER is omitted here. For a 5000 mile long link (the longest terrestrial link), the sustainable error rate works out to be around 3.0e-5 for a 64 Kb/sec link, 1.0e-5 for a 512 Kb/sec link, and 2.0e-6 for a 4 Mb/sec link. Since these error rates are rather high compared with the typical link quality found in practice; it is okay (and perhaps even desirable) to design the error monitor assuming a lower SER. In this sense, these numbers should be regarded as upper bounds on the error rates at which the link could be left into service for a long time.

(Kant 1995) shows an analytic model of SSCOP performance, which can be used to obtain SER as a function of various network parameters. Clearly, the SER depends on some key link parameters such as link length, link speed, offered load, and message size. This tends to make the error monitor parameters dependent on the link parameters, which is undesirable from an administrative and operations point of view. This issue is particularly important in the ATM signalling context because the link parameters may vary over a wide range. In particular, the following ranges must be considered:

- Link Length: 0 to 5000 miles for terrestrial links. Links to geo-synchronous satellite are about 30,000 miles long.
- Link Speed: Currently specified as 64 Kb/sec to 4 Mb/sec.
- Offered Load: No filler PDUs ⇒ Load may drop to almost zero.
- Message sizes: 1-4 cells with current services. Up To 20 cells in future.

This paper presents an error monitor for in-service SAAL links that meets all the challenges described above. Before inventing this new algorithm, the authors examined the existing error monitoring mechanisms, namely SUERM (signal unit error rate monitor) of low speed (56/64 Kb/sec) CCS links (BCR 1991), EIM (errored interval monitor) of high-speed (1.5 Mb/sec) CCS links, and the error monitoring provided by the ATM/SONET layers. None of these were found entirely satisfactory for SSCOP environment.

2 ALGORITHM DESCRIPTION

Although the algorithm is not inherently tied to SSCOP, we describe it here specifically for SSCOP. A description of SSCOP may be found in (Kant 1995) and (Quinn 1993). SSCOP assigns a sequence number to all the user PDUs to keep track of retransmitted PDUs and to ensure that the PDUs are delivered in the correct order to the higher layer on the receive side. When the receiver detects a new gap in the received sequence numbers, it alerts the transmitter via a *ustat* message, which contains a list of missing sequence numbers. In response, the transmitter retransmits those PDUs. In addition, the transmitter periodically sends a *poll* message, to which the receiver responds via a *stat* message, which lists *all* existing gaps in received sequence numbers. The transmitter retransmits these messages. Unnecessary retransmissions are avoided by the protocol using some mechanisms detailed in (Quinn 1993). The poll-stat combination also provides the "I-am-alive" functionality between peers. In particular, there is a timer known as the *no-response timer* which times the gaps between successive stat arrivals on the transmit side. If this timer expires, the link (or the virtual circuit) is reset.

The proposed error monitoring algorithm resides in the SAAL *Layer Management* and needs the following pieces of information from SSCOP for its operation:

1. End of a polling interval, i.e., sending of a poll by SSCOP, and the following auxiliary information:
   - The current VT(S) value (i.e., the sequence number to be given to the next user PDU that is to be transmitted for the first time).
   - Indication of any retransmissions in the last polling interval (rexmit_flag).
   - Number of stats received in the last polling interval (n_stats).

2 Indication of a credit rollback by the peer receiver.

Currently, SSCOP only reports total number of retransmissions during a polling interval, which can be used to extract "rexmit-flag". Other information is, however, available within SSCOP, and can be easily reported to the layer management.

The basic idea behind the error monitoring algorithm is as follows. At the end of each polling interval, a "penalty" factor is computed for the polling interval. The algorithm uses penalty factors over a block of N_{blk} consecutive polling intervals and computes a quality of service (QoS) measure for the block. This QoS is simply the arithmetic average of the penalty factors, clipped at 1.0. That is, if the average penalty over the block exceeds 1.0, the QoS is made 1.0. Finally, the block QoS is used to compute an overall (or running) QoS by using exponential smoothing over consecutive block QoS's. That is, if Q denotes the running QoS, and Q_b denotes the QoS from the current block, Q is updated as follows:

$$Q = (1 - \alpha)Q + \alpha Q_b$$

where α is the exponential smoothing factor in the range (0,1). Whenever the running QoS Q exceeds a threshold *thres*, the link is taken out of service. The penalty factor over an interval is computed as follows: If no stat has been received for the last N_{gap} polling intervals, set penalty to $N_{blk} - 1$; otherwise, set it to 1 or 0 depending on whether "rexmit_flag" is set.

At moderate to low error rates, the behavior of this *core algorithm* is primarily governed by retransmissions of user PDUs. Consequently, the take-down time will increase as the offered user load drops. However, the error monitor should be capable of detecting unacceptable error rates even at zero offered load, since all links may experience long idle periods. Links that are used as alternate routes would almost always be idle. This motivated the addition of the following mechanism to the core algorithm:

The auxiliary mechanism counts the number of stats received (denoted N_{sr}) over a "super-block" of size N_{sup} polling intervals. At the end of the super-block, if the number of missing stats (given by $N_{sup} - N_{sr}$) exceeds a threshold N_{loss}, the link is taken out of service. Otherwise, we repeat the process over the next super-block.

This mechanism is almost independent of the load and enforces a minimum link quality depending on the choice of N_{sup} and N_{loss}.* It also complements the core monitor at low loads by limiting the failure time.

The algorithm also includes provision to handle situations where the receiver causes retransmissions by withdrawing credit for already transmitted PDUs. This is done by ignoring retransmissions after such a *credit rollback* until a poll with a higher VT(S) value is sent out.

Although it is convenient to describe the algorithm in terms of counting variables N_{gap}, N_{blk}, N_{sup}, and N_{loss}, their use makes the algorithm parameterization heavily dependent on the polling interval T_{poll}. To remove this dependence, it is better to use time-unit versions of these quantities, henceforth denoted by replacing "N" by "T" in the notations. The relationship between the two is simply $N_x = T_x/T_{poll}$, for $x = gap, blk, sup, loss$.

The primary parameters of the algorithm and their recommended values are listed in

*A slight dependence on load exists because SSCOP resets the keep-alive timer whenever a stat is received during an idle period.

Table 1 Primary parameters of the error monitor and their default values

Parm	Meaning	value	Parm	Meaning	value
T_{gap}	Max gap between stats	0.3 sec	T_{blk}	Block size	0.3 sec
T_{sup}	Super-block size	120 sec	T_{loss}	Stat loss limit	1.3 sec
α	Smoothing factor	0.1	β	Clipping threshold	1.0
thres	Threshold for failure	0.191			

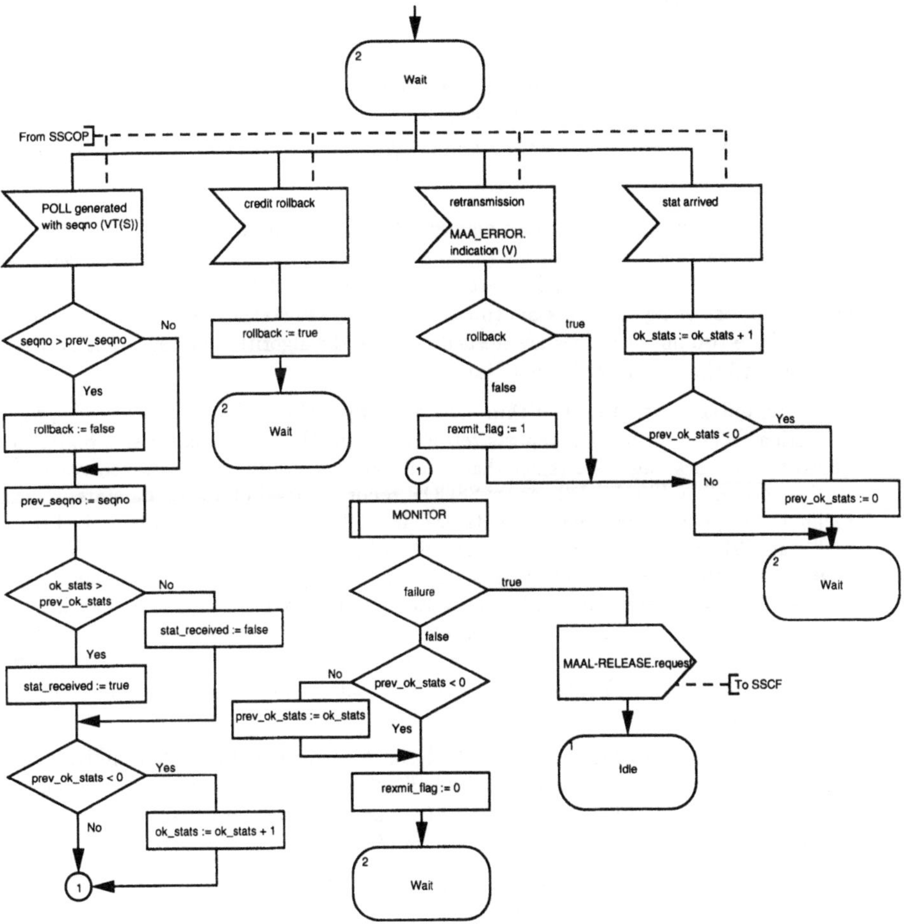

Figure 1 Top Level of the Error Monitor

Table 1. Figures 1 and 2 show the flow chart of the entire algorithm. The algorithm uses a number of variables, which are described below:

- N_{gap}, N_{blk}, N_{sup}, N_{loss}: counter versions of T_{gap}, T_{blk}, T_{sup}, T_{loss}.
- PI_count: A counter for polling intervals (PIs).
- ok_stats: A counter for number of received stats in a superblock.
- prev_ok_stats: Remembers ok_stats value for the previous PI.
- stat_log[N_{gap}]: A boolean array to hold stat-arrival indication over last N_{gap} PI's.
- stat_received: A boolean indicating stat reception in the last polling interval.
- tot_penalty: Running total of penalties over PIs of a block.
- block_qos: QoS measure over a block.
- tot_qos: Overall running QoS measure.
- seqno: VT(S) at the time of most recent sending of a poll.
- prev_seqno: VT(S) value from the previous polling interval.
- rollback: A boolean to indicate a credit rollback by the peer receiver.

Figure 1 shows the interactions between the error monitor and SSCOP. The conversion of time parameters to counting parameters and auxiliary variable initializations are not shown. Most auxiliary variables are initialized to 0 (or false in case of booleans); the only exceptions being prev_ok_stats (initialized to -1) and stat_log array (initialized to all true). These two initializations are needed to make the algorithm pretend that no PDUs are lost until the arrival of the first stat.

After initialization, the algorithm waits for a trigger from SSCOP and takes appropriate action. Whenever SSCOP reports retransmissions, the error monitor sets rexmit_flag unless the rollback flag is set (since that condition indicates that the retransmission may be due to credit rollback.) Whenever SSCOP reports arrival of a stat, ok_stats is incremented. Also, if prev_ok_stats is negative, it is set to zero to indicate the arrival of the first stat. When SSCOP reports a credit rollback, the flag "rollback" is set. When SSCOP reports the expiry of poll timer, seqno, prev_seqno, and rollback are updated. Also, stat_received is set to indicate whether a stat was received in the last polling interval. It can be verified that this boolean will evaluate to true until the first stat is received. Incrementing of ok_stats on each poll reception before the first stat has arrived (i.e., while prev_ok_stats is negative) ensures that stat and poll counters run in sync until the first real stat arrives.

Following these updates, the main body of the error monitor, called MONITOR, is run. In case of a normal exit from the MONITOR (i.e., if a changeover is not declared), prev_ok_stat is updated if it is nonnegative (i.e., only after the very first stat has arrived). The rexmit_flag is also reset upon return from MONITOR.

The MONITOR routine, shown in Figure 2, works as follows. The tot_penalty is incremented by $N_{blk} - 1$ if no stat has arrived for the last N_{gap} intervals. Otherwise, it is incremented by rexmit_flag. If the current polling interval ends a block, tot_qos is computed and compared against *thres*. If the current polling interval ends a superblock, a check is made to see if more than N_{loss} stats were lost. The flow-chart assumes that the superblock size is an integer multiple of the block size.

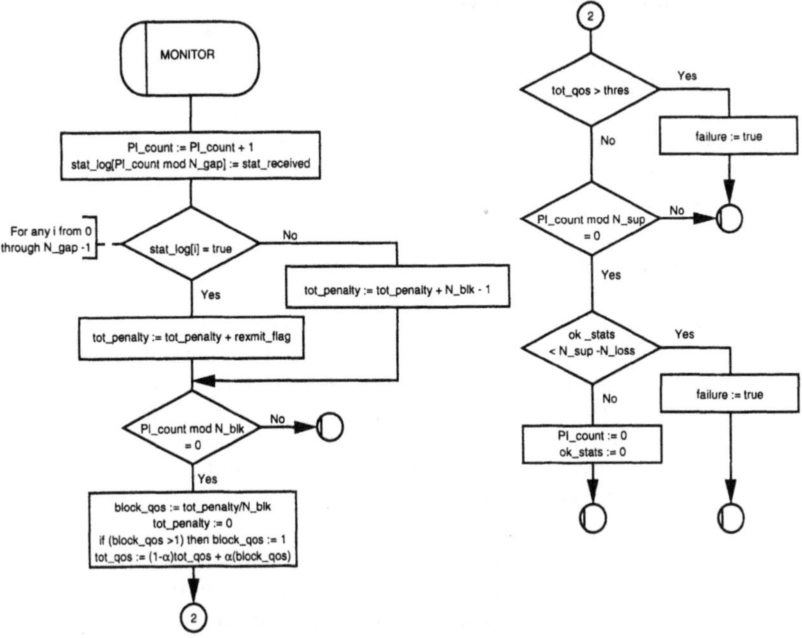

Figure 2 MONITOR routine of Error Monitor

3 RATIONALE FOR PARAMETER CHOICES

The most critical parameter for an error monitor is t_b, the duration of the error burst that must be tolerated. From experience with CCS networks, it is known that a suitable value for t_b is 0.3–0.4 secs. Since riding over a longer burst entails higher transmit congestion, the design here enforces $t_b = 0.4$ only for bit error ratio (BER) approaching 0.5. Consequently, the overall algorithm rides over 0.3 sec error burst in nearly all cases, but may not always ride-over 0.4 sec error bursts (although it does so in most cases).

In order to minimize transmit congestion at changeover time, it can be shown that N_{blk} should be about one-half of t_b/T_{poll}. (The factor 1/2 comes from the fact that an error burst can start anywhere within a block, and thus can always span two blocks irrespective of how large the block is.) To minimize link take-down time under severe errors, T_{gap} should be set as small as possible subject to the constraint that normal jitter in stat reception should not result in nonzero penalty. This leads to the choice of $T_{gap} = 0.3$ secs assuming that the SSCOP implementation gives the highest priority to polls/stats/ustats as stipulated in the SSCOP specification (i.e., assuming that the queuing delays suffered by polls and stats are always less than 0.1 sec).

The need for clipping and the choice of $\beta = 1.0$ is tied to the use of penalty > 1 in connection with excessive gap between successive stats. The purpose of clipping is simply to limit blk_qos to the same value as the one without the use of penalty> 1. In this sense, $\beta = 1.0$ is the only possible choice. The main reason for not altering the limit on blk_qos is that excessive loss of stats can occur only at very high error rates; at other error rates the algorithm should behave as if the stat-gap mechanism was not present. The choice of $\beta = 1.0$ can be justified in other ways too; in fact, a β other than 1 is unlikely to be useful; for this reason, the implementation in Figure 2 does not even regard β as an algorithm parameter.

The main motivation for monitoring the gap between successive stats comes by considering the algorithm performance for long messages at high error rates. In such cases, the message error probability could be almost 1 whereas the cell error probability is still not too high. In this case, almost no ustats will be generated and the error reporting happens almost entirely via stats. This tends to increase link take-down times substantially in a very narrow range of error rates. By using a nonzero penalty when stats also do not arrive, one can partially compensate for the lack of ustats and thereby bring down this peak in take-down time. Nevertheless, the congestion can still increase substantially in narrow BER range for very long messages.

The rationale for choosing penalty=$N_{blk} - 1$ when the inter-stat gap exceeds T_{gap} is as follows: It is desirable to set penalty to the highest possible value to minimize link take-down time in the situation described above. Given $\beta = 1.0$, the largest useful value of penalty is N_{blk}. This value is fine from an error burst ride-over perspective; however, if stats were lost during the error burst, and the first stat to arrive following the error burst takes more than 200 ms, the link will be taken out of service. Setting penalty=$N_{blk} - 1$ eliminates this problem.

The two remaining parameters of the core algorithm, namely α and *thres*, can be interrelated by the error burst ride-over requirement. Suppose that a clean link is hit by an error burst of length t seconds with BER approaching 0.5, after which the link again becomes clean. It is then easy to compute the maximum value of QoS, say q, as a function of α for any given scenario of PDU losses due to the error burst. Let this event. Let $q_{\max}$ denote the maximum q over all the scenarios. Then, it is required that $q < thres$ iff $t \le t_b$. Such an analysis indicates that $thres = \alpha(2 - \alpha)$. At lower BERs, such a choice does not necessarily ride over error bursts of length t_b at all BER values. However, the region of vulnerability is rather small as the results show later.

The design approach is now to choose α and then compute *thres* from the above equation. With such a design, the maximum transmit congestion does not change substantially as α is decreased. By using an analytic model, it is possible to characterize the behavior the algorithm rather accurately at low error rates (i.e., when the take-down time is at least a few seconds). This model shows that a small α is preferable for two reasons: (a) it results in a sharper knee, and (b) the variability in the take-down time goes down. However, since *thres* also decreases with α, the knee tends to move towards a lower BER. Thus, to ensure that the knee BER for a 4 Mb/sec link does not become too small, α is chosen as 0.1.

To parameterize the auxiliary mechanism, it is best to set T_{loss} to the minimum possible value, and then determine the minimum value of T_{sup} needed to enforce a desired link quality at zero load. Anticipating that T_{sup} will not be much more than a few minutes, it suffices to allow for stat losses due to one $t_b = 0.4$ secs error burst and 2 random errors. The error burst can wipe out at most $t_b/T_{poll} + 2$ stats. The two random errors can wipe out 2 more stats. Stats may also get delayed due to queuing delays suffered by polls or the stats generated by the polls. We assume that this cumulative delay, say T_{qd}, does not exceed 0.5 secs. It follows that $T_{loss} = t_b + 4T_{poll} + T_{qd}$, which works out to be 1.3 secs.

Given T_{loss}, it is then easy to estimate T_{sup} to enforce a given link quality, say θ under random errors. Here θ is defined as the BER at which the auxiliary mechanism should leave the link in service with a high probability. It is adequate to choose θ as the sustainable error rate of a 4 Mb/sec link at normal engineered load. The idea is that an idle in-service link should be good enough to suddenly take normal engineered load without causing excessive delays. From these considerations, θ=5.0e-6, which, in turn, gives $T_{sup} \approx 120$ secs. (A precise setting of T_{sup} is unnecessary; for implementation ease, it helps to make T_{sup} a multiple of T_{blk}.)

4 EVALUATION OF THE ALGORITHM

The evaluation of the algorithm was done via a detailed simulation of SSCOP. Most results were obtained under "double-sided errors", i.e., when both forward and backward directions experience synchronized error bursts simultaneously. This scenario is more appropriate since it results in worse performance (because the loss of ustats/stats yields less information about the existing error conditions.)

A tolerable burst length of $t_b = 0.4$ sec can wipe out at most 6 stats, which creates a gap of 0.7 secs between successive stats. Allowing another 0.1 sec for jitter, the minimum value of the SSCOP no-response timer T_{nr} is 0.8 secs, and was used in all the experiments. At very high error rates, it is the no-response timer that will take the link out of service. The algorithm, however, is not totally dependent on a low setting of no-response timer. In particular, without a no-response timer, the stat-gap mechanism will make the tot_qos exceed the *thres*hold in 3 to 4 blocks (i.e., the link will always be taken out of service in 1.2 secs or less).

The evaluation used three sample link speeds: lowest (64 Kb/sec), highest (4 Mb/sec), and their geometric mean (512 Kb/sec). The default link length was 5000 miles, although the behavior of the algorithm for other link lengths was also explored. The default (or "normal") message size distribution was (0.5, 0.2, 0.1, 0.1, 0.1), i.e., 50% of the PDUs are 1 cell long, 20% 2 cells long, etc. This distribution is a rough estimate for initial BISUP traffic. Several other distributions were also used in the experiments, a few of which are reported here.

For ease of reference, the offered loads are expressed as percentage of failure-mode engineered load. The latter is chosen as 0.8 Erlang (including polls and stats). With a constant polling interval of 100 ms, the corresponding user load is 0.668, 0.783, and 0.798 Erlangs respectively, for a 64 Kb/sec, 512 Kb/sec, and 4 Mb/sec link. These loads are referred to as 100% loads. The normal engineered load, which is the default, is half as much, or 50%. The major performance parameters for evaluation are as follows:

Take-down time: Time to declare changeover (i.e., to send a COO message) from the time the error burst starts.

Transmit Congestion: Total number of unacknowledged messages (measured in cells) in the system at the time of changeover order. In order to avoid the question of how full the cells are, congestion was measured in terms of cells rather than PDUs or bytes.

PDU delay: Total delay (time of delivery to the higher layer on receive end minus the time of arrival on the transmit side) experienced by all PDUs that get through before changeover.

Burst ride-over: The probability that the algorithm will fail to ride-over an error burst of a given duration.

For the first 3 parameters, we report the mean and "m+2s" (mean plus twice the standard deviation). All graphs show two different scenarios; for each scenario, the upper curve is for "m+2s" value, and the lower one is for the mean value.

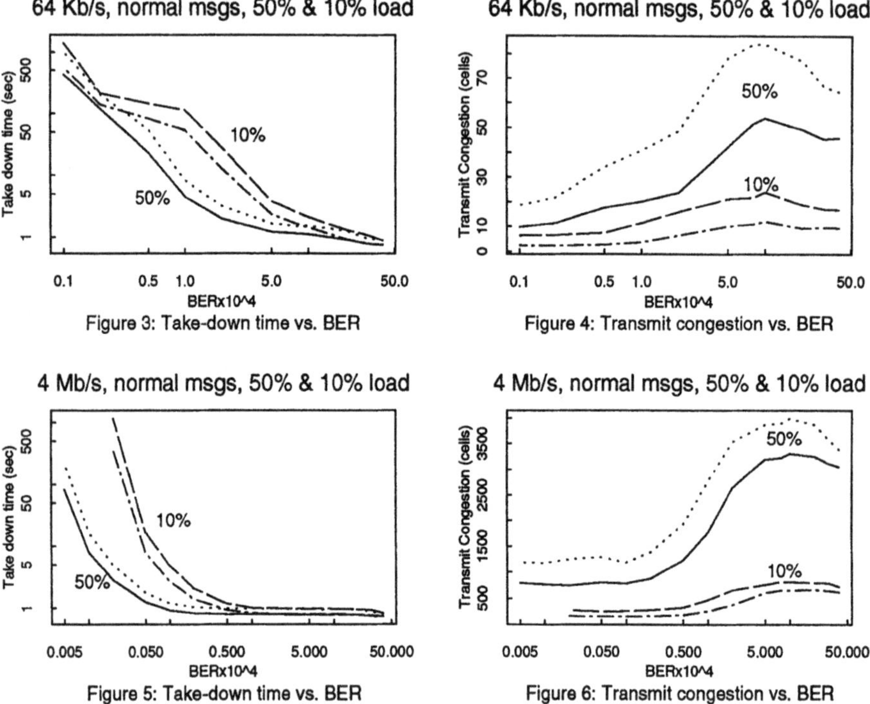

Figure 3: Take-down time vs. BER

Figure 4: Transmit congestion vs. BER

Figure 5: Take-down time vs. BER

Figure 6: Transmit congestion vs. BER

Due to space limitations, only a few graphs are included here, however, the discussion draws upon a large number of other cases as well. Figures 3 and 5 show the link take-down time as a function of BER for a 5000 mile carrying "normal" messages. These graphs show performance for two different link speeds (64 Kb/sec and 4 Mb/sec) and two different offered loads (50% and 10%). In the low BER region the simulation was terminated if the link survived for more than 50 minutes (an arbitrarily chosen cut off point based on the available computing resources). It is seen that the curves have a rather sharp knee that moves appropriately with the link speed. (The 64 Kb/sec curve also rises very sharply below a BER of 1.0e-5.) For normal messages at 100% load, the cutoff BERs were 5.0e-6, 1.0e-6, and 1.0e-7 respectively for 64 Kb/s, 512 Kb/s and 4 Mb/s links. For 50% load, these BERs were 5.0e-6, 2.0e-6, and 2.0e-7. These BERs are about 1/10th of the sustainable error rates (SERs) for those links. This is perhaps a bit too conservative, but should be satisfactory in all cases except when a 4 Mb/sec link is implemented in copper (as opposed to fiber).

As the offered load decreases, the core algorithm has fewer and fewer messages to work with, which tends to increase the take-down time. (See Figure 5). To an extent this is a desirable behavior since the SER increases as the load drops; however, when the load becomes too small, the core algorithm may keep the link in service for a long time at error rates that are generally considered too high. Furthermore, as the offered load approaches zero, the core algorithm will become totally ineffective since there are no retransmissions to monitor. This is where the auxiliary algorithm involving superblocks comes into play. This auxiliary algorithm takes the link out in 2 minutes or less irrespective of the load or link speed up to an error rate of about 2.0e-5. At a BER of 1.0e-5, the algorithm takes about 10 minutes, and at a BER of 5.0e-6, it takes more than 50 minutes. For a 64 Kb/sec link, the SER is higher than the θ for the auxiliary algorithm; therefore, the latter comes into play even at 50% offered load, as seen by the rather slow ascent of take-down time below a BER of 5.0e-5. At 10% load, the auxiliary algorithm comes into play at an even higher BER. This explains the somewhat irregular shape of curves in Figure 3. In contrast, at 4 Mb/sec, the SER is much less than θ even at 10%, and thus one sees only the core algorithm operating in Figure 5.

Figures 4 and 6 show the transmit congestion for the two scenarios discussed above. It is seen that the transmit congestion shows a unimodal behavior, with the maximum typically occurring around 1.0e-3. This error rate is just below the error rate when most of the take-downs will be due to no-response timer. At higher error rates, the congestion trails off because the no-response timer will consistently take the link out of service in about 0.8 seconds with very little variability. If the no-response timer is set to a high value, this effect will not occur and the congestion will continue to increase beyond the BER of 1.0e-3. However, this further increase cannot be too much, since if a stat is not received for more than 2 blocks, the threshold will be crossed at the end of third block. That is, without the no-response timer, the link will be taken out of service in 3 to 4 blocks at very high error rates.

A more careful examination of link take-down times shows that the algorithm adjusts itself automatically in the right direction as the link-speed, message size, or offered load change. Figures 3 and 5 show this effect with respect to link speed and offered load. As the message size increases, the sustainable error rate decreases, which means that the take-down time should decrease at BERs around the knee and lower. Figure 7 shows this effect; although, the difference is almost indiscernible. At very high BERs, long messages increase take-down time somewhat because there are much fewer messages in a block.

As the message size increases, the peak congestion increases, however, the nature of increase depends on the PDU size distribution. The mechanism at work here is the

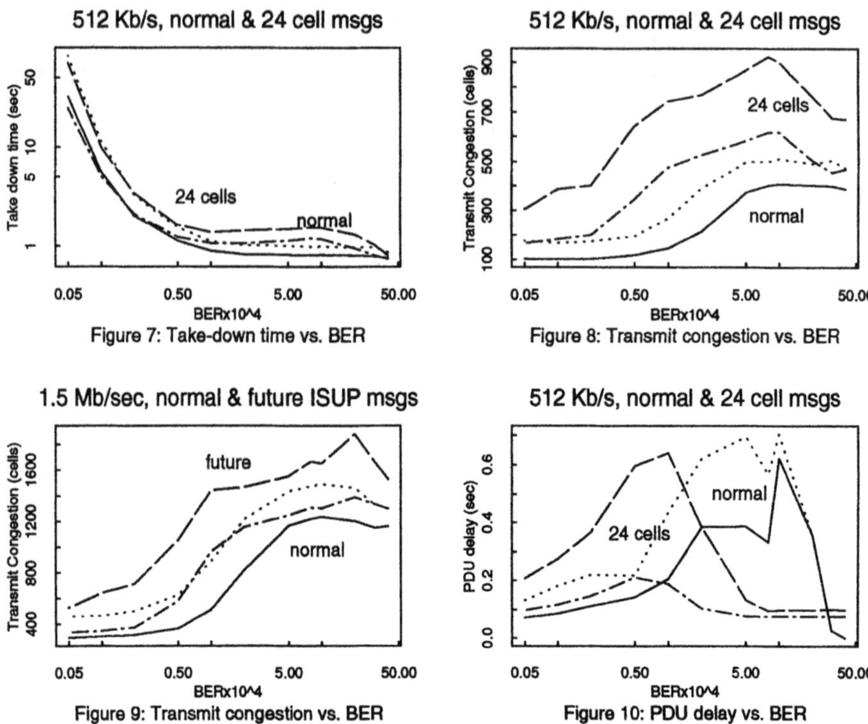

Figure 7: Take-down time vs. BER

Figure 8: Transmit congestion vs. BER

Figure 9: Transmit congestion vs. BER

Figure 10: PDU delay vs. BER

lack of ustat generation when a string of long messages is transmitted. Consequently, if most PDUs are very long, very few ustats are generated in the BER range 4.0e-4 and 2.0e-3, which leads to a peak in congestion which may go as high as 25–40% depending on the message size. The BER region over which the congestion remains high also expands accordingly. This is shown in Figure 8 which compares the transmit congestion for normal and 24 cell messages. However, if a significant percentage of PDUs (say 20–30%) are short, peak congestion does not increase significantly. For example, the projected BISUP size distribution over 10-20 year window should not be too far from the following distribution that was used for experimenting with a 1.5 Mb/sec link:

(4, 0.75), (8, 0.05), (12, 0.05), (16, 0.05), (20, 0.10)

Figure 9 compares transmit congestion for normal messages against the one for this distribution. It is seen that the increase in congestion is no longer very substantial. Therefore, in summary, the algorithm should work with most of the anticipated traffic mixes; however, if the link were to carry very long messages exclusively, it could result in a significant peak in transmit congestion.

Figure 10 shows PDU delays as a function of BER for a 512 Kb/sec link for normal and 24 cell PDUs. The PDU delays show a jump at BERs around 5.0e-5 or higher, but remain small overall. It is to be noted that at BERs of 5.0e-5 or more, the link will be

Table 2 Probability of not riding over a 0.4 sec error burst

BER	short,ss	long,ss	short,ds	long,ds
2.0e-4	0	0	0.005	0.01
5.0e-4	0.006	0	0.015	0.037
8.0e-4	0.025	0.02	0.21	0.17
1.0e-3	0.063	0.14	0.25	0.30
2.0e-3	0.25	0.19	0.37	0.29
4.0e-3	0.35	0.37	0.20	0.14
6.0e-3	0.20	0.18	0.023	0.015
8.0e-3	0.13	0.089	0.005	0.006
1.0e-2	0.053	0.057	0	0
1.4e-2	0.005	0.005	0	0
1.6e-2	0	0	0	0

taken out of service in a matter of seconds (see Figure 7); therefore, a large PDU delay in this range is inconsequential.

By analyzing various error burst scenarios, it can be concluded that the burst ride-over will be poorest for a heavily loaded high speed link. Consequently, the ride-over tests were performed using a fully-saturated 512 Kb/sec and 4 Mb/sec link. The algorithm was able to ride-over a 300 ms error burst in all cases. With a 400 ms error burst, however, there were very narrow regions of BER, where the algorithm did not perform too well. This is to be expected, as stated earlier. Table 2 shows some sample results for 400 ms ride-over. The four columns after the BER column are for the four combinations of short/long link, and single-sided (ss)/double-sided (ds) errors. A short link is 50 miles, whereas a long link is 5000 miles. Each entry in the table shows the probability that the algorithm will not ride-over a 400 ms error burst. It is seen that the probabilities become rather large in the range 8.0e-4 to 4.0e-3 and drop off to zero very quickly outside this very narrow range. For example, for BERs above 1.6e-2, the algorithm never fails to ride over the error burst. In practice, the errors can be expected to cover a large BER range, say from 1.0e-3 (or lower) to 0.5. Assuming a uniform distribution, the probability that the algorithm will fail to ride-over the error burst will be negligible.

In evaluating the algorithm, it is important to examine its sensitivity with respect to various parameters. For example, Figures 11 and 12 compare the take-down time and transmit congestion for a 50 mile (terrestrial) link against a 30,000 mile (satellite) link at 512 Kb/sec. It is seen that the take-down times are almost identical in the two cases. However, the transmit congestion is much higher and rolls off much more slowly for the satellite link. These characteristics result from the fact that a large number of messages reside in the pipeline for a satellite link. With the same parameterization, the burst tolerance for a satellite link will go down significantly; however, available data indicates that error bursts are typically much shorter for satellite links. Therefore, the same parameterization should work well for satellite links as well.

Figures 13 and 14 compare the performance for a 4 Mb/sec link when the polling interval is reduced from 100 ms to 25 ms. (Reduction in polling interval can be beneficial only at very high speeds — see (Kant 1995).) It is easy to show that in this case the no-response timer can be decreased to 0.65 secs (from 0.8 secs) without compromising the burst tolerance of the error monitor. Lowering the polling interval has the following two effects on the algorithm performance:

1. The knee of the take-down curve shifts towards higher BER by the same factor as the decrease in the poll timer. Thus, the knee will now occur at a BER of 8.0e-7 instead of 2.0e-7. This is fine (and even desirable) since the sustainable error rate for a 4 Mb/sec link is 2.0e-6.
2. The congestion remains almost unchanged up to a BER of 1.0e-3, but continues to increase and attains a somewhat higher maximum of 3800 cells at a BER of 3.0e-3. This phenomena can be explained by the fact that 4 times as many stats are being sent, which makes their survival more likely.

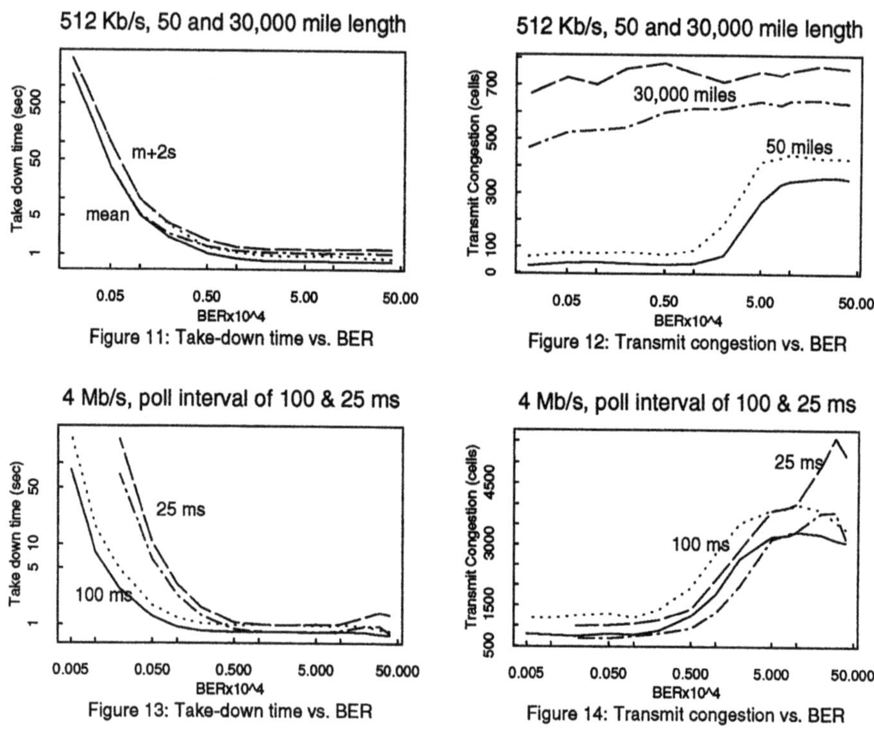

Figure 11: Take-down time vs. BER

Figure 12: Transmit congestion vs. BER

Figure 13: Take-down time vs. BER

Figure 14: Transmit congestion vs. BER

All experiments until now assumed that the signalling traffic is Poisson. In certain applications (e.g., user to signalling, distance learning, point to multipoint connections, etc.), the offered traffic may be considerably burstier. By using a 2-state MMPP arrival model, it is found that the mean take-down time and mean congestion are not very sensitive to burstiness, but the variance increases noticeably (as expected).

5 CONCLUSIONS

In this paper we presented an algorithm for error monitoring of in-service SAAL links. It was shown that the algorithm adapts itself well over a wide range of link parameters, including link lengths, link speeds, PDU sizes, and offered loads without any changes to the algorithm parameters. Consequently, a single set of error monitor parameters is adequate for almost all links likely to be used in signalling applications. Furthermore, the algorithm works well without any change in parameters even for satellite links. The algorithm also works well without any parameter changes if the polling interval is decreased for high speed links, although this results in a slightly higher peak congestion. Finally, the algorithm is simple, easy to implement, and does not need much interaction with SSCOP. These are important considerations because (1) with high speed links, processors often become a bottleneck, which makes expensive algorithms undesirable, and (2) an algorithm requiring substantial changes to SSCOP standard would be unattractive.

The only negative aspect of the algorithm is a sharp peak in transmit congestion in a narrow BER range when the link carries long messages primarily. Perhaps, the algorithm can be improved further in this area.

6 REFERENCES

"Bellcore Specification of the Signalling System Number 7", TR-NWT-246, Issue 2, Vol 2.

Quinn, S. (ed.), "BISDN ATM Adaptation Layer – Service Specific Connection Oriented Protocol", TD PL/11-20C Rev1, 1993.

Kant, K., "Analysis of Delay Performance of ATM Signalling Links", Proc of INFOCOM 95, Boston, MA, April 1995.

Hou, V.T., Kant, K., Ramaswami, V., and Wang, J.L., "Error Monitoring Issues for Common Channel Signalling", IEEE Journal of Selected Areas in Communications, 12, 3, April 1994, pp 456-467.

Ramaswami, V., and Wang, J.L., "Analysis of Link Error Monitoring Protocols in Common Channel Signalling Network", IEEE Trans on Networking, 1, 1, pp 31-47, 1993.

7 BIOGRAPHY

Krishna Kant received his Ph.D. degree in Computer Science from The University of Texas at Dallas in 1981. From 1981-1984 he was with Northwestern University, Evanston, IL. From 1985-1991, he was with the Pennsylvania State University, University Park, PA. In 1988 he served with the Teletraffic Theory division at AT&T Bell Labs, Holmdel, NJ and in 1991 with the Integrated Systems Architecture division at BellCore, Piscataway, NJ. Since Jan 1992, he has been with Network services performance and control group at BellCore, Red Bank, NJ. His research interests include performance modeling of computer and communication systems and fault-tolerant software design. He is the author of a book titled *Introduction to Computer System Performance Evaluation* (New York: McGraw-Hill, 1992).

25

Virtual Path Assignment in ATM Networks

Metin Aydemir and Yannis Viniotis
Department of Electrical and Computer Engineering
North Carolina State University, Raleigh, NC 27695-7911, USA.
Phone: (919) 515-5148. Fax: (919) 515-2285.
email: {metin,candice}@eceyv.ncsu.edu

Abstract

Virtual Path (VP) Assignment is an important resource management activity in ATM networks. By using the VP concept, the logical layout of the network can be administered to minimize the total network cost to the provider. We formulate the VP Assignment problem as a constrained optimization problem; the objective function to be minimized reflects processing/control costs at nodes and transmission costs between nodes. The quality of service requirements and overall throughput of the network are taken into consideration in the solution process as constraints.

Since the solution space is complex, we develop a heuristic algorithm based on descent methods to reach a "near-optimal" VP configuration for a given physical network and traffic demand. The Multi-Rate Loss Network model is used in the calculation of connection-level blocking probabilities. Results for static connection-routing are presented, for various forms of the objective function and constraints.

Keywords

ATM, virtual path, optimization, loss networks, quality of service, high-speed networks.

1 INTRODUCTION

The Asynchronous Transfer Mode (ATM) is accepted as the transport technology for Broadband Integrated Services Digital Networks (B-ISDN), due to its flexibility in accommodating a wide range of services with different traffic characteristics, and due to its efficient use of transport capacity by means of statistical multiplexing, among other advantages (ITU-T Recommendation I.150, 1991). The Virtual Path concept in ATM Networks is developed in order to achieve lower total network costs by reducing the processing and control costs (Kanada *et al.*, 1987) (Tokizawa *et al.*, 1988) (Addie *et al.*, 1988).

The Virtual Path concept is accommodated into the ATM Specifications by the ITU-T standards body (ITU-T Recommendation I.311, 1991). The Virtual Paths are logical paths defined at the transmission path level (Physical Layer) of ATM. In ITU-T terminology, a Virtual Path is defined as a labeled path (bundled multiplexed virtual channels or connections) between its end-nodes (also called virtual path terminators). The virtual path concept and implementation issues are addressed in detail in Sato *et al.* (1990), and Burgin and Dorman (1991).

The definition of VPs is refined by associating a deterministic capacity with the virtual path. This reservation of capacity for VPs has two important consequences: i) the statistical multiplexing is limited to the traffic within a virtual path; ii) a logical network (VP Network) emerges on top of the physical network topology.

Assuming that all connections are made by using VPs, this logical VP Network is a very flexible tool that can be employed in the management of network resources. The configuration of the VP Network has a major impact on the processing/control costs and transmission costs of a network.

VPs impact the network processing and control costs in three areas: call acceptance and call set-up, cell routing, and adaptability to changes. Employing VPs reduces the processing and delay associated with the call acceptance control (CAC) functions. Since VPs have guaranteed bandwidth along their path, these functions have to be performed only at the beginning node of a VP. When a call is active, only the VPI field of each ATM cell is processed in intermediate nodes of the VP. The cell processing time is reduced compared to VC Switching, which uses both VPI and VCI fields. Varying traffic conditions and network failures can be managed by adjusting the VP Network to accommodate the changing conditions and to maintain network performance.

VPs increase network transmission costs and decrease network throughput. The VP concept is based on reserving capacity in anticipation of expected traffic, and this results in decreased utilization of available transmission capacity. Allocating the capacity of a physical link to VP's using it reduces the sharing of the transmission capacity resource. The application of VP concept reduces the link utilization and overall throughput of the network, compared to total sharing (no VP) case.

We can consider two possible VP Assignment cases to illustrate the effect of the VP Network on the network performance. First, consider a fully-connected VP Network. A VP Network can be formed such that every node has one or more logical link (VP) to every other node. In this network, connections are established using only one VP, thus minimizing control costs. Applicability of this scenario decreases as the number of nodes are increased. For a sufficiently large network, the VPI field in ATM Cell header renders this solution impossible, due to the fact that the size of this field imposes a limit on the number of VPs associated with one node. The proliferation of VPs has a negative effect on the network utilization. For most network topologies, the reduced throughput will probably be prohibitive, even before any other factor comes into play. Second, consider the case where the VP Network is the same as the physical ATM Network. In this case, the VP Network is such that each physical link in the ATM Network contains only one VP, which carries all connections in the link. This is equivalent to not using the VP feature at all. In this case, the network utilization is maximized, and traffic will observe the lowest connection blocking probability. However, the connection establishment activity and processing at the intermediate nodes are at their highest cost. In general, an optimum VP Network solution lies between the two extreme VP Assignment cases considered above. Its exact location is a function of how we define optimality, the physical network under study, and the traffic flows in that network. This paper contributes to the existing work in this subject by proposing a new heuristic search method that is intuitively simple, for the solution of VP assignment problem.

The rest of the paper is organized as follows. In Section 2, we define the details of the problem definition and methods chosen on attacking the problem. Section 3 addresses the computationally challenging blocking probability calculation in a VP Network. In Section 4, we describe our proposed search method, used in the VP Assignment Algorithm. Initial results from test runs of the algorithm and some conclusions are presented in Sections 5 and 6.

2 PROBLEM FORMULATION

We consider the problem of finding a VP Network (i.e. assigning VPs and allocating their capacities) which minimizes an objective function subject to a set of QoS related constraints, for a given network topology, link capacities, and traffic demands.

The traffic demand is defined in terms of *traffic streams.* A *traffic stream* is a distinct traffic flow between two nodes, with defined call arrival rate, call holding time, and cell arrival rate. The cell arrival rate can be the average cell arrival rate if a constant bit rate (CBR) source is considered, or it can be an equivalent rate for bursty sources based on their quality of service requirements. Users can be categorized into different classes with distinct holding time statistics and bandwidth requirements. This way, different traffic types can be modeled with different *traffic streams.* This multi-class traffic representation provides us with a method to define the traffic demand in detail.

The solution of the VP Assignment problem is highly dependent on the solution of the connection routing problem and vice versa. The following argument gives insight into this interdependency. One needs to know, at least statistically, the traffic offered to various paths in order to accurately judge the merit of allocating a VP on the path. This requires selection of physical routes for traffic streams before the VP Assignment Algorithm proceeds. On the other hand, the connection routing problem can be solved most effectively when the final logical VP Network is known and available.

Thus, the problem can be approached in three ways:

1. Assign VP end nodes, and determine traffic streams that belong to each VP, then select routes for VPs (Chlamtac *et al.*, 1993). This method precludes the possibility of using multiple VP's on a connection.
2. Assign the VP's and connection routes simultaneously within the same procedure (Lin and Cheng, 1993). This formulation yields a complex problem, which cannot be solved directly (Cheng and Lin, 1994).
3. Select physical routes for each traffic stream, then form the VP Network based on this traffic layout.

In this paper, we assume that physical routes for traffic streams are provided (last approach in the above list). This approach is chosen as a first attempt to tackle the problem, due to its simplicity.

We make certain assumptions regarding the network structure, which we summarize below. No restrictions are imposed on the topology of the network. VPs can be established between any two nodes of the network. The nodes of the network are functionally identical. All nodes can be the source or destination of the network traffic. Nodes are not classified as ATM Cross-Connect or ATM Switching nodes. The assumptions on the node characteristics are adopted in order to provide an unrestricted solution space for the VP assignment problem. The physical links of the network are bidirectional; the capacity available on a link can be used in either direction. These assumptions give us a realistic and flexible model, which can be applied to a wide range of practical networks.

We also make the following assumptions on the connection establishment/handling: i) each connection can use one or more VPs. Allowing more than one VP on a single physical route of a connection results in a larger solution space, ii) the same route is used by connections of a traffic stream. A fixed routing scheme is assumed for simplicity. This assumption does not preclude other possible routing schemes at the call acceptance time, iii) each connection request has an associated equivalent bandwidth based on the traffic stream it belongs to. This bandwidth is allocated to the connection deterministically, when the connection is accepted, iv) if a connection request cannot be satisfied, due to capacity constraints, then the connection request is rejected. This event is called connection blocking, and the probability of call blocking is the same for all connections of a certain traffic stream.

The attributes of Virtual Path j are represented by the triplet $\mathbf{V}_j = \{V_j^{pa}, V_j^{ts}, V_j^c\}$. In this triplet, V_j^{pa} represents the set of physical links in the path of the virtual path j. V_j^{ts} denotes the set of traffic streams using the virtual path j. V_j^c denotes the capacity allocated to the virtual path j. The set of all virtual paths is denoted by $\mathbf{V} = \{\mathbf{V}_j; j = 1, 2, \ldots, J\}$. The size J of $\mathbf{V}$ is an output variable, since the number of virtual paths in the system will depend on the solution of VP Assignment problem.

The following two VP characteristics are also assumed : 1) VP's are unidirectional logical links. Since the traffic streams are also defined as one-way traffic, the flow of traffic streams of a VP should match the direction of that VP. 2) Virtual Paths reserve the capacity allocated to them deterministically (i.e., statistical multiplexing of VP's is not considered). Statistical multiplexing within a VP is of course allowed. This is implicit in the equivalent bandwidth concept used in the characterization of traffic streams.

We assume that the VP Assignment Algorithm will be used off-line, at a central point for the whole network. We also assume that there will not be real-time constraints, thus the algorithm can be computationally intensive.

Based on the assumptions and choices made above, we can now define the input and output variables of the VP Assignment Algorithm. Traffic streams are input variables, defined by their connection arrival rate λ_r, bandwidth requirement b_r, and mean connection holding time $1/\mu_r$, where $r = \{1, 2, \ldots, R\}$. The physical route T_r of stream r is also an input parameter, since we assume that traffic routes are already selected. T_r is a set of links that connect the source to the destination node. The network topology is specified by the node pairs that are connected by physical link i, and the capacity C_i of physical link i, where $i = \{1, 2, \ldots, I\}$. The output of the VP Assignment Algorithm is the set of all VPs ($\mathbf{V}$).

In the next subsection, we concentrate on the selection of the objective function and constraints for the optimization problem.

2.1 Objective Function and Constraints

The objective function should reflect all parameters that are taken into account when considering the network cost. In this stage, we opted to use very simple objective functions due to the need to calculate their value repetitively. First, we choose to minimize the number of logical links used by a connection as a measure of control and processing costs. Secondly, since the network topology and transmission capacity is fixed and given, we try to balance maximizing the utilization of the network versus minimizing the control and processing costs.

Before considering variations of objective functions further, let us define the constraint parameters of this problem. The physical limitations of the network are obvious constraints in the optimization procedure. The sum of capacities allocated to VP's that are using physical link i must be less than or equal to the physical link capacity C_i. For link i, this constraint can be formulated as:

$$\sum_{\forall j: i \in V_j^{pa}} V_j^c \leq C_i \tag{1}$$

where V_j^{pa} is the set of physical links used in virtual path j, and V_j^c is the capacity assigned to virtual path j. Inequality (1) is enforcing the capacity limitations of the network. It is obvious from (1) that we consider policies where the allocated VP capacities are guaranteed to be available (i.e. statistical multiplexing at the VP level is not performed).

There is a set of VPs for each traffic stream r, such that this set covers the physical path T_r, of stream r. For stream r, this constraint can be written as:

$$\bigcup_{\forall j: r \in V_j^{ts}} V_j^{pa} = T_r, \tag{2}$$

where V_j^{ts} is the set of traffic streams using the virtual path j. Equation (2) provides that there is a path in the VP Network to carry each defined traffic stream, and no traffic stream is left out in the VP Network.

The quality of service requirements of various classes can be implemented either as constraints, or as part of the objective function. We follow both approaches here. Of course, a variety of QoS measures can be defined in a network. For simplicity, we deal only with the following QoS measures: 1) *Blocking of Calls:* The blocking probability of traffic stream r, call it $\beta_r(\mathbf{V})$, must be less than or equal to ε_r, the maximum blocking probability allowed. In general, $\beta_r(\mathbf{V})$ for a given stream r will only depend on the subset of virtual paths used by traffic stream r, and the other traffic carried on these virtual paths. 2) *Number of Hops:* H_r, the number of VPs traversed by a connection of traffic stream r between its source and destination must be less than K, where K is a given positive number. By imposing this limitation, we bound the connection set-up time for all traffic streams. Note that this constraint does not put a bound on the maximum delay a connection will encounter, since a VP may traverse more than one physical link.

Based on the above, the following three objective functions are considered in this paper:

1. **Minimum hops with connection blocking as a constraint**

$$\min_{\{\mathbf{V}\}} \sum_{r=1}^{R} H_r, \tag{3}$$

subject to:

$$\sum_{\forall j: i \in V_j^{pa}} V_j^c \leq C_i, \qquad \forall i \in \{1,2,\ldots,I\}, \tag{4}$$

$$\bigcup_{\forall j: r \in V_j^{ts}} V_j^{pa} = T_r, \qquad \forall r \in \{1,2,\ldots,R\}, \tag{5}$$

$$\beta_r(\mathbf{V}) \leq \varepsilon_r, \qquad \forall r \in \{1,2,\ldots,R\}, \tag{6}$$

$$H_r(\mathbf{V}) \leq K, \qquad \forall r \in \{1,2,\ldots,R\}, \tag{7}$$

where H_r is the number of VPs used by the traffic stream r.
In the objective function of equation (3), the total number of hops experienced by all traffic streams is minimized. The integer-valued sum of equation (3) is an indicator of processing costs over the entire network, which is minimized while blocking probabilities are kept at acceptable level by means of inequality (6).

2. **Minimum hops (weighted) with connection blocking as a constraint**

$$\min_{\{\mathbf{V}\}} \sum_{r=1}^{R} \lambda_r \cdot H_r, \tag{8}$$

subject to:

$$\sum_{\forall j: i \in V_j^{pa}} V_j^c \leq C_i, \qquad \forall i \in \{1,2,\ldots,I\}, \tag{9}$$

$$\bigcup_{\forall j: r \in V_j^{ts}} V_j^{pa} = T_r, \qquad \forall r \in \{1,2,\ldots,R\}, \tag{10}$$

$$\beta_r(\mathbf{V}) \leq \varepsilon_r, \qquad \forall r \in \{1,2,\ldots,R\}, \tag{11}$$

$$H_r(\mathbf{V}) \leq K, \qquad \forall r \in \{1,2,\ldots,R\}, \tag{12}$$

where λ_r is the connection arrival rate of traffic stream r.
Equation (8) is a measure of number of hops experienced by an arbitrary connection. Since the intensity of traffic is taken into account in equation (8), it is a more precise minimization compared to equation (3). Equation (8) is a real-valued sum, thus computationally more expensive. Since the parameters $\{\lambda_r\}$ are fixed for a given network, objective function (8) is not a special case of objective function (13). The VP networks obtained from the two formulations will in general be different.

3. **Balancing control and transmission costs**

$$\min_{\{\mathbf{V}\}} \left\{ \alpha \sum_{j=1}^{J} D_j + \sum_{r=1}^{R} H_r \right\}, \tag{13}$$

subject to:

$$\sum_{\forall j: i \in V_j^{pa}} V_j^c \leq C_i, \qquad \forall i \in \{1,2,\ldots,I\}, \tag{14}$$

$$\bigcup_{\forall j: r \in V_j^{ts}} V_j^{pa} = T_r, \quad \forall r \in \{1, 2, \ldots, R\}, \tag{15}$$

$$\beta_r(\mathbf{V}) \leq \varepsilon_r, \quad \forall r \in \{1, 2, \ldots, R\}, \tag{16}$$

$$H_r(\mathbf{V}) \leq K, \quad \forall r \in \{1, 2, \ldots, R\}, \tag{17}$$

where D_j is the number of VPs in physical link j, and H_r is the number of VPs used by traffic stream r. As we discussed earlier in this section, J is an output variable, the value of which is determined as part of the solution. The value of the input parameter α is chosen based on the network characteristics. Constraint (16) guarantees an acceptable level of blocking, and constraint (17) makes sure the call set-up delays are bounded. Equation (13) is a simple balancing formula for two reasons: i) Maximize the sharing of the VP's, so that network utilization is maximized and blocking of connections is minimized (first sum); ii) Minimize the number of hops in a connection so that the set-up costs are low (second sum).

Problems (3), (8), and (13) are hard optimization problems, due to their nonlinearity and complexity of the constraint spaces. There are two reasons for this complexity. First, some of the optimization variables are integer-valued. Second, the constraints in inequalities (6), (11), and (16) are not convex, and therefore traditional solution techniques are not applicable.

The calculation of blocking probability $\beta_r(\mathbf{V})$ is difficult in most cases, and an exact method is not available. In the next section, we describe two approximations for this calculation, as a function of VP Network selection.

3 MODEL FOR CONNECTION-LEVEL BLOCKING PROBABILITY CALCULATION

In this section, we describe the product-form model for blocking probability calculation, and two approximate methods that are employed in the VP Assignment algorithm.

For a given VP configuration in an ATM Network, we can view each VP as a logical link in the network. This new logical network topology is referred to as the *VP Network*. Let's assume that a total of R traffic streams exist. The connection requests of stream r are assumed to arrive following a Poisson process with rate λ_r. The connection holding time distribution is assumed to be of phase-type, with mean $(1/\mu_r)$. A connection request that cannot be satisfied (not enough capacity in one of the links on its path), will be cleared without any other consequences. This type of networks has been named *multi-rate loss networks* (Chung and Ross, 1993), and analyzed in the context of circuit-switched telephony networks (Kelly, 1986). Multi-Rate Loss Networks have been shown to fit into a generalized BCMP Network model (Baskett *et al.*, 1975) (Lam, 1977). The equilibrium state probability distribution of this model has a product-form solution.

We next provide the state definition, other parameters, and the set of feasible states as an introduction to the approximation schemes that will follow. The logical link capacities are denoted by $\mathbf{V^c} = (V_1^c, V_2^c, \ldots, V_J^c)$, where J is the total number of logical links (VPs). The matrix $\mathbf{A}$ is constructed such that the element A_{jr} of the matrix denotes the bandwidth or cell arrival rate required by the traffic stream r, on logical link j. The dimension of $\mathbf{A}$ is $J \times R$. Typically, a traffic stream has a constant bandwidth demand, c, on all of its links, thus a row of matrix $\mathbf{A}$ takes values of zero or c. The logical link capacities and connection bandwidth requirements are integer multiples of a unit capacity.

The state of the system is defined as $\mathbf{n} = (n_1, n_2, \ldots, n_R)$, where n_r is the number of active connections of stream r. The set of feasible states $\mathcal{S}(\mathbf{V^c})$ is given by:

$$\mathcal{S}(\mathbf{V^c}) = \{\mathbf{n} : \mathbf{A} \cdot \mathbf{n} \leq \mathbf{V^c}\}.$$

This representation encapsulates the constraints imposed on the state space due to link capacities. The jth row of the vector inequality $\mathbf{A} \cdot \mathbf{n} \leq \mathbf{V^c}$ provides the constraint imposed by the capacity of link j:

$$A_{j1}.n_1 + A_{j2}.n_2 + \ldots + A_{jR}.n_R \leq V_j^c.$$

The state probability distribution is in product-form:

$$P(\mathbf{n}) = \frac{1}{G(\mathbf{V^c})} \cdot \prod_{i=1}^{R} \frac{a_i^{n_i}}{n_i!},$$

where $G(\mathbf{V^c})$ is the normalization constant for a network with capacity vector $\mathbf{V^c}$ and $a_i \stackrel{\text{def}}{=} \lambda_i/\mu_i$. $G(\mathbf{V^c})$ is given by:

$$G(\mathbf{V^c}) = \sum_{\mathbf{n} \in \mathcal{S}(\mathbf{V^c})} \left\{ \prod_{i=1}^{R} \frac{a_i^{n_i}}{n_i!} \right\}.$$

The blocking probability, $\beta_r(\mathbf{V})$ of connections belonging to traffic stream r can be represented by:

$$\beta_r(\mathbf{V}) = 1 - \frac{G(\mathbf{V^c} - \mathbf{A_r})}{G(\mathbf{V^c})},$$

where $\mathbf{A_r}$ is the r-th column of matrix $\mathbf{A}$.

Although an explicit formula is available, the calculation of the normalization constant proves difficult as the network size increases. An efficient method to compute the exact probability distribution of large multi-rate loss networks is not known. In the following subsections, two methods to calculate *approximate* blocking probabilities are presented. These approximations are all variations of the *reduced load approximation* principle, first introduced by Whitt (1985) for single-rate loss networks. They are presented in order of increasing complexity.

3.1 Erlang Fixed Point (EFP) Approximation

If one assumes that blocking between links and between streams of a link, occurs independently, then the blocking probability $\beta_r(\mathbf{V})$, can be approximated by :

$$\beta_r(\mathbf{V}) = 1 - \prod_{j=1}^{J}(1 - L_j(\mathbf{V}))^{A_{jr}}. \tag{18}$$

The $L_j(\mathbf{V})$'s can be considered as the blocking probability of link j for a unit bandwidth request. With the simplifying independence assumption, the traffic offered to a link will be Poisson. The rate of the Poisson traffic will be thinned by blockings occurred in the other links. This view of the offered load being reduced by blocking was first employed by Whitt for single-rate loss networks (Whitt, 1985). The following equations are used to obtain the $L_j(\mathbf{V})$ values :

$$L_j(\mathbf{V}) = E\left[V_j^c, \frac{1}{1 - L_j(\mathbf{V})}\sum_r A_{jr}a_r\prod_i(1 - L_i(\mathbf{V}))^{A_{ir}}\right], \tag{19}$$

where

$$E[C, v] = \frac{v^C}{C!}\left(\sum_{n=0}^{C}\frac{v^n}{n!}\right)^{-1}. \tag{20}$$

Equation (20) is the Erlang Loss Formula for a link with capacity C and Poisson traffic rate v. Kelly (1986) has shown that these fixed-point equations have a unique solution in the set $[0, 1]^J$. The Erlang Fixed Point Approximation has been shown to be asymptotically correct in a natural limiting regime (as the link capacity and offered loads are increased simultaneously, while their ratio remains the same).

3.2 Knapsack Approximation

Extensions to the Erlang Fixed Point Approximation were developed by S-P. Chung and K.W. Ross (1993), among others. One approach is to incorporate a one-dimensional iteration (Kaufman, 1981) into fixed point equations.

Consider a single-link system with link capacity V_j^c, and traffic streams $r \in V_j^{ts}$. The bandwidth of stream-r connections is A_{jr}, and the offered load of stream-r connections is a_r. Define

$$K_{jr}[V_j^c; a_l, l \in V_j^{ts}] = 1 - \frac{\sum_{n=0}^{V_j^c - A_{jr}} w(n)}{\sum_{n=0}^{V_j^c} w(n)}, \quad r \in V_j^{ts}$$

where

$$w(0) = 1,$$

$$w(n) = \frac{1}{n} \sum_{r \in V_j^{ts}} A_{jr} a_r w(n - A_{jr}), \quad n = 1, 2, \ldots, V_j^c.$$

If we assume that connection blocking is independent from link to link, then we can write the following "thinning" approximation for the offered load (a_{lj}) of stream-l connections on link j. Let T_l be the set of links in the path of stream-l. Then we have that:

$$a_{lj} = a_l \cdot \left(\prod_{i \in T_l - \{j\}} (1 - L_{il}) \right). \tag{21}$$

The original offered load a_l is thus reduced due to the blockings experienced on the other links (i.e., the set $T_l - \{j\}$) of stream-l.

Thus, we have the following set of equations, corresponding to each stream r of each link j:

$$L_{jr}(\mathbf{V}) = K_{jr} \left[V_j^c; a_{lj}, l \in V_j^{ts} \right], \qquad \forall j, r : A_{jr} \neq 0. \tag{22}$$

$L_{jr}(\mathbf{V})$ is defined as the approximate probability that the available bandwidth on link j is less than A_{jr}. The same argument applied in EFP approximation can be used to show that there is always a solution to the set of equations in (22). However in this case unlike EFP, the solution does not have to be unique. Finally, the blocking probability of stream r, is approximated by:

$$\beta_r(\mathbf{V}) = 1 - \prod_{j \in T_r} (1 - L_{jr}(\mathbf{V})).$$

This approximation takes into account the multi-class characteristics of the links, thus improving on the EFP approximation which lumps all classes into one flow of unit bandwidth.

3.3 Comparison of Approximations

The fastest approximation procedure is the Erlang Fixed Point (EFP) Approximation. Its iteration involves the computation of Erlang-B formula for each link. The experimentation with EFP Approximation reported in Chung and Ross (1993) shows that EFP tends to underestimate the blocking probabilities in some cases. The Knapsack Approximation computes one dimensional Kaufman Iteration for each traffic stream of each link. This approximation takes into account the multi-class characteristics of the links above the EFP approximation which lumps all classes into one flow of unit bandwidth. The drawback of the Knapsack Approximation is the non-uniqueness of the solution. By choosing the initial values of blocking probabilities realistically, we can insure converging to the correct solution, in most cases. In all executions of the Knapsack Approximation, we have never experienced a non-convergent case.

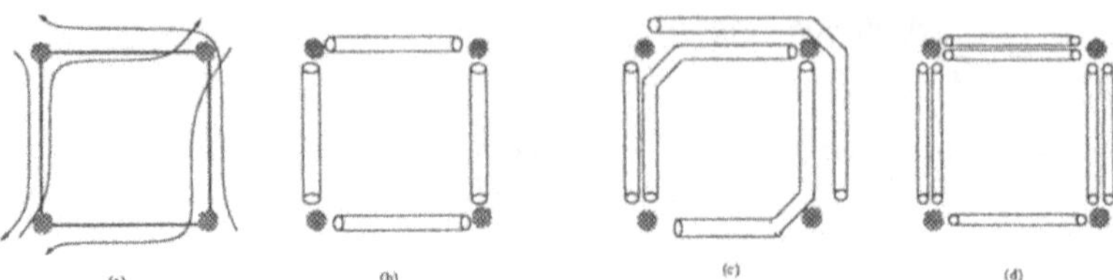

Figure 1 The three extreme solutions: a) Physical network and traffic streams, b) **1-VP/LINK** case, c) **1-VP/TS** case, d) **1-VP/LINK/TS** case.

4 VP ASSIGNMENT ALGORITHM

In this section, we present algorithm VPA, a heuristic VP Assignment algorithm to calculate optimal or near-optimal VP Networks. The current version of the algorithm outputs the following: the final VP allocations and their capacities, the objective function value, the blocking probabilities of traffic streams, and the unused capacities in all physical links.

To achieve this, we define a search method to move around in the domain of valid assignments (VP Networks) and check for optimum solution. Our search method consists of the following three basic steps:

1. Finding a starting point which is a valid solution (a valid VP Network) that satisfies the constraints,
2. Finding incremental changes in the VP Network that achieve a lower value for the objective function, and satisfy the constraints,
3. Repeating the previous step until no more improvement can be found and a termination criterion has been met.

Before describing our method of finding an optimum solution out of all possible VP Networks, let us consider three extreme solutions that give us "boundaries" on the range of possible solutions. These three extreme cases are illustrated in figure 1 for a very simple network, and are described below:

1. Each physical link is a VP. This case is referred to as **1-VP/LINK**. The **1-VP/LINK** corresponds to the maximum sharing case. It yields the lowest blocking probabilities for a fixed connection routing. However, it results in the maximum number of VPs along the path of a connection.
2. Each traffic stream has its own unique VP from source to destination. This case is referred to as **1-VP/TS**. The **1-VP/TS** corresponds to the minimum processing cost, since each connection is using only one VP from source to destination. Conversely, the link sharing is minimized. Each traffic stream has its own reserved capacity along the

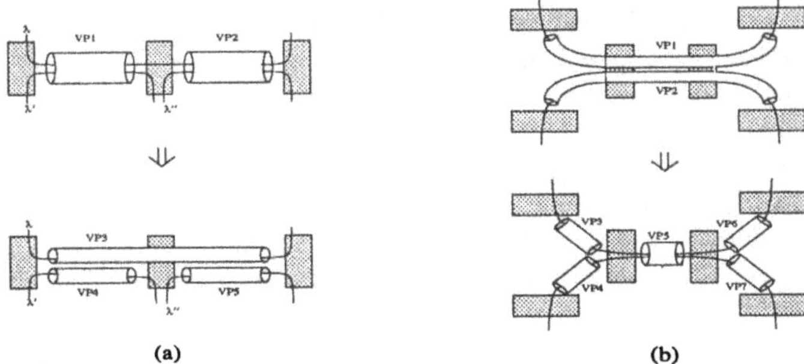

Figure 2 (a) Extending, and (b) Contracting Operations

physical path. The VPs can be shared by the connections of the same traffic stream only.

3. Each traffic stream has a unique VP on each physical link of its path from source to destination. This case is referred to as **1-VP/LINK/TS**. The **1-VP/LINK/TS** results in the maximum number of VPs over the whole network. This corresponds to minimum sharing and maximum processing.

An initial solution (a valid VP assignment) has to be chosen as the starting point of the VPA Algorithm. Any VP Network that satisfies the constraints can be the initial solution. We have used the **1-VP/LINK** solution (or a "neighboring" one) as a starting point, depending on the objective function and constraints.

4.1 Finding a Neighboring Assignment

The following two operations (moves) are used to find a neighbor VP Network that is slightly different than the VP Network at hand. Both of these moves involve two VPs at a time.

Extending In the extending operation (see figure 2.a), traffic streams common to two VPs (stream λ in VP1 and VP2) are used to form a new VP (VP3). Non-common traffic streams on the original VPs (λ' and λ'') are carried on separate VPs (VP4 and VP5). In order to apply *extending* to two VPs, the following conditions must be met: i) one VP must start where the other ends, and, ii) two VPs must have at least one common traffic stream.

Contracting In the contracting operation (see figure 2.b), two VPs (VP1 and VP2) are dismantled, and a new merged VP (VP5) is formed on the common physical path of the original VPs. Other VPs (VP3, VP4, VP6, and VP7) are formed for the remaining physical paths of traffic streams if needed. In order to apply *contracting*, the two VPs must have a common physical link or links that form a path.

Next, we introduce the notion of reachability and show that the contracting and extending operations constitute a sufficient set to traverse the solution space and reach all possible solutions.

Reachability If the topological layout of a valid VP Network, called VPN1, can be obtained by applying predefined operations to another valid VP Network, called VPN2, then VPN1 is *reachable* from VPN2.

Note that only the topological layout of the VP's is considered, while the capacities allocated to VP's are not.

Complete Reachability A set of operations S_{op}, has the *complete reachability property* over a solution set S_{vp}, if any element of S_{vp} is reachable from any other element of S_{vp} by applying a finite number of operations in S_{op}.

In order to show the complete reachability property, we have to show the Reciprocity property, define two special cases of above-mentioned moves, and refine the valid VP Network solution set.

Reciprocity One can easily observe that *extending* and *contracting* are reciprocal operations. From an assignment that has been reached by *extending*, one can go back to the original assignment by one or more *contracting* operations and vice versa.

If an *extending* operation generates only one VP from the original two VPs, it is called a *simple extending* operation. This is possible when the original VPs have the same traffic stream set. Similarly, if a *contracting* operation generates only one VP from the original two VPs, it is called a *simple contracting* operation. This is possible when the original VPs have the exact same path in the physical network.

Restricted VP Network Set The Restricted VP Network set is defined as the set of valid VP Networks where simple contracting and/or simple extending can not be applied to any VP pair.

We will assume that our VP Network solution will be in the Restricted VP Network set. This assumption is justified, because the simple contracting and simple extending operations are desirable in all circumstances. The simple contracting operation improves blocking without affecting the processing costs. The simple extending operation improves processing costs without degrading the blocking probabilities. We can show that if a VP Network solution which may undergo simple extending/contracting operation(s) is chosen, then we will definitely improve the VP Network solution by executing the simple extending/contracting operation(s).

Let us assume now that we apply the simple extending operation to the **1-VP/LINK** solution wherever possible. The resultant VP Network is called "modified **1-VP/LINK**" case. Similarly, the "modified **1-VP/TS**" is the VP Network reached by applying all possible simple contracting operations to **1-VP/TS** case. These modified extreme cases are within our Restricted VP Network set.

Theorem (Complete Reachability). Any element of *Restricted VP Network Set* can be reached by a finite number of successive applications of *extending* and *contracting* operations, starting from an arbitrary *Restricted VP Network Set* element (i.e., the *extending* and *contracting* operations has complete reachability property over the *Restricted VP Network Set*).

Outline of Proof. The following points outline the skeleton of our proof of reachability in Restricted VP Network Set. The complete proof is given in Aydemir and Viniotis (1995).

- Starting with any valid assignment, one can reach one of the modified extreme cases by applying extending or contracting continuously. If extending is applied, the final VP Network will be the "modified **1-VP/TS**" case. If contracting is applied, the final VP Network will be the "modified **1-VP/LINK**" case.
- Due to the reciprocity property of *extending* and *contracting*, one can reach any valid assignment from one of the modified extreme cases.
- Finally, from a valid assignment one can reach all other valid assignments by successively applying extending and/or contracting. We have shown two possible paths between any two VP Network that pass through the modified extreme cases. □

Once a new VP Network is found, the capacity of new VPs should be assigned and the capacity of existing VPs should be revised. The next section explains our approach to the capacity assignment task.

4.2 Capacity Allocation to VPs

The capacity allocation algorithm is applied to reallocate VP capacities after each move in the VP Assignment Algorithm. The algorithm keeps the blocking probabilities of traffic streams in balance by applying the max-min fairness approach (Bertsekas and Gallager, 1992). This approach can be summarized as trying to improve the blocking probability of the worst-blocked traffic stream by adding capacity to VPs used by this traffic stream.

The algorithm starts with a minimal initial capacity allocation for each VP. The traffic stream with the highest blocking is always chosen and its bandwidth is added to the capacity of all VPs along its path. If physical link capacity limits are reached, no further improvement can be achieved for the VPs involved. The algorithm continues to equally improve the blocking for all traffic streams, until one of the following conditions is met: i) the blocking probability constraints are satisfied for all traffic streams, or ii) no more improvement in blocking is possible, due to link capacity constraints.

5 RESULTS

Preliminary runs with the Virtual Path Assignment (VPA) Algorithm have been executed for various objective functions and constraints for two sample networks (see figure 3). The 10-node network, and the 23-node network topologies are originally presented in Arvidsson (1994) and Siebenhaar (1994), respectively. For each network, we defined traffic streams with fixed routes between every node pair in both directions (90 traffic streams for 10-node, and 506 traffic streams for 23-node network).

The primary outcome of the VPA Algorithm is the VP Network (the set of VPs with assigned capacity and assigned physical path). Additionally, we output the blocking probabilities and number of hops for traffic streams. The unused capacity of physical links at the end of the algorithm is also monitored, even though we do not report it here (see Aydemir and Viniotis, 1995).

We have tested the response of the VPA Algorithm by varying the blocking probability

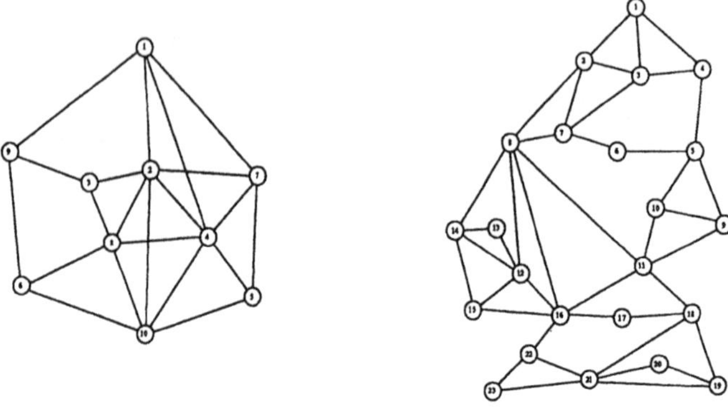

Figure 3 The 10-node and 23-node networks used in the test runs.

constraint, ε_r, and maximum hop constraint (K). We have also varied the total offered load on a fixed network topology, and the proportionality constant α in objective function (13). The blocking probability constraint ε_r value of 0.01 is applied in all of the experiments except for figure 6.

In figure 4, we demonstrate the accuracy of the EFP and Knapsack Approximations with respect to simulation values. The points in the plot for EFP and Knapsack Approximations are mean error percentages of approximate stream blocking probabilities with respect to a simulation estimate of stream blocking probabilities. The points for the simulation are the average of error percentages of upper bound of confidence interval (95 %) with respect to simulation estimate. The Knapsack Approximation is almost always within the confidence interval of simulation results; the EFP approximation underestimates the blocking probability of connections. In the rest of the runs, the knapsack approximation is employed for blocking probability calculations.

In figure 5, we present the effect of maximum hop constraint (K) on the minimum hop and weighted min hop objectives. Limiting the number of hops experienced, has resulted in achieving better optimization results for the same load conditions. Starting solutions for each K value were different, and this may have caused the algorithm to reach better solutions for lower K values. The variation in the parameter α and its effect on the optimal value is also illustrated in figure 5. As the α value increases, the emphasis on the minimization of number of hops decreases, resulting in the gradual increase of average number of hops.

Figure 6 illustrates the response of the algorithm to changing blocking probability constraints in both 10-node and 23-node networks. The same ε_r value is applied to all traffic streams. In the 10-node network, the weighted and non-weighted versions of the objective function that minimizes the number of hops were used and both generated the same results. As the blocking probability constraint is relaxed, the number of hops used

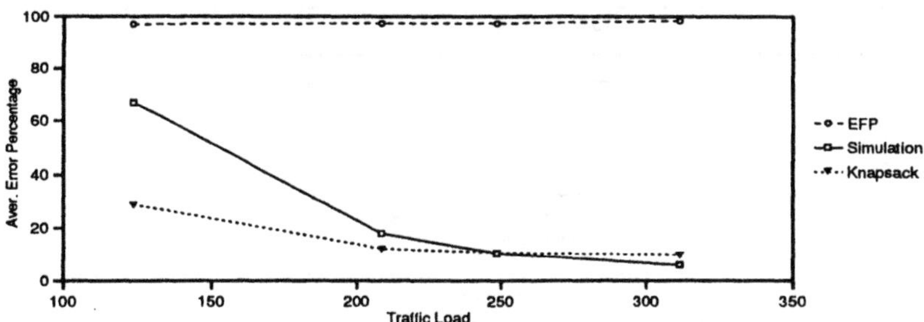

Figure 4 Comparison of Connection Blocking Probability Approximation Methods.

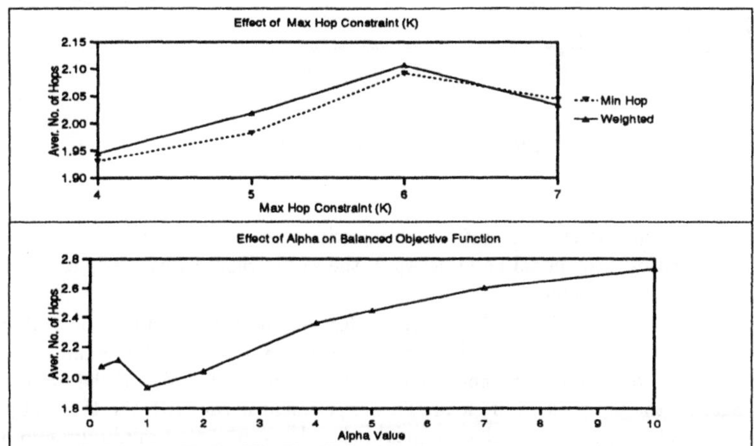

Figure 5 Effects of K (Max Hop Constraint) and α in Balanced Objective.

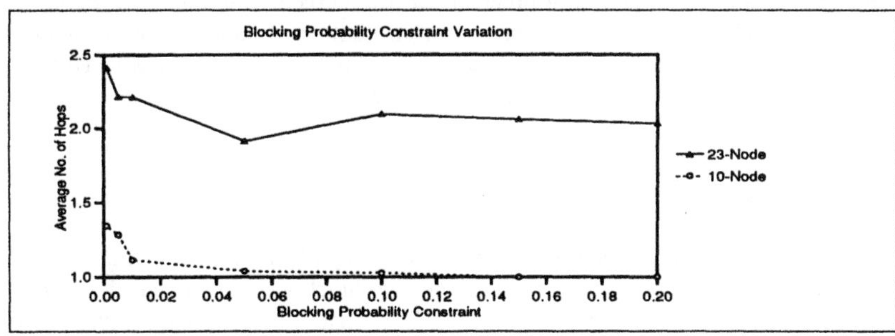

Figure 6 Blocking Probability Constraint (ε_r) Variations.

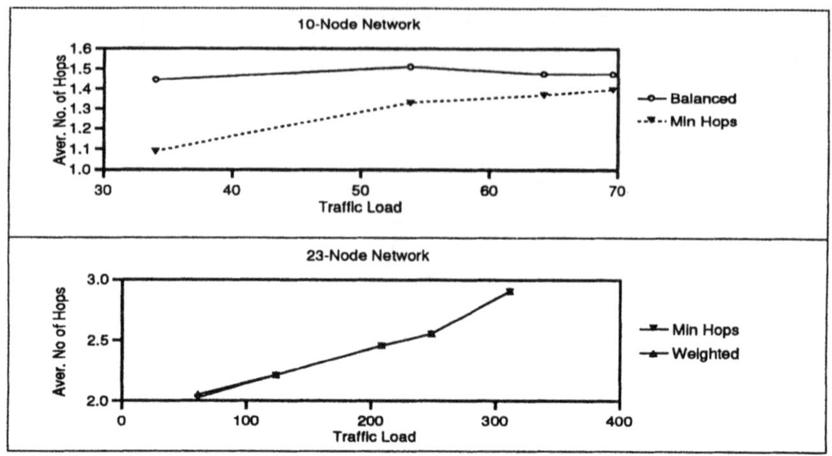

Figure 7 Traffic Load Variations.

by a connection also decreases, until all connections use one VP (**1-VP/TS** case), in the 10-node network. In the 23-node network, the same trend is observed, except for an unexpected dip at $\varepsilon_r = 0.05$.

Figure 7 shows the results of the algorithm for varying traffic loads. The traffic load for the whole network is calculated by using the formula $\sum \lambda_r \cdot b_r$. The mean number of hops experienced by connections and the overall blocking probability of the network are plotted against the network traffic load. This experiment is run with the three objective functions of Section 2.1. The results for weighted and non-weighted minimum hop objective functions were the same. The balanced objective function is run with α set to 0.5. As the traffic load increases on a fixed network, the blocking probability constraint plays a bigger role in all cases. Thus the difference in the outcome diminishes with increasing traffic load. The slope of the increase of mean number of hops is more pronounced in the 23-node network. The minimum hop and weighted minimum hop objectives have performed almost the same, in the large network case too.

6 CONCLUSIONS

In this paper, we presented a heuristic optimization algorithm for the Virtual Path Assignment function in ATM networks. The objective functions chosen reflect processing and transmission costs on the whole network. The constraints reflect quality of service requirements. The procedure is flexible and robust, since it allows multi-class traffic and poses no restrictions on the VP Network structure to be obtained.

One of the most formidable tasks in the search of the optimal VP Assignment is the calculation of blocking probabilities observed by the users. Since exact methods are practically not available, a set of approximation schemes is studied. The approximation schemes are all inspired by the reduced load approximation principles, and have varying computational complexities. The experience gathered from several comparisons with simulations indicate that the Knapsack approximation performs accurately for our purposes. The results of the VPA Algorithm applied to a sample network show that the procedure runs successfully with different constraints and parameters. Increasing traffic loads under a constant blocking constraint forced the VPA Algorithm to use shorter VPs and increase sharing of links. Thus, the number of hops traversed by a connection has increased. Similarly, increasing the blocking constraint over same network conditions resulted in an increase on the number of hops traveled. The minimum hops and weighted minimum hops objectives have resulted in very close VP assignments. This may be due to the fairly balanced traffic in the network we have tested. The results of the experiments with maximum hop constraint (K) suggests the existence of local optima, and sensitivity of the result to the initial starting solution.

Areas for improvement and further study include incorporation of connection routing function into the optimization process, and the verification of the optimality of the resultant VP Network by investigating other search methods.

7 ACKNOWLEDGEMENTS

We would like to thank Dr. G. Bilbro for his helpful suggestions on the algorithm of section 4.

8 REFERENCES

Addie, G. and Burgin, John and Sutherland, S. L. (1988) B-ISDN Protocol Architecture, in *Proceedings of IEEE GLOBECOM*, 22.6.1, Hollywood,FL.

Arvidsson, A. (Oct. 1994) Real Time Management of Reconfigurable Virtual Path Networks in B-ISDN, *TRC Report No. 15/94*, Teletraffic Research Centre, The University of Adelaide.

Aydemir, M. and Viniotis I. (1995) Virtual Path Assignment Problem in ATM Networks, *CACC Report*, NCSU, Raleigh, NC.

Baskett, F. and Chandy, K. M. and Muntz, R. R. and Palacios, F. G. (1975) Open, Closed and Mixed Networks of Queues with Different Classes of Customers, in *Journal of the ACM*, **22**, 248-60.

Bertsekas, Dimitri and Gallager, Robert. (1992) *Data Networks, 2nd Edition*. 524-8. Prentice Hall.

Burgin, John and Dorman, Dennis. (Sep. 1991) Broadband ISDN Resource Management: The Role of Virtual Paths, in *IEEE Communications Magazine*, 44-48.

Cheng, Kwang-Ting and Lin, Frank Yeung-Sung. (1994) On the Joint Virtual Path Assignment and Virtual Circuit Routing in ATM Networks, in *Proceedings of the IEEE GLOBECOM*, 777-82.

Chlamtac, Imrich and Faragó, András and Zhang, Tao. (May 1993) How to Establish and Utilize Virtual Paths in ATM Networks, in *Conference Record of the IEEE International Conference on Communications (ICC)*, Geneva, Switzerland.

Chung, Sun-Ping and Ross, Keith W. (1993) Reduced Load Approximations for Multirate Loss Networks, in *IEEE Transactions on Communications*,**41(8)**, 1222-31.

Kanada, T. and Sato, K. and Tsuboi T. (1987) An ATM Based Transport Network Architecture, in *IEEE COMSOC Int. Workshop on Future Prospects of Burst/Packetized Multimedia Communications*, Osaka, JAPAN.

Kaufman, Joseph S. (1981) Blocking in a Shared Resource Environment, in *IEEE Transactions on Communications*,**COM-29(10)**, 1474-81.

Kelly, F. P. (1986) Blocking Probabilities in Large Circuit-Switched Networks, in *Advances in Applied Probability*, **18**, 473-505.

Lam, S. S. (Jul. 1977) Queueing Networks with Population Size Constraints, in *IBM Journal of Research and Development*, 370-8.

Lin, Frank Yeung-Sung and Cheng, Kwang-Ting. (Dec. 1993) Virtual Path Assignment and Virtual Circuit Routing in ATM Networks, in *Proceedings of the IEEE GLOBECOM*, 436-41. Houston, Texas.

Sato, Ken-Ichi and Ohta, Satoru and Tokizawa, Ikuo. (1990) Broad-Band ATM Network Architecture Based on Virtual Paths, in *IEEE Transactions on Communications*, **COM-38(8)**, 1212-22.

Siebenhaar, Rainer (1994) Optimized ATM Virtual Path Bandwidth Management Under Fairness Constraints, in *Proceedings of the IEEE GLOBECOM*, 924-8.

Tokizawa, I and Kanada, T. and Sato, K. (1988) A New Transport Network Architecture Based on Asynchronous Transfer Mode Techniques, in *Proceedings of ISSLS'88*, 11.2.1-5, Boston, MA.

Whitt, Ward. (1985) Blocking When Service is Required From Several Facilities Simultaneously, in *AT&T Technical Journal*, **64(8)**, 1807-56. Prentice Hall Inc., NJ.

9 BIOGRAPHY

Metin Aydemir received the B.S.E.E. degree from Middle East Technical University, Ankara, Turkey in 1981. He continued to study towards a master's degree at Electrical Engineering Dept. in Florida Institute of Technology, Melbourne, Florida (1985), and worked in telecommunications software design area in ITT and SIEMENS. Since 1992, he is a PhD candidate in Electrical and Computer Eng. Department at NCSU, Raleigh, NC. His research interests are in high-speed network analysis and modeling, queueing networks, and optimization algorithms.

Yannis Viniotis received the B.Sc. degree from the University of Patras, Greece, and M.S. and Ph.D. degrees in Electrical Engineering from the University of Maryland, College Park. He is currently an Associate Professor of Electrical and Computer Engineering at North Carolina State University, Raleigh, NC. His research interests are in computer communication system design and analysis, with particular emphasis on Quality of Service, multimedia, multicasting and adaptive network control algorithms. He has published over 30 articles and lectured extensively on these topics. He was a guest editor of the *Performance Evaluation* Journal, in 1995, on High Speed Networks. He has chaired two international conferences on networking, in 1992 and 1993.

26

Enhancing ATM Network Performance by Optimizing the Virtual Network Configuration

A. Faragó[2], S. Blaabjerg[1], W. Holender[1], B. Stavenow[1], T. Henk[2], L. Ast[2], S. Székely[2]

[1]Ellemtel Telecommunications Systems Laboratories
S-22370 Lund, Sweden

[2] Dept. of Telecommunications and Telematics
Technical University of Budapest

XI. Sztoczek u. 2, Budapest, Hungary H-1111
Phone: +36 1 463 1861
Fax: +36 1 463 3107
E-mail: farago@ttt-atm.ttt.bme.hu

Abstract

Virtual or logical subnetworks are expected to play an important role in large B-ISDN configurations. This gives an additional degree of freedom to ATM network architectures, since even for a fixed physical network the logical configuration can still vary depending on particular demands and conditions. This new degree of freedom calls for new solutions to utilize the opportunity for the potential enhancement of network performance by optimizing the logical configuration, as part of optimizing the distributed network architecture. In this paper a framework and model, along with efficient solution algorithms, are presented to dimension virtual ATM networks on top of the same physical infrastructure network, such that the virtual networks share the infrastructure, while the total network revenue is optimized. The algorithms are tried on various network scenarios and a trade-off between the quality of the result and running time is exhibited.

Keywords: ATM network, virtual subnetwork, network partitioning, network revenue.

1 Introduction

The trend towards service integration in telecommunications has been steadily more profound and at the time when ATM was introduced the expectations regarding the range of services that could be integrated had practically no limitations. During the last few years, however, it has been recognized that it is not at all easy to integrate services with very different demands regarding e.g. bandwidth, grade of service (GoS) or congestion control functions. The latest development within the ITU SG13 with the agreement of the definitions of 4 ATM bearer capabilities clearly demonstrate this. In some cases it turns out to be better to support different services by offering separate logical networks, and limiting the degree of integration to only partial, rather than complete, sharing of physical transmission and switching resources.

A second context in which separation into logical networks may take place is virtual leased networks. Large business customers realizing e.g. LAN interconnections may require *guaranteed* resources. Furthermore, virtual paths are special cases of logical networks and peak rate allocation of VP's can be seen as a (virtual) separation of resources.

Resource separation for segregation of ATM layer bearer capabilities, for offering different GoS classes, virtual leased networks with guaranteed resources and peak rate allocated virtual paths are examples of a new feature in the design, dimensioning and management of ATM networks. On top of the physical infrastructure a number of logical or virtual subnetworks can co-exist, sharing the same physical transmission and switching capacities.

The complete partitioning of the infrastructure network is an important component in network optimization and can be seen as the contrast to complete sharing. It is evident that in many circumstances a solution somewhere between the two extremes with a minimum guaranteed resource and possibility to use a certain number of extra resources will be applied. However, since complete partitioning is an important landmark in the context and since so far most work has been done only for isolated links, an investigation of the issue on the network level is of importance.

The problems with full service integration are also realized in [12], and an optimization model taking into account transmission, switching and set up cost is developed in which different strategies like complete partitioning, complete sharing and sharing with trunk reservation can be evaluated. In [1] a greedy algorithm for Dynamic Capacity Management is used to design

VP networks and the problem of gathering reliable measurement data is also considered. Concerning multi-rate models, [13] gives an excellent overview.

The objective of this paper is to formulate an optimization model of the partitioning problem for a network and to present three efficient algorithms that solve the problem. The objective function is the *total network revenue,* and together with the physical constraints also grade of service constraints are taken into account in the simplest model.

In Section 2 the B-ISDN network environment in which the model is to be applied is described and further motivation for and consequence of the partitioning approach is given. Section 3 presents the optimization model, while Section 4 describes three algorithms that find solutions.

2 B-ISDN and Network Partitioning

In this section the B-ISDN scenario considers a situation that is somewhat more developed than it is today.

2.1 B-ISDN and Overlay Networks

A fully developed B-ISDN will have a very complex structure with a number of overlay networks. Conceptual tools to simplify the description are needed and one such conceptual model suitable of describing overlay networks is the *Stratified Reference Model,* [7]. This model adopts the general layering of the OSI model, it works with the three lowest layers, but allows for recursion by a generalization of the physical layer.

Adopting the concept of the Stratified Reference Model, the B-ISDN will consist of the following *strata,* see Fig. 1. A transmission stratum based on SDH at the bottom, a cross connect stratum based on either SDH or ATM on top of that, which acts as an infrastructure for the ATM VP/VC stratum with switched connections. Finally, the large set of possible applications uses the ATM or cross connect stratum as an infrastructure.

2.2 Cross Connect and Partitioning

In this paper the focus will be on the ATM and the cross connect stratum in general and on partitioning of the ATM stratum network in particular.

Whether the cross connect stratum is realized by SDH or ATM will have important implications for the partitioning, see Fig. 2. If the cross connect stratum is based on SDH and the ATM network is realizing different QoS classes by resource separation, the partitioning can only be done in integer portions of the STM modules of the SDH structure, see Fig. 2.a. On the other hand, if the cross connect is realized by ATM virtual paths (VP), then no integrality restriction exists and the partitioning can be done in any real portions, see Fig 2.b. The SDH cross connect solution will therefore give rise to a model that is discrete in the ATM link capacities while the ATM cross connect solution gives rise to a continuous model. It should be emphasized that if the partitioning is made in order to support different QoS classes, then either the integer solution should be adopted or the ATM switches will have to be designed in a way to support partitioning at the individual input and output ports.

2.3 Relevant QoS Parameters

Since ATM has similarities with both packet switched and circuit switched networks it is not *a priori* obvious which properties should have the greatest impact to an optimization model. However, it is the data transfer phase when the similarities with packet switched networks are the largest. At the connection setup phase the similarities to circuit switching dominate, especially if a preventive connection control concept with small ATM switch buffers has been adopted together with the equivalent bandwidth concept.

In an optimization model that models the call scale phenomenon it is natural to take the view that an ATM network can be modeled as a multi-rate circuit switched network in which the most important quality of service parameter is the connection blocking probability. This is the basis for the model to be considered in the next section.

3 The Optimization Model

A fixed infrastructure network with a number of nodes and a number of trunk groups connecting the nodes in an arbitrary way is considered. This network will model the cross connect stratum in the B-ISDN overlay network.

On top of this network a number of logically separated ATM networks are to be carried. The topology of these *virtual* or *logical* ATM networks will, in general, differ from the topology of the underlying network and an ATM link may use more than one trunk group. An ATM network can be a virtual leased network for a large business customer, it can be the part of an overall ATM network realizing a QoS class by resource separation, or it can simply be a virtual path.

The first set of *constraints*, expressible by linear inequalities in the model, comes from the natural fact that the sum of ATM link capacities on a physical trunk group cannot exceed the capacity of the trunk group.

The traffic demand between any two nodes in any of the ATM networks is assumed given, and a number of fixed routes in each of the ATM networks is also assumed given in advance. In general, there exists a number of routes between a given node pair. One objective of the optimization model is to distribute the traffic offered to the node pair optimally between the routes which can realize the communication. The distribution of the offered traffic between the possible routes is termed *load sharing* and the parameters according to which it takes place are called *load sharing coefficients.*

A second set of constraints in the model comes from the fact that the traffic demand between any pair of nodes should be equal to the sum of traffic demands on the routes realizing communication between the origin-destination pair. This means, the sum of load sharing coefficients is one.

To explain the idea in the simplest case, each virtual ATM network is assumed to operate under fixed non-alternate routing. A connection attempt is assigned at random to a single route, and if insufficient resources are available, the connection request is blocked and disappears from the system. Since alternate routing complicates the analysis, we restrict ourselves to the simplest fixed routing case in this paper.

The bandwidth demand of any ATM connection is characterized by a single parameter that can be interpreted either as equivalent bandwidth in case of adoption of the sustainable cell rate, or as peak rate of the connection in case of peak rate allocation. It is assumed that the one-parameter characterization of the ATM connections together with an adequate admission control algorithm ensures that cell level GoS degradations, like cell loss, cell delay and cell delay variation are well under control, and need not be considered at this level. Instead, the main GoS parameter in this context is the *call scale* parameter: *connection blocking probability.*

For reasons of fairness it is natural to impose a third set of constraints, namely the GoS constraint, that the connection blocking probability on each route should not exceed a given maximum value.

The *objective function* is the total network revenue summed up over all virtual ATM networks. It is assumed that each accepted connection will generate revenue at a given rate depending on the traffic type and route, and the total network revenue can then be seen as a weighted sum of carried traffic values. For a complete mathematical formulation see [2].

To summarize, the task of the optimization model is to find the partitioning of the infrastructure network into virtual subnetworks and to find the load sharing coefficients for each ATM node pair which maximizes the total network revenue subject to the physical constraints listed above and optionally also the GoS constraints.

4 Solution Approaches

The objective function to be optimized is inherently difficult to deal with, since it requires the knowledge of the carried traffic and thereby route blocking probabilities that can be computed in an exact way only for very small networks. A common feature of the methods to be applied is that they make use of the natural and well known reduced load and link independence assumptions, see [9], [10], [5]. Based on this approximation the partial derivatives of the network revenue can be found in a tractable form suitable for a gradient based hill climbing. Since this method has quite high running time, therefore, a somewhat simpler method based on a sequence of linear programming tasks is presented. Finally, a very simple alternative approach is suggested. All the three presented solution approaches assume that the ATM link capacities can take any real number, thus implying that the underlying cross connect structure is based on ATM.

4.1 Gradient Based Hill Climbing

Applying an extension of the arguments given in [8], see [2] for details, the partial derivatives of the network revenue can be computed. The computation requires that the fixed point equations associated with the reduced load approximation are solved, and in addition a set of linear equations has to be solved. The size of these equation systems equals to the product of the number of links in the network and the number of traffic types.

Using these derivatives, the algorithmic solution is reduced for a standard gradient based hill climbing over a convex feasibility domain defined by linear inequalities. Here one can use either commercial or freely available software, see e.g. [3]. Unfortunately, however, the speed of the algorithm is greatly reduced by the large amount of side computations, mentioned in the preceding paragraph.

The algorithm will in general converge towards a local optimum. However, by a careful choice of the initial values, the risk of getting stuck in a local optimum that falls far from a global optimum can be reduced.

4.2 Solution by Sequential Linear Programming

In general, the complexity of computing blocking probabilities is much larger in the case where many different bandwidth demands (traffic types) co-exist. An efficient way to avoid the increased complexity is to approximate the blocking probability of a traffic type requiring d units of capacity by grabbing d times one unit independently. It is proven that this approximation asymptotically leads to the correct blocking probabilities, see [9] and [11].

Adopting this approximation and assuming all revenue coefficients are the same, independently of traffic types, it is shown in [4] that a slightly changed objective function and the same constraints can be accurately approximated by a sequence of linear programming tasks. The accuracy of the approximation increases in the regime where capacities are large.

In the following the total number of virtual links over all virtual networks is denoted by J, and the capacity of virtual link j is denoted by C_j.

The incidence of physical and virtual links is expressed by a $K \times J$ zero-one structure matrix $\mathbf{S}$ (K is the number of physical links) in which the entries are given by

$$s_{k,j} = \begin{cases} 1 & \text{if physical link } k \text{ is used by virtual link } j \\ 0 & \text{otherwise} \end{cases} \tag{1}$$

Let us denote the blocking probability of logical link j by B_j, and let A_{ir} be the amount of capacity used by a route r call on link j and ν_r the offered traffic to route r.

$E(\rho, C)$ is Erlang's B-formula, and a_j is a logarithmic blocking measure, defined by the procedure, related to link j.

The algorithm is as follows, see [4] for details.

Step 1 Set $B_j := 0$, $a_j := 1$ for each j.

Step 2 Solve the linear programming problem

$$\text{Maximize} \quad \sum_j a_j C_j$$

$$\text{Subject to} \quad SC \leq C_{phys} \text{ and } C \geq 0.$$

(C_{phys} is the physical capacity vector.)

Step 3 Compute new values for the blocking probabilities by

$$B_j := E\left((1 - B_j)^{-1} \sum_r A_{jr} \nu_r \prod_i (1 - B_i)^{A_{ir}},\ \tilde{C}_j\right)$$

where $\tilde{C}_1, \ldots, \tilde{C}_J$ come from Step 2 as a solution of the linear programming problem.

Step 4 Set

$$a_j := -\log(1 - B_j), \qquad j = 1, \ldots, J.$$

Step 5 If all variables differ from their previous value by less then a given error treshold, then STOP, else repeat from Step 2.

The above iterative procedure has an intuitively appealing interpretation. If a link j at a certain iteration has large blocking probability B_j, then the corresponding a_j coefficient in the objective function will be large. That will inspire the linear programming to increase the value of C_j, since a variable with larger objective function coefficient can contribute more to the maximum. This conforms with the intuition that a link with high blocking probability needs capacity increment.

Numerical experience shows that the algorithm converges typically within a few iterations [4].

4.3 Simplified Partitioning Based on Equivalent Link Blocking

The approach to be presented here assumes that the traffic demand to each *route* is known in advance, implying that no optimization with respect to load sharing can take place. Besides having the traffic demand matrix as input, we also assume here that for each route r a maximum allowed route bloking probability $B(r)$ is given.

Now some additional notations are introduced. Let $\mathcal{R}$ be the set of (fixed) routes in the network, taking all logical subnetworks into account. We want to design the logical link capacities such that the blocking probability of any route r is at most $B(r)$.

Let $L(r)$ be the set of logical links used by route r and denote by $l(r)$ the length of route r, that is, the number of links on the route.

Now, assuming link independence and using the reduced load approximation, the prescriptions on route blocking can clearly be satisfied if

$$1 - B_j \geq (1 - B(r))^{1/l(r)}$$

holds for each route r and for each logical link $j \in L(r)$. The idea here is that we distribute the route blocking probability evenly among the links used by the route. If $\mathcal{R}_j$ denotes the set of routes that use link j then we obtain the condition

$$1 - B_j \geq \max_{r \in \mathcal{R}_j} (1 - B(r))^{1/l(r)}$$

Now let B_j^0 be the maximum possible value for the blocking probability on link j that follows from the above model, based on the idea of evenly distributed blocking. From the above considerations we have

$$B_j^0 = 1 - \max_{r \in \mathcal{R}_j} (1 - B(r))^{1/l(r)}$$

Applying the B_j^0 values, we can approximate the link offered traffic as

$$\rho_j^0 = \sum_{r \in \mathcal{R}_j} A_{jr} \nu_r \prod_{i \neq j} (1 - B_i^0)^{A_{ir}}.$$

Once we know the value of B_j^0 and ρ_j^0 from the above explicite formulas, the capacity C_j of logical link j can now be calculated by inverting numerically Erlang's formula

$$B_j^0 = E(\rho_j^0, C_j).$$

If the sum of the obtained C_j logical capacity values exceed the available physical capacity on a physical link, then we normalize them such that they satisfy the physical capacity constraints. That is, we multiply all of them by a constant, smaller than 1, such that their sum becomes equal to the physical capacity, while keeping the ratio of the logical capacities.

It should be noted that this procedure does not optimize the network revenue, it only dimensions the logical links, taking into account the requirements on route blocking probabilities. Thereby, it considers network revenue in an indirect way.

The computational complexity of this model is *very small*, since only the ATM-link capacities are varied and no linear programming is performed.

5 Numerical Example

The algorithms are being tested on various network examples. Due to space limitations, we show just very briefly the results of the comparison of two algorithms: the sequential linear programming approach, called Fixpoint and the equivalent link blocking approach, called ELB.

The two algorithms were compared on a 6-node physical network that carried five different virtual ATM networks, each with four traffic classes. 16 different traffic scenarios were considered, characterized by two parameters. The parameter α defines the distribution of traffic among the classes. The higher the value of α, the more unbalanced is the distribution. The parameter β defines the homogenity of bandwidth demands: the higher the value of β, the more inhomogeneous are the bandwidth demands.

The results are shown in Fig. 3. For balanced and homogeneous traffic scenarios the two methods show the same performance. For unbalanced and/or inhomogeneous traffic demands, however, the more sophisticated Fixpoint algorithm performs significantly better, according to the expectations.

On the other hand, when running time is considered, we find a converse relationship, see Fig. 4. Although for very small networks there is no significant difference in running times, but if the network size gets larger, such as a 6-node physical network with 5 virtual subnetworks carriyng 4 traffic classes, as in the example, then there is already a dramatic difference in running times. This shows a clear trade-off between quality and speed: if rougher estimates suffice, then ELB is recommended. On the other hand, if more accuracy is needed, then the slower but more accurate Fixpoint algorithm is favourable.

6 Conclusion

A framework and model, along with efficient solution algorithms, are presented to dimension virtual ATM networks on top of the same physical infrastructure network, such that the virtual networks share the infrastructure, while the total network revenue is optimized. The algorithms are tried on various network scenarios and a trade-off between the quality of the result and running time is exhibited. The algorithms are being implemented as part of an integrated system for ATM network planning, simulation and management called PLASMA [6].

7 Acknowledgment

The research was performed as part of a joint project between Ellemtel Telecommunication System Laboratories, Sweden and the Department of Telecommunications and Telematics, Technical University of Budapest, Hungary. The authors are grateful to Miklós Boda and Géza Gordos for stimulating discussions and for their continuous support.

References

[1] Å. Arvidsson, *Management of Reconfigurable Virtual Path Networks*, ITC-14, France, 1994, pp. 931-940.

[2] A. Faragó, S. Blaabjerg, L. Ast, G. Gordos and T. Henk, "A New Degree of Freedom in ATM Network Dimensioning: Optimizing the Logical Configuration", to appear in *IEEE Journal on Selected Areas of Communications,* Spec. Issue on the Fundamentals of Networking.

[3] C. Lawrence J.L. Zhou, A.L. Tits, "Users Guide for CFSQP version 2.2: A C Code for Solving (Large Scale) Constrained Nonlinear (Minimax) Optimization Problems, Generating Iterates Satisfying All Inequality

Constraints", E.E. Department, Institute for System Research, University of Maryland, TR 94-16r1.

[4] A. Faragó, S. Blaabjerg, W. Holender, T. Henk, L. Ast, Á. Szentesi, Zs. Ziaja, "Resource Separation - an Efficient Tool for Optimizing ATM Network Configuration", Proc. *NETWORKS'94, Sixth International Network Planning Symp.*, Budapest, Hungary, Sept. 1994, pp. 83-88.

[5] A. Girard, *Routing and Dimensioning in Circuit Switched Networks*, Addison Wesley, 1990.

[6] Zs. Haraszti, I. Dahlqvist, A. Faragó and T. Henk, "PLASMA - An Integrated Tool for ATM Network Operation", *International Switching Symp. (ISS'95)*, Stockholm, 1995.

[7] T. Hadoung, B. Stavenow, J. Dejean, "The Stratified Reference Model: An Open Architecture to B-ISDN", *International Switching Symp. (ISS'90)*, Stockholm, 1990.

[8] F.P. Kelly, "Loss Networks", *The Annals of Applied Probability*, 1(1991/3), pp. 319-378.

[9] F.P. Kelly, "Blocking Probabilities in Large Circuit Switched Networks", *Adv. Appl. Probab.*, 18(1986), pp. 473-505.

[10] F.P. Kelly, "Routing in Circuit Switched Networks: Optimization, Shadow Prices and Decentralization", *Adv. Appl. Probab.*, 20(1988), pp. 112-144.

[11] J.-F. P. Labourdette and G.W. Hart, "Blocking Probabilities in Multitraffic Loss Systems: Insensitivity, Asymptotic Behavior and Approximations, *IEEE Trans. Communications,* 40(1992/8) pp. 1355-1366.

[12] M. Menozzi, U. Mocci, C. Scoglio, A. Tonietti, *Traffic Integration and Virtual Path Optimization in ATM Networks*, Networks'94, pp. 71-76.

[13] M. Ritter, P. Tran-Gia (Eds.), *Multi-Rate Models for Dimensioning and Performance Evaluation of ATM Networks*, COST 242 Interim Report, Commission of the European Communities, June 1994.

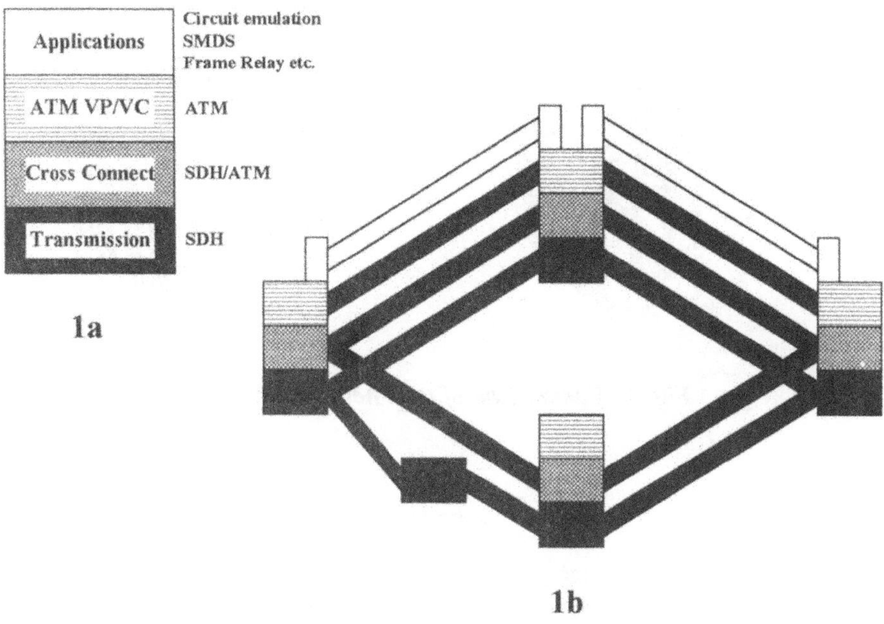

Fig. 1 B-ISDN from the viewpoint of the stratified reference model 1a) the protocol viewpoint, 1b) the network viewpoint

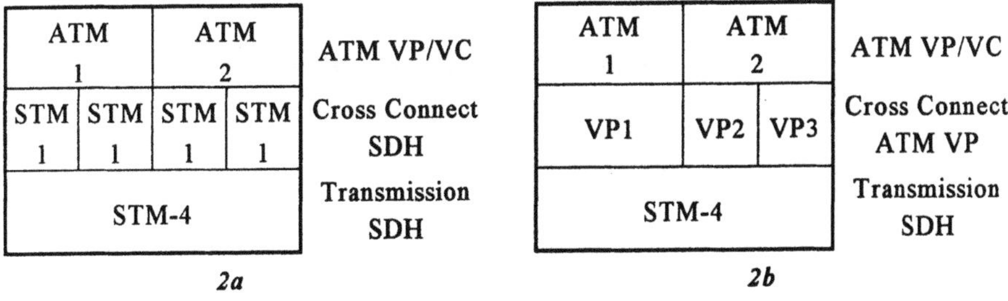

Fig. 2 Partitioning of the Cross Connect infrastructure. In case where it is based on SDH whole blocks of STM-1 must be chosen, while in case of ATM VPs any continuous values are feasible.

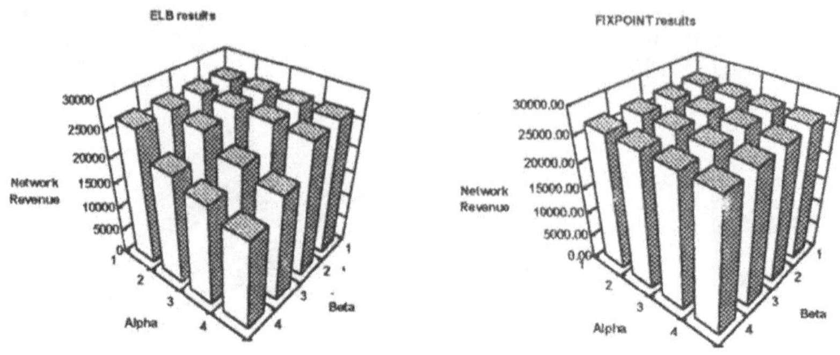

Fig. 3 Comparison of total network revenue

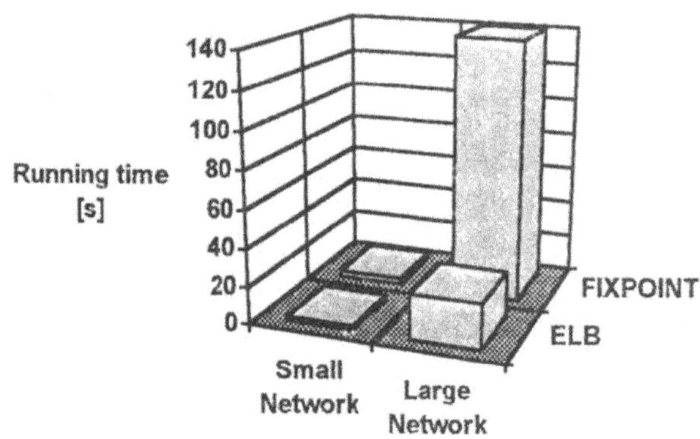

Fig. 4 Comparison of running times

INDEX OF CONTRIBUTORS

KEYWORD INDEX

GPSR Compliance
The European Union's (EU) General Product Safety Regulation (GPSR) is a set of rules that requires consumer products to be safe and our obligations to ensure this.

If you have any concerns about our products, you can contact us on

ProductSafety@springernature.com

In case Publisher is established outside the EU, the EU authorized representative is:

Springer Nature Customer Service Center GmbH
Europaplatz 3
69115 Heidelberg, Germany

www.ingramcontent.com/pod-product-compliance
Ingram Content Group UK Ltd.
Pitfield, Milton Keynes, MK11 3LW, UK
UKHW021902260726
13966UKWH00006B/187

* 9 7 8 1 4 7 5 7 4 9 0 7 6 *